Technologies for Near-Zero-Emission Gasoline-Powered Vehicles

Other SAE titles of interest:

Advanced Developments in Ultra-Clean Gasoline-Powered Vehicles
By Fuquan (Frank) Zhao
(Order No. PT-104)

Automotive Gasoline Direct-Injection Engines
By Fuquan (Frank) Zhao, David L. Harrington, and Ming-Chia Lai
(Order No. R-315)

Vehicular Engine Design
By Kevin L. Hoag
(Order No. R-369)

For more information or to order a book, contact SAE International at
400 Commonwealth Drive, Warrendale, PA 15096-0001;
phone (724) 776-4970; fax (724) 776-0790;
e-mail CustomerService@sae.org;
website http://store.sae.org.

Technologies for Near-Zero-Emission Gasoline-Powered Vehicles

Fuquan (Frank) Zhao

Warrendale, Pa.

All rights reserved. No part of this publication may be reproduced, stored in a retrieval system, or transmitted, in any form or by any means, electronic, mechanical, photocopying, recording, or otherwise, without the prior written permission of SAE.

For permission and licensing requests, contact:
SAE Permissions
400 Commonwealth Drive
Warrendale, PA 15096-0001 USA
E-mail: permissions@sae.org
Tel: 724-772-4028
Fax: 724-772-3036

Library of Congress Cataloging-in-Publication Data

Zhao, Fuquan.
　　Technologies for near-zero-emission gasoline-powered vehicles / Fuquan (Frank) Zhao.
　　p. cm.
　　Includes bibliographical references and index.
　　ISBN-13: 978-0-7680-1461-7
　　ISBN-10: 0-7680-1461-1
　　1. Automobiles--Motors--Technological innovations.
　　2. Internal combustion engines, Spark ignition--Technological innovations. I. Title.

TL210.Z454 2007
629.25'04--dc22　　　　　　　　　　　　　　　　2006040905

SAE International
400 Commonwealth Drive
Warrendale, PA 15096-0001 USA
E-mail: CustomerService@sae.org
Tel: 877-606-7323 (inside USA and Canada)
　　　724-776-4970 (outside USA)
Fax: 724-776-0790
Copyright © 2007　SAE International
ISBN-10 0-7680-1461-1
ISBN-13 978-0-7680-1461-7
SAE Order No. R-359
Printed in the United States of America.

Contents

Preface xv

Chapter 1   Transient Engine Startup and Shutdown Processes 1
              By Wai K. Cheng

    1.1   Introduction 1
    1.2   The Engine Shutdown Process for Port Fuel Injected Engines 2
        1.2.1   General Behavior During the Shutdown Process 2
        1.2.2   Impact of the Shutdown Process on Hydrocarbon Emissions 4
    1.3   The Engine Startup Process for Port Fuel Injected Engines 5
        1.3.1   Initial Conditions for Engine Startup 5
        1.3.2   General Behavior During the Startup Process 6
        1.3.3   Mixture Preparation in the Startup Transient 10
        1.3.4   Combustion in the Startup Transient 20
        1.3.5   Hydrocarbon Emissions in the Startup Transient 20
    1.4   The Engine Startup Process for Direct Injection Spark Ignition Engines 22
    1.5   Summary 25
    1.6   References 26

Chapter 2   Mixture Formation Processes 31
              By Ron Matthews and Matt Hall

    2.1   Introduction 31
    2.2   Liquid Fuel as a Source of Hydrocarbon Emissions 32
    2.3   Fuel Injection Hardware and Controls 44
        2.3.1   Injector Types 44

		2.3.2 Injection Timing 47
		2.3.3 Other Injection Parameters 51
	2.4	Flow Field Effects 56
	2.5	Strategies to Improve Fuel Preparation and Reduce Liquid Fuel Effects During Cold Start and Warm-Up 60
	2.6	Summary 64
	2.7	References 69

Chapter 3 Cold-Start Hydrocarbon Emissions Mechanisms 79
By James A. Eng

	3.1	Introduction 79
	3.2	Global Engine Behavior During a Cold Start 80
		3.2.1 Required Fueling Levels 83
		3.2.2 Fuel Accounting 84
	3.3	Hydrocarbon Storage Mechanisms 86
		3.3.1 Storage in Crevices 88
		3.3.2 Absorption in Oil Layers and Deposits 90
		3.3.3 Liquid Fuel 92
		3.3.4 Quench Layers 96
		3.3.5 Partial Burns 98
		3.3.6 Rich Air/Fuel Operation 100
	3.4	Hydrocarbon Transport Mechanisms 100
		3.4.1 Transport Mechanisms at Warm Conditions 101
		3.4.2 Transport Mechanisms at Cold Conditions 103
	3.5	Hydrocarbon Oxidation 104
		3.5.1 Hydrocarbon Oxidation Mechanisms 105
		3.5.2 Hydrocarbon Consumption Rates 106
		3.5.3 Post-Flame Hydrocarbon Consumption in Engines ... 109
	3.6	Summary 112
	3.7	References 114

Contents

Chapter 4 Characterization of Cold Engine Processes 121
By Choongsik Bae

4.1 Introduction 121
4.2 Fuel Injection Characteristics and Fuel Delivery into the Engine Cylinder 122
    4.2.1 Fuel Sprays 122
    4.2.2 Wall Wetting 127
    4.2.3 Fuel Delivery into the Engine Cylinder 130
4.3 Mixture Distribution and Its Interaction with Flow 134
4.4 Combustion Processes and Pollutant Formation 142
4.5 Summary 145
4.6 References 146

Chapter 5 Spark Retardation for Improving Catalyst Light-Off Performance 157
By Stephen Russ

5.1 Introduction 157
5.2 Calibration Actions for Improving Catalyst Light-Off 157
5.3 Engine Operation with Retarded Ignition 160
5.4 Approaches for More Robust Operation with Ignition Retard 163
    5.4.1 Enhanced Charge Motion 163
    5.4.2 Dual Spark Ignition 167
5.5 Summary 168
5.6 References 169

Chapter 6 Secondary Air Injection for Improving Catalyst Light-Off Performance 173
By Fuquan (Frank) Zhao and Mark Borland

6.1 Introduction 173
6.2 Principle and System Layout of Secondary Air Injection 174

vii

6.3	Thermal Oxidation Versus Catalytic Oxidation		178
6.4	Role of Temperature and Mixing in Enhancing the Thermal Oxidation Process		181
6.5	Requirements for Engine Enrichment and Secondary Air Injection Quantity		191
6.6	Secondary Air Injection Control and Onboard Diagnostics		193
	6.6.1	Open-Loop Control	194
	6.6.2	Closed-Loop Control	194
	6.6.3	Sensors for Feedback Control	196
6.7	Other Application Considerations for Secondary Air Injection		197
	6.7.1	Application of Secondary Air Injection to Vee Engines	197
	6.7.2	Application of Secondary Air Injection to Turbocharged Engines	198
	6.7.3	Other Application Issues	199
6.8	Summary		200
6.9	References		201

Chapter 7 Effects of Fuel Properties and Fuel Reforming on Cold-Start Hydrocarbon Emissions and Catalyst Light-Off 205
By James A. Eng

7.1	Introduction		205
7.2	Gasoline Properties		207
	7.2.1	Composition	207
	7.2.2	Volatility	208
	7.2.3	Driveability	211
	7.2.4	Reformulated Gasoline	214
7.3	Fuel Effects on Hydrocarbon Emissions		215
7.4	Onboard Fuel Reformers		220
	7.4.1	Steam Reforming	222

	7.4.2	Partial Oxidation Reforming 223
	7.4.3	Autothermal Reforming... 225
	7.4.4	Cold-Start Performance Improvements..................... 226
	7.4.5	Improved Catalyst Light-Off... 227
	7.4.6	Cold-Starting Alcohol-Fueled Vehicles 228
7.5	Onboard Fuel Distillation... 229	
7.6	Summary 233	
7.7	References 234	

Chapter 8   Advanced Catalyst Design...241
*By Paul J. Andersen, Todd H. Ballinger, and
David S. Lafyatis*

8.1	Introduction 241
8.2	Advanced Three-Way Catalyst Concepts and Design.............. 247
8.3	Catalyst System Design Principles for Meeting Partial Zero Emissions Vehicle Emissions Standards......................... 255
8.4	Summary 263
8.5	References 264

Chapter 9   The Hydrocarbon Trap 269
By Kimiyoshi Nishizawa

9.1	Introduction 269	
9.2	Functions of the Hydrocarbon Trap 269	
	9.2.1	Hydrocarbon Trap System... 269
	9.2.2	Materials 270
9.3	Factors to Control Efficiency 272	
	9.3.1	Selecting and Developing Trapping Material............ 273
	9.3.2	Selecting and Developing the Catalyst Coating 275
	9.3.3	Selecting the Shape of the Catalyst Substrate 276
9.4	Measures for Improving System Efficiency............................. 277	

		9.4.1	Actively Controlled Systems	277

	9.4.2	Improved Passive Systems	279
9.5	Summary		281
9.6	References		281

Chapter 10 Three-Way Catalytic Converter System Modeling 283
By Tariq Shamim

10.1	Introduction		283
10.2	Modeling Approaches		284
	10.2.1	Single-Channel-Based One-Dimensional Modeling	284
	10.2.2	Multidimensional Modeling	286
10.3	Chemical Reaction Mechanisms		289
	10.3.1	Three-Step Chemical Reaction Mechanism	290
	10.3.2	Four-Step Chemical Reaction Mechanism	291
	10.3.3	Modified Four-Step Chemical Reaction Mechanism	292
	10.3.4	Five-Step Chemical Reaction Mechanism	292
	10.3.5	Six-Step Chemical Reaction Mechanism	294
	10.3.6	Thirteen-Step Chemical Reaction Mechanism	296
	10.3.7	Multistep Chemical Reaction Mechanism with Elementary Reaction	297
	10.3.8	Influence of Catalyst Deactivation on Reaction Mechanism	298
10.4	Oxygen Storage Mechanism		299
	10.4.1	Simple Single-Step Oxygen Storage Capacity Mechanism	299
	10.4.2	Detailed Nine-Step Oxygen Storage Capacity Mechanism	301
10.5	Heat and Mass Transfer Phenomena		302
10.6	Inlet Flow Distribution		304
	10.6.1	Flow Distribution Index	308
	10.6.2	Improvement of Flow Uniformity	309

	10.7	Modeling of Catalyst Dynamic Behavior	309
	10.8	Summary	313
	10.9	Mathematical Nomenclature	317
	10.10	References	320
	10.11	Appendix	328

Chapter 11 Evaporative Emissions Reduction 333
By Jenny Spravsow and Christopher Hadre

	11.1	Introduction	333
		11.1.1 Overview of Evaporative Emissions Standards	333
		11.1.2 Types of Evaporative Emissions	334
		11.1.3 Evaporative Emissions Test Procedures	334
	11.2	Types of Evaporative Emissions Control Systems	336
	11.3	Reducing Evaporative Emissions	337
		11.3.1 Seals	339
		11.3.2 Connectors	340
		11.3.3 Materials	340
		11.3.4 Canister and Engine Control Technology	342
	11.4	Summary	345
	11.5	References	345

Chapter 12 Onboard Diagnostics 347
By Glenn Zimlich, Kathleen Grant, and Timothy Gernant

	12.1	Introduction	347
		12.1.1 Emissions Failure Thresholds for Diagnostic Monitors	348
		12.1.2 Proper Identification of Diagnostic Failures	350
	12.2	Catalyst System Monitor	351
		12.2.1 Theory, Application, and Regulatory Implications	351
		12.2.2 Catalyst Monitor Operation	353
	12.3	Comprehensive Component Monitor	354

12.4	Cold-Start Emissions Reduction Control Strategy Monitor		354
12.5	Engine Misfire Monitor		355
	12.5.1	Theory, Application, and Regulatory Implications	355
	12.5.2	Misfire Monitor Operation	356
12.6	Evaporative System Monitor		356
	12.6.1	Theory, Application, and Regulatory Implications	356
	12.6.2	Initial Vacuum Decay-Based Method for Leak Detection	356
	12.6.3	Positive Pressure Decay Leak Detection	358
	12.6.4	Natural Vacuum-Based Leak Detection	358
12.7	Exhaust Gas Recirculation System Monitor		360
12.8	Fuel System Monitor		360
	12.8.1	Theory, Application, and Regulatory Implications	360
	12.8.2	Fuel System Monitor Operation	361
12.9	Oxygen Sensor Monitor		363
	12.9.1	Theory, Application, and Regulatory Implications	363
	12.9.2	Oxygen Sensor Monitor Operation	363
12.10	Secondary Air System Monitor		364
12.11	Variable Valve Timing/Control System Monitor		364
12.12	In-Use Performance Tracking		365
12.13	Summary		366
12.14	References		367

Chapter 13 Emissions Measurements 369
By Michael Akard

13.1	Introduction		369
13.2	Exhaust Emissions		370
13.3	Constant Volume Sampler		372
	13.3.1	Dilution Air	372
	13.3.2	Exhaust Dilution	375
	13.3.3	Dilution Ratio Optimization	376

	13.3.4	Bag Sampling 376
	13.3.5	Bag Materials.. 379
	13.3.6	Flow Rate Measurement and Control 382
	13.3.7	Sample Transfer from Bags to Analyzers 383
13.4	Bag Mini-Diluter 384
	13.4.1	Bag Mini-Diluter Dilution Gas 389
	13.4.2	Oxygen Interference 394
	13.4.3	Modeled Performance for Sampling Systems 396
	13.4.4	Differences Among Bag Mini-Diluters 396
	13.4.5	System Verification 397
13.5	Analyzer Accuracy 398
	13.5.1	Calibration Gas Requirements 399
	13.5.2	Utilities and System Components 400
13.6	Summary 401
13.7	References 402

Chapter 14 Near-Zero-Emission Gasoline-Powered Vehicle Systems 407
By Fuquan (Frank) Zhao

14.1	Introduction 407
14.2	System Requirements for a Near-Zero-Emissions Gasoline-Powered Vehicle 408
14.3	BMW Partial Zero Emissions Vehicle System 410
14.4	Ford Partial Zero Emissions Vehicle System 411
14.5	Honda Ultra-Clean Gasoline-Powered Vehicle System 413
14.6	Nissan Partial Zero Emissions Vehicle System 416
14.7	Toyota Partial Zero Emissions Vehicle System 424
14.8	Toyota Ultra-Clean Hybrid Vehicle System 428
14.9	Summary 432
14.10	References 432

Acronyms 435

Index 443

About the Contributing Authors 457

About the Editor 463

Preface

During the last several years, significant efforts have been directed toward the development of near-zero-emissions gasoline-powered vehicles in the automotive industry to meet increasingly stringent emissions legislation. Several vehicle manufacturers have already successfully launched these types of vehicles that are powered by conventional internal combustion engines and are certified in California to meet the most difficult emissions standard of partial zero emissions vehicle (PZEV) in the world. Accompanying this extensive development effort is the continuing generation of a large volume of technical information, with a growing need for systematic organization, description of fundamental processes, sorting of insights on technical issues, and identification of key trends and future research and development (R&D) directions. This book was planned to serve this essential purpose.

Subjects related to the development and certification of near-zero-emissions gasoline-powered vehicles are covered in this book. Experts in the field were invited to address a broad spectrum of key R&D issues in the rapidly progressing area of near-zero-emissions gasoline-powered vehicles. The book presents nearly all topics that a reader in the field may wish to comprehend, and future technology directions and R&D needs are outlined. The purpose is to provide the reader with a concise, brief introduction to the state of the art of technology developments in near-zero-emissions gasoline-powered vehicles. The material reflects the latest global technical initiatives that are being incorporated or investigated within the automotive and research communities. Engineers and researchers in this field will find the book invaluable in developing optimum systems for near-zero-emissions gasoline-powered vehicles and in understanding and interpreting test data. Corporate managers who are responsible for product decisions in this area will benefit significantly from the clear statements of the advantages and disadvantages of a spectrum of subsystems and from the discussions of current global best practices.

This book is organized into 14 chapters. The complex processes associated with engine startup and shutdown, mixture formation, unburned hydrocarbon (HC) emissions formation, and characterization of cold engine processes are presented in Chapters 1 through 4 in a clear and organized manner, and are placed within a framework of current state-of-the-art hardware and processes. Chapters 5 through 7 describe in detail the key technologies for improving cold-start catalyst light-off performance, such as spark retardation, secondary air injection, fuel properties, and fuel reforming. Chapter 8 describes state-of-the-art catalyst developments in meeting the most stringent emissions regulations such as PZEV. Chapter 9 is dedicated to the development of HC traps for controlling cold-start HC emissions. Chapter 10 introduces approaches developed to systematically model three-way catalytic converters, and Chapter 11 outlines the measures recently developed to control evaporative emissions. Issues related to the development of onboard diagnostics (OBD) for low emissions vehicles are addressed in Chapter 12, and Chapter 13 thoroughly reviews the important considerations in certifying near-zero-emissions gasoline-powered vehicles. Finally, Chapter 14 provides a comprehensive summary of the current best practices of near-zero-emissions gasoline-powered vehicle systems in the field, to give the reader a general understanding of all key elements in near-zero-emissions gasoline-powered vehicles.

The compiled information in this book would constitute an excellent base for someone who recently started working in the field, for someone who is already familiar with the area but wants opinions on certain issues from other experts in the field, or for someone who simply wants a general idea about recent technology developments in near-zero-emissions gasoline-powered vehicles. Anyone in industry or academia can expect this book to provide a rapid learning curve, and it will serve as a valuable desk reference on technologies related to the development of near-zero-emissions gasoline-powered vehicles.

In closing, I would like to extend my thanks to all the chapter authors for their professional contributions to the successful compilation of this book. I also would like to express my gratitude to Martha Swiss of SAE International for her excellent support in publishing this book. Finally, I would like to sincerely acknowledge my wife Dannie and my son James, for their understanding, patience, and encouragement throughout the lengthy process of preparing this book. They tolerated the time I spent writing and editing during weekends and holidays, which was time I could have spent with them. Without my family's unflagging support, it is difficult to conceive how this book project would have been carried out as an enjoyable and fruitful experience.

Fuquan (Frank) Zhao
April 25, 2006

CHAPTER 1

Transient Engine Startup and Shutdown Processes

Wai K. Cheng
Massachusetts Institute of Technology

1.1 Introduction

The startup and shutdown processes are important engine transients that substantially affect the overall vehicle emissions. In cold start, the catalyst is not effective until light-off. Because the conversion efficiency of a fully warmed-up modern catalyst is in the upper 90%, the startup emissions become a significant part of the trip total. Hydrocarbon (HC) emissions are of the most concern because of strict regulatory requirements. Technologies for reducing startup emissions include the development of fast light-off and HC-trapping catalysts, of engine operating strategies that would provide fast light-off, and of means to lower the engine-out emissions in the startup process [Nishizawa *et al.* (2001); Kidokoro *et al.* (2003)]. Precisely controlling the engine during the startup period is difficult because there are significant uncertainties about the state of the engine (e.g., the amount of leftover fuel inside the engine from the previous stop, the piston starting positions, the properties of the gasoline). Also, at low temperatures, the thermal environment for mixture preparation is not favorable. In addition, the engine should provide smooth operation during this period. Many efforts have been made to understand the startup process and to improve the emissions of that process.

The role of the engine shutdown process in emissions is largely due to its impact on the next startup. As will be discussed in Section 1.2.2 of this chapter, the exhaust system usually is large enough to trap all engine effluent in the few cycles during which the engine coasts to a stop. This trapped gas will be pumped out during the next startup and contributes substantially to emissions if the catalyst has cooled. For port fuel injected (PFI) engines, substantial fuel also remains in the cylinder and in the port. This fuel will impact the next startup behavior and emissions.

Many physical and chemical processes are involved in engine startup and shutdown: the mixture preparation process, the HC emissions mechanisms, and the catalyst behavior. Because these processes will be reviewed in both general as well as detailed manners in subsequent chapters of this book, only behaviors specific to the startup and shutdown transients are discussed in this chapter. Furthermore, most of the materials discussed in this chapter are relevant to PFI engines. The startup and shutdown processes for gasoline direct injection (GDI) engines are discussed in Section 1.4.

1.2 The Engine Shutdown Process for Port Fuel Injected Engines

1.2.1 General Behavior During the Shutdown Process

The engine behavior during shutdown depends on the engine configuration, the operating conditions, and the details of the process. Figure 1.1 illustrates a typical shutdown of a fully warmed-up four-cylinder engine at idle [Klein and Cheng (2002)]. The power shutoff to the engine control unit (ECU) in the key-off process (which was not synchronous with the engine operation) is indicated by the baseline voltage drop from 12 to 0V in the injector coil and in the ignition coil signals (at approximately 2.2 seconds on the time axis). The power-off occurred immediately after ignition in cylinder 3. For cylinders 2 and 4, fuel was injected, but there was no ignition. A major part of the unburned mixture in these cylinders thus escaped to the exhaust system.

For unsynchronized power-off, depending on the phasing between the power-off and the crank timing, there could be two or three cylinders that will have fuel injection but no ignition in a four-cylinder engine. For a six-cylinder engine, there could be three or four such cylinder events. Thus, a substantial amount of unburned mixture is delivered to the exhaust system during the shutdown process.

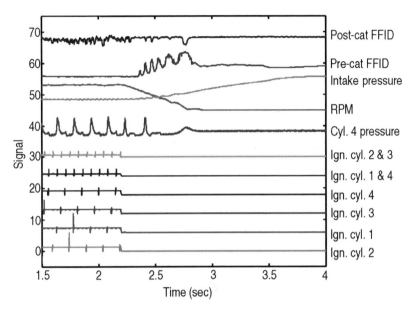

Figure 1.1 Events in a normal four-cylinder 2.4-L engine shutdown process from idle. Power-off is indicated by the baseline voltage drop from 12 to 0V in the injector and ignition signals. Last fired cycle: cylinder 3; in cylinders 2 and 4, fuel was injected, but there was no ignition [Klein and Cheng (2002)].

After power-off, the engine took approximately 0.6 second (four revolutions) to coast from the idle speed (800 rpm) to rest. The intake pressure, however, took longer to stabilize because of the filling process of the manifold; it did not recover to atmospheric pressure until approximately 1.4 seconds from power-off. Thus, during the engine coast-down period, the intake pressure is sub-atmospheric.

The engine gas flow in the coast-down process is complex. Because the intake manifold pressure is still quite low (0.3 to 0.4 bar) during this period, there is substantial backflow from the exhaust to the intake in the valve overlap period. In addition, because the cylinder pressure at the exhaust valve opening (EVO) of a throttled nonfiring cylinder is below atmospheric, there is a backflow from the exhaust to the cylinder for a brief period after EVO before the normal exhaust displacement flow. These flow processes resulted in significant cycle-to-cycle and cylinder-to-cylinder mixing of the exhaust gas.

1.2.2 Impact of the Shutdown Process on Hydrocarbon Emissions

The HC mole fractions at the entrance and exit of the catalyst were measured by Klein and Cheng (2002) using fast-response flame ionization detectors (FFIDs) and are shown in Figure 1.1 as the pre-catalyst (pre-cat) and post-catalyst (post-cat) FFID signals. The peaks in the pre-cat FFID signal show the pumping out of the unburned HCs during the four revolutions in the shutdown process. There were eight cylinder events: two firing events that had comparatively little HC emissions, and six non-firing events that contributed to the peaks in the trace. The gradual buildup of the HC level in the pre-cat signal was a result of the incremental filling of the exhaust volume by the HCs coming out of the cylinders and the cycle-to-cycle/cylinder-to-cylinder mixing processes. The sources of HCs included both the unburned mixture from the cycles with fuel injection but no spark ignition, and the subsequent delivery of the residual fuel in the intake port and cylinder after the injectors were switched off. The peak values of HCs were ~18 kppm C_1, which were significantly lower than those of a stoichiometric mixture (125 kppm C_1, excluding residual gas dilution) but much higher than the normal engine-out values (a few thousand ppm C_1). Thus, the exhaust system was filled with a diluted mixture of unburned fuel-air and burned gas in the shutdown process.

The exhaust HCs in the engine coast-down process were not detected in the post-cat FFID signal (top trace of Figure 1.1; the "noise" on the trace before the engine stop was due to the mechanical vibration of the engine). This observation resulted from the following. Because of the low intake pressure, the total exhaust volume flow (approximately 3.6 L) of the three non-firing revolutions (of the four) in the coast-down process was less than the catalyst volume (5 L); hence, all the exhaust gas from the process was trapped. Therefore, the shutdown process alone does not contribute to the tailpipe emissions. The trapped HCs, however, will contribute to the emissions of the next startup.

Klein and Cheng (2002) measured the mass of HCs trapped in the catalyst in the coast-down process to be approximately equal to the fuel mass of one injection event at idle. A simple and effective way to lower this value is to shut off the fuel first while the ignition remains on; then there is no prepared charge without ignition. Klein and Cheng (2002) reported a factor of 2 reduction in the trapped HC mass by this method. Thus, in a normal key-off shutdown, the injection-without-ignition events and the residual fuel in the port each contributes approximately half of the trapped HCs in the catalyst during the engine coast-down process.

1.3 The Engine Startup Process for Port Fuel Injected Engines

1.3.1 Initial Conditions for Engine Startup

The engine startup behavior depends on the state of the engine from the last shutdown and the cranking/startup strategy. Important considerations are the catalyst temperature, which determines whether the catalyst is effective in the startup, and the coolant temperature (essentially the metal temperature of the engine), which affects the fuel evaporation process.

Figure 1.2 shows the catalyst and coolant temperatures of an engine after shutdown. These data were taken from an engine on a dynamometer stand and at an ambient temperature of 25°C. The cool-down times to ambient (times for temperature difference to decrease by $\frac{1}{e}$) were 40 minutes for the catalyst and 60 minutes for the engine coolant. The actual cool-down time for an enclosed engine in a vehicle will be somewhat slower. Because the threshold temperature for effective catalyst operation is approximately 300°C, the catalyst remains

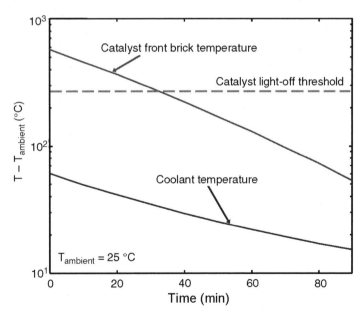

Figure 1.2 Catalyst and coolant temperatures after engine shutoff for an engine on a dynamometer stand [data from Klein and Cheng (2002)].

effective for half an hour or so after shutdown. Thus, if the engine is shut off for a short period of time (e.g., in a hybrid vehicle), the catalyst will still be effective when the engine is restarted.

The camshaft position when the engine stops determines the initial condition of cranking. The relative phasing of the valve timing and the injection events affects the mixture preparation process in startup. When an engine coasts to a stop, the final piston position is determined by the gas load, inertia load, and friction [Henein *et al.* (1995); Castaing *et al.* (2000)]. When the gas load is the dominant term, simulation results show that the engine stops with one of the pistons at 88° crank angle (measured from bottom dead center [BDC] compression) for a four-cylinder engine and 118° for a six-cylinder engine [Castaing *et al.* (2000)]. These values are consistent with observations. Thus, there are four or six different normal stopping cam positions for a four- or six-cylinder engine, respectively.

1.3.2 General Behavior During the Startup Process

Henein *et al.* (1995) and Santoso and Cheng (2002) recorded the cycle-to-cycle behavior of the engine startup process. Figure 1.3 shows the general behavior for a 2.4-L four-cylinder engine calibrated as a 1998 Model Year low emissions vehicle (LEV). The firing order was 1-3-4-2. The crank position before starting was at the mid-stroke of compression of cylinder 2. The starting and ambient temperatures were 25°C. (Note that calibration strategies are changing rapidly; thus, the detailed behavior shown in Figure 1.3 may not be representative for future engines. Nevertheless, the basic features will still be pertinent.)

Referring to Figure 1.3, the startup behavior may be described as follows:

1. Key-on cranking started at (a). The starting motor accelerated the crank speed to 250 rpm quickly (in tens of milliseconds). The ECU was powered up at (b), which is ~0.2 second after key-on.

2. The first injection happened at (c). Fuel was supplied by all four injectors simultaneously at ~70 mg/cylinder. For comparison, the fuel vapor mass required to produce an in-cylinder stoichiometric mixture at the starting condition was 45 mg/cylinder. The first injection happened with the intake valve closed for cylinders 1 through 3, and with the intake valve open for cylinder 4.

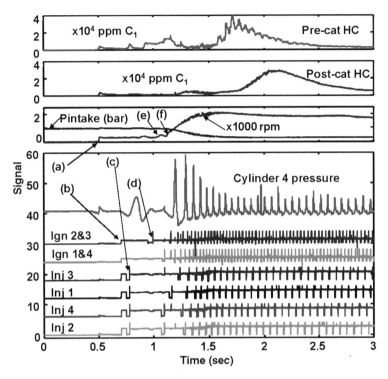

Figure 1.3 Hydrocarbon emissions and engine events in the startup process of a four-cylinder spark ignition (SI) engine. The firing order was 1-3-4-2. The initial cam position was with cylinder 2 in mid-stroke compression [Santoso and Cheng (2002)].

3. Ignition was first applied to cylinder 2 at (d), and then to cylinder 1. The firing torque resulted in accelerations at (e) and (f). Note that in this period, the engine speeds were quite low, and there was substantial change in speed even within one cycle. Therefore, the conventional set of values of spark advance for optimum combustion phasing would not apply. This point will be discussed later.

4. Subsequent firing accelerated the engine to 2200 rpm in a period of 0.5 second. At the same time, the intake manifold pressure dropped from atmospheric to 0.3 bar. Subsequently, the engine speed decreased to the idle speed of 800 rpm (not shown in Figure 1.3). The speed surge in the startup process is sometimes referred to as the speed flare.

5. At the clock time of 1.1 second, the ECU began to inject fuel synchronously with the cam events. As a result, cylinder 4 received two injection pulses before the first ignition. Note that the first injection started at roughly the middle of the cylinder 4 intake valve opening (IVO) period; thus, part of this fuel entered the cylinder, never experienced ignition, and was exhausted as HC emissions. Such incidents are common when the first injection pulses are timed asynchronously.

6. Because the catalyst had not reached light-off temperature, the HC mole fraction at the exit of the catalyst was the same as that at the entrance, with the appropriate delay and dispersion.

7. The HC flow into the catalyst consisted of two peaks. Klein and Cheng (2002) had identified that the first peak was due to the flushing of the residual HCs from the previous engine shutdown. These HCs comprised both the HCs trapped in the exhaust system and those from the residual fuel in the port and cylinder. The second peak was a result of the current crank-start procedure.

8. The cumulative HC mass flow associated with each of the two peaks was 16 and 55 mg, respectively.

9. Klein and Cheng (2002) switched off the fuel with the ignition remaining on in the previous shutdown process so that there was no misfired cycle. With this procedure, the cumulative HC mass in the first peak decreased from 16 to 11 mg. The latter value was the contribution of the residual fuel in the engine (in the port and inside the cylinder).

This discussion leads to a division of the engine startup process into the following events (Figure 1.4):

1. **First round of firing of all cylinders**—In these firings, there is little residual gas in the cylinders, and the intake pressure is close to atmospheric. As such, the torque output is high, and the engine accelerates quickly.

2. **The speed flare**—The extent of the speed run-up is determined by the process of drawing down intake pressure to the idle value by the engine. At first, the net torque (gross indicated torque minus friction torque) is positive, and the crankshaft accelerates. As intake pressure decreases, the torque decreases. At the peak speed point, the net torque is zero; beyond that, the net torque becomes negative, and the crankshaft decelerates until idle speed is reached. For good driveability, the engine speed should

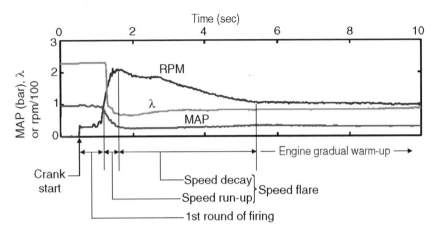

Figure 1.4 Events in a startup engine transient.

increase in every cycle in the run-up period [Samenfink *et al.* (2003)]. Care also must be taken to limit the intake manifold absolute pressure (MAP) undershoot in the speed flare. The low MAP results in high residual gas fractions that could cause engine stability problems.

3. **The gradual warm-up of the engine**—The speed flare usually takes place in a few seconds. After that, the engine idles at more or less constant speed and intake pressure until the driver demands the first acceleration. The thermal state of the engine, however, is far from equilibrium (see Section 1.3.3). Strategies such as fast idle speed [Acke and Marsh (2001)] and retarded spark timing [Russ *et al.* (1999a and 1999b)] are usually used at startup to create a high-enthalpy flow to the catalyst to facilitate catalyst light-off.

Note that the engine behavior during the speed flare period is highly dynamic. The air/fuel (A/F) ratio depends not only on the fuel vapor delivered to the charge, but also on the amount of air inducted in the cycle. The latter, however, depends on the MAP, which is affected by the engine acceleration, which, in turn, is affected by the A/F ratio and the amount of inducted air. Thus, the engine behavior is governed by a complex feedback system with a high level of interdependency between the fuel vapor preparation and the airflow. Furthermore, the engine must be calibrated with sufficient margin to provide smooth operation with the substantial variations of fuel properties in the market (e.g., regional, seasonal, and brand variations). Thus, substantial empiricism is employed in the calibration procedure.

To provide good cyclic stability, engines were usually calibrated to achieve a slightly rich A/F ratio in the warm-up period (Figure 1.4). However, Kidokoro *et al.* (2003) reported substantial advantage in operating lean in this period because the catalyst activation temperature threshold was lower in an oxidizing environment. By changing the A/F ratio from 14.5 to 15.5, there was a 50% reduction in the supplied exhaust gas enthalpy required to reach the 50% HC conversion point.

1.3.3 Mixture Preparation in the Startup Transient

During engine cold start, the thermal environment is not favorable to mixture preparation. (The factors affecting mixture preparation will be discussed in Chapter 2.) For conventional injectors, the drop size is of the order of 150 μm, so that the drag force is small compared to the inertia in the period of flight. Thus, most of the fuel first lands on surfaces (i.e., port wall and valves). Fuel vapor is formed by the following processes [Shin *et al.* (1995)]:

- Direct evaporation from the injected droplets in flight

- Evaporation from the wall film in the intake port area

- Evaporation from the fuel drops strip-atomized by the reversed flow when the intake valve opens, and by the flow through the valve induced by the piston motion

- Evaporation from the cylinder and piston surfaces

As such, the factors that influence the mixture preparation in the startup transient are as follows:

1. **Port wall/intake valve temperatures**—The port wall temperature is essentially the coolant water temperature, which typically does not change appreciably in the first 10 or so seconds from startup (Figure 1.5). The temperature of the intake valve is determined by the thermal inertia of the valve, and the balance between the energy addition from the combustion chamber, the heat loss through the valve seat contact, and the cooling due to the vaporization of the fuel [Cowart and Cheng, 1999]. (The heat exchanges to the gas in both the backflow and forward flows are relatively less important; the heat loss through the valve stem is relatively

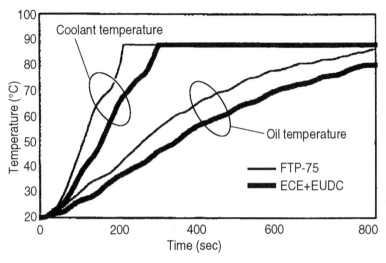

Figure 1.5 Coolant and oil temperature history in the U.S. Federal Test Procedure (FTP-75) and the European ECE+EUDC drive cycles; 1.8-L engine in a 1300-kg vehicle [Shayler et al. (1997)].

unimportant.) The temperature rise in the first few seconds of cranking depends much on the amount of liquid fuel (which cools the valve) that lands on the valve; the rate of rise is typically of the order of 10°C/sec.

2. **Fuel properties**—A more volatile fuel will facilitate mixture preparation in the startup transient. However, the effects have not been quantified in a satisfactory manner. Part of the difficulty is due to the substantial differences in the engine-specific responses to the fuel characteristics and to the lack of a direct relationship between fuel properties and engine cold-start calibration. The fuel effects are traditionally measured in terms of total weighted demerits (TWD) in a cold driveability test procedure. The TWD is a subjective score based on the driver's perception of the vehicle malfunctions (e.g., stalling, hesitation, stumble, rough idle, surge, backfire) that are associated with mixture preparation problems. The most accessible fuel properties are the distillation curve and the Reid vapor pressure (RVP). The information in the distillation curve is usually summarized by the 10, 50, and 90% distillation temperatures (T_{10}, T_{50}, and T_{90}; values are in degrees Fahrenheit [°F]). Note that normally there is a strong T_{10} and RVP correlation. The overall evaporative behavior of the fuel is further

characterized by a single number, the driveability index (DI), which is defined as

$$DI = (1.5 \times T_{10}) + (3.0 \times T_{50}) + (1.0 \times T_{90})$$

The range of DI varies from 1100 to 1300; for example, the California Phase II calibration gasoline has a DI of ~1115; high-DI-calibration fuel has a DI of ~1275. Fuels with lower DI values are more volatile. For a history of the development of the DI measure, see Barker *et al.* (1988). The DI definition is based on TWD and distillation temperature correlations for vehicles in the 1970s and early 1980s. There was substantial spread in the vehicle-to-vehicle responses, and the correlation was sensitive to ambient temperature. In the 1989 Coordinating Research Council (CRC) Driveability Program, it was shown that T_{10} and RVP had almost no effect on cold-start TWD for PFI vehicles; the major factor was T_{50} [Graham *et al.* (1991)]. In spite of all the uncertainties, the DI is still used as the gauge for the overall fuel evaporative properties. However, no direct relationship has been established between DI (or the distillation temperatures) and the cold-start calibration requirement of an engine.

3. **Fuel injection characteristics**—The important parameters are droplet size, fuel targeting, and injection timing. The injected fuel droplets comprise a distribution of diameters (d). Typically, the size distribution is summarized in terms of a single parameter: the Sauter mean diameter (SMD), which is the ratio of the average of d^3 to that of d^2. Hence, the SMD is proportional to the ratio of the average volume to average surface area of the drops; it affects the overall evaporation time of the fuel spray. Typical low-pressure injectors used in PFI engines employ a pressure differential of 4 bar, which produces a spray with an SMD of the order of 150 μm.

For droplet diameters larger than ~100 μm, most of the liquid fuel lands on the metal surfaces, and the mixture preparation process is not sensitive to the drop size. For a smaller drop size, Kidokoro *et al.* (2003) reported improvement of mixture preparation by reducing the SMD from 85 to 75 μm. To have the droplets follow the airflow and avoid significant wall wetting, the drop size must be of the order of 20 μm or less. Injectors with heated tips have been developed to create flash boiling in the fuel spray by which a drop in SMD of 20 to 50 μm was obtained [Zimmerman *et al.* (1999); Samenfink *et al.* (2003)].

Fuel normally is targeted at the back of the intake valve, which is the hottest surface and is the surface that warms the fastest in the intake port. Furthermore, fuel film collected there is effectively strip-atomized by the reversed blowdown flow when the intake valve opens and by the forward flow induced by the piston. These processes facilitate evaporation.

For timing, the usual practice is to inject during the closed-valve period. Higher HC emissions were observed for open-valve injection, especially at low temperatures, because of the significant direct flight of liquid fuel from the spray into the cylinder [Alkidas and Drews (1996)]. At low temperatures, because of the large amount of injected fuel required to produce a combustible charge in the first round of firing, spark-plug wetting often happens for open-valve injection [Castaing *et al.* (2000)].

Meyer *et al.* (1998) measured the liquid droplet flow into the cylinder for open- and closed-valve injections by integrating the droplet flux (measured by a phase Doppler particle anemometer) across the valve curtain. Figure 1.6 shows the normalized volume flow for the first 90 seconds of warm-up. Initially, the closed-valve injection produced a lower volume

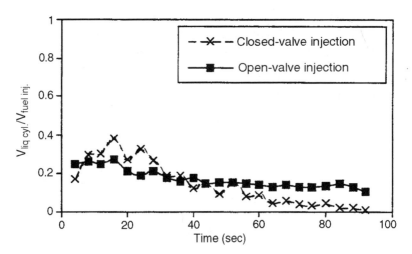

Figure 1.6 Volume of liquid entry (as a fraction of the total injected volume) into the cylinder. Engine was motored at 1100 rpm, 0.5-bar MAP before fuel injection and firing. Fuel pulse width was fixed at a value that would provide $\Phi = 0.9$ at steady state except for the first pulse, which was six times longer [Meyer et al. (1998)].

of liquid flow into the cylinder. Then for the period of approximately 10 to 40 seconds after firing, the reverse was true. After 40 seconds, the closed-valve injection again produced a lower value. This behavior was attributed to the fact that for the open-valve injection, a substantial part of the droplet flow into the cylinder was directly from the injector spray; thus, the liquid volume flow into the cylinder did not change appreciably as the engine warmed up. For closed-valve injection, the fuel film in the port/back-of-valve areas took multiple injections to build up. Initially, the film was thin, and the volume flow of the droplets strip-atomized into the cylinder was lower than that in open-valve injection. As a thick fuel film built up, this droplet volume flow increased and became larger than that in the open-valve injection. As the engine warmed up, surface evaporation significantly reduced the film thickness. Then the stripped droplet flow decreased, and the volume flow was less than that in open-valve injection.

Although there may be a period in the warm-up process in which open-valve injection produces a lower droplet volume flow into the cylinder than closed-valve injection, the HC emissions using the latter could still be lower. That is because the in-cylinder distribution of liquid film is different in the two cases. The droplets from open-valve injection mostly impinged on the cylinder wall opposite to the intake valve; those from closed-valve injection mostly landed on the piston [Stanglmaier *et al.* (1997)]. Because of the colder wall temperature, the liner and head fuel films survive the combustion process, evaporate in the exhaust stroke, and contribute to exhaust HCs [Takeda *et al.* (1995); Shin *et al.* (1995)]. Because of the higher temperature, the fuel on the piston tends to evaporate and burn and has less impact on emissions. Hence, the HC emissions are lower with closed-valve injection.

4. **Port flow characteristics**—Strip-atomization by the reversed blowdown of the hot burned gas is advantageous to mixture preparation. This process can be enhanced by increasing the valve overlap period. The high residual fraction resulting from a large overlap is detrimental to engine idle stability, but the tradeoff may be achieved by a variable valve timing (VVT) system. Using a cam phase shifting VVT system, Kidokoro *et al.* (2003) reported a 20% reduction in FTP nonmethane hydrocarbons (NMHCs) with a 6° valve overlap at starting. Increasing the local port flow velocity judicially also can enhance fuel film evaporation and strip atomization. The strategy usually is implemented by using a flap-type air control valve in the intake

port. The valve restricts the flow to a small opening to produce a directed flow of high velocity. The scheme provides not only a better mixture preparation, but also an increase of cylinder charge motion that enhances combustion stability [Kidokoro et al. (2003); Nishizawa et al. (2000)].

5. **Intake valve timing**—The strategy of opening the intake valve early to increase overlap backflow has already been discussed here. Another strategy is to have a late intake valve opening so that the cylinder pressure is substantially below atmospheric when the valve opens. As such, (a) there is no reversed blowdown, but the initial forward flow driven by the manifold/cylinder pressure difference is much stronger and thereby helps strip-atomization; (b) the initial low cylinder pressure enhances evaporation; and (c) the temperature of the in-cylinder charge is higher due to the work done by the manifold/cylinder pressure difference on the gas flow during pressure equilibration.

Roberts and Stanglmaier (1999) studied the IVO timing effect by shifting the timing chain in steps of one tooth on the cam sprocket in an engine controlled by a stock ECU, while the cylinder pressure and exhaust HCs were measured in a normal cold-start period. As such, it was difficult to study the confounding effects of fuel metering, mixture preparation, and volumetric efficiency on the indicated mean effective pressure (IMEP) and HC emissions. The overall results, however, were encouraging, in that retarding the IVO by 19° CAD (crank angle degrees) (one tooth), the coefficient of variation (COV) of the IMEP in the first 200 cycles was reduced from the original engine configuration while the HC emissions were unchanged.

Lang and Cheng (2004) studied in detail the effects of IVO timing on the first cycle fuel delivery during cranking. Their engine simulation results showed that the high velocity and shear rate at the valve curtain occurred only for a brief period (a few milliseconds) during which the lift is small (fraction of a millimeter). Thus, the fuel transport to the cylinder would be limited by the upstream feed, and only a small amount of liquid fuel located at the valve rim and seat is benefited from the high-shear-rate atomization. In addition, in the first cycle, because there is no residual burned gas in the cylinder, the expansion of the charge from exhaust valve closing (EVC) to IVO can significantly cool the charge so that the first fuel entering the cylinder is exposed to cold air, which inhibits evaporation. (The charge temperature at IVO dropped by 50°C when the IVO timing was changed from top dead center [TDC] to 30° after TDC [ATDC].) Furthermore, with fixed cam duration, a late IVO implies a late intake valve closing (IVC). Thus, the compression time and the effective compression ratio are reduced, thereby reducing the in-cylinder evaporation of the fuel. These effects may negate the previously stated benefits of delaying the IVO timing.

The measured in-cylinder fuel equivalence ratio Φ for the first cycle of cranking as a function of the IVO timing and the amount of injected fuel is shown in Figure 1.7 [Lang and Cheng (2004)]. The IVO timing for the stock engine was at $-1°$ CAD. (All timings were referenced from TDC–exhaust.) Values for Φ increased when IVO was retarded from TDC–exhaust. The trend, however, continued when IVO was advanced from TDC–exhaust (i.e., Φ decreased.) This behavior contradicts the explanation that the IVO timing effect is due to the forward flow when the valve opens, because then such flow no longer exists. (The EVC for this engine was at $8°$ after TDC; there was no backflow because the MAP was atmospheric.) A better explanation that covers both the retarded and advanced IVO timing is shown in Figure 1.8 [Kapus and Poetscher (2000);

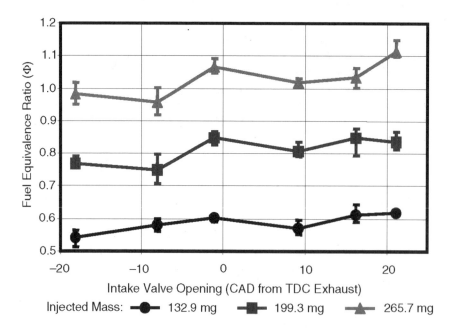

Figure 1.7 In-cylinder fuel equivalence ratio for the first cycle of cranking as a function of IVO timing for different injected fuel mass [Lang and Cheng (2004)].

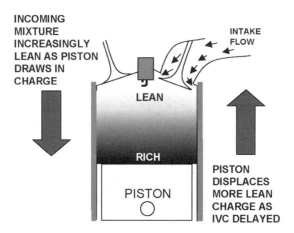

Figure 1.8 Schematic illustrating charge stratification and enrichment by displacing the leaner charge in the upper part of the cylinder with delayed IVC [Lang and Cheng (2004)].

Lang and Cheng (2004)]. During the intake stroke, the in-cylinder charge is likely to be stratified with a richer mixture at the bottom. (This fact was confirmed by in-cylinder FFID measurements.) In Figure 1.7, the IVO timing changed from -18 to $+21°$, which corresponded to an IVC change from 214 to $253°$. For the timing change, then, relatively more air than fuel would be displaced because of the mixture stratification as the piston came up before IVC. Hence, a richer mixture resulted.

Preparing a combustible charge in the first round of firing is especially difficult. The intake port surfaces are cold, there is no reversed blowdown of hot burned gas, and the forward flow is not strong because of the low engine speed. As such, a substantial amount of fuel is injected for these cycles to achieve a combustible mixture. Figure 1.9 shows the resulting first cycle in-cylinder fuel equivalence ratio and the delivery ratio (defined as the fraction of the injected fuel that comprises the combustible charge). The most sensitive parameter is the engine coolant temperature. The fuel delivery to the charge improves with cranking speed, but the effect is modest. An important observation is that the delivery efficiency decreases with the increase of injected fuel mass. This is because the injection footprint does not increase appreciably with the injected mass, so that a thick fuel film is formed at the vicinity of the intake valve seat. In the first cycle, there

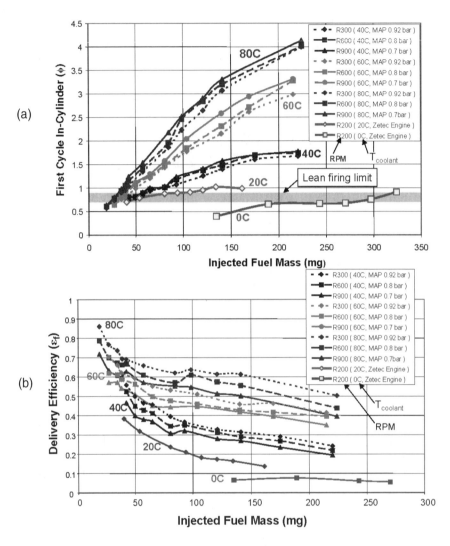

Figure 1.9 (a) In-cylinder fuel equivalence ratio and (b) delivery efficiency of first injection pulse as a function of coolant temperature and revolutions per minute (rpm). The intake pressures were adjusted according to the cranking speed [Santoso and Cheng (2002)].

is no reversed blowdown to strip-atomize and spread out the liquid fuel. When the forward flow commences, because of the low gas velocity at the valve curtain (due to the low engine speed), atomization is not efficient, and a large part of the liquid fuel enters the cylinder as film flow [Shin et al. (1995)]. The situation is especially severe at low temperatures.

Takeda et al. (1995) did an accounting of the injected fuel mass by trapping the HCs in the port and cylinder using a set of hydraulically controlled valves. The minimum fuel injected for stable combustion in the first three cycles in Figure 1.10(a) was interpreted in Figure 1.10(b) as contributing to part of the "required" amount of fuel, with the remaining part coming from

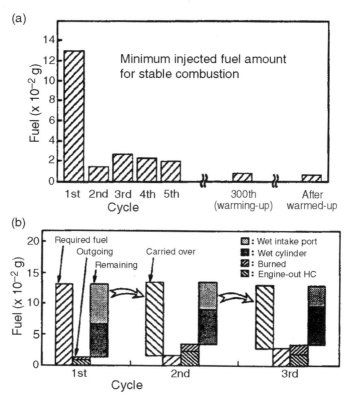

Figure 1.10 (a) Minimum fuel calibration for stable combustion, cycle-by-cycle value. (b) Fuel accounting for the first three cycles; the required amount for the cycle consisted of the injected amount and the carried-over amount that comprised the remaining liquid fuel in the intake port and in the cylinder from the previous cycle [Takeda et al. (1995)].

the wetted port and cylinder from the previous cycle. Note that in the first three cycles, the mass of fuel burned is a small fraction of the "required" amount, and a large amount of fuel is carried over from the previous cycle in both the intake port and the cylinder wall. A significant part of the latter ended up in the lubrication oil.

1.3.4 Combustion in the Startup Transient

Referring to Figure 1.4, after the first few firings, the engine speed goes beyond 700 rpm or so. Then, if a satisfactorily flammable mixture is prepared, the combustion characteristic in each cycle is not materially different from that under regular engine operation. In the first few firings, especially for the first cycle, however, the engine speed is low, and there is significant speed change during the cycle. As such, the usual notions of proper combustion phasing (e.g., minimum spark advance for best torque [MBT] timing would have peak pressure at ~15–18° ATDC or 50% burned point at 7° ATDC) are not valid. The slow engine speed means that the combustion duration is shorter in terms of crank angles; thus, the combustion process may be overly advanced. This fact is illustrated in Figure 1.11, which compares the heat release analysis of a first round cycle at a higher speed (Case A) to one at a lower speed (Case C) at the same ignition timing [Castaing et al. (2000)]. The fuel equivalence ratios of the trapped charge in both cases were approximately the same, so that the combustion duration should be comparable in terms of time. In terms of crank angles, however, combustion in Case C is much faster, as evidenced by the much steeper pressure rise. The effectively advanced combustion phasing leads to a lower gross indicated mean effective pressure (GIMEP) mainly because of the higher heat loss associated with the advanced timing.

1.3.5 Hydrocarbon Emissions in the Startup Transient

The HC emissions formation mechanisms during cold start will be assessed in Chapter 3. Referring to Figure 1.4, if the engine is calibrated properly (i.e., with no misfired or partial burn cycles), the engine-out HC emissions mechanisms in the flare speed decay and engine gradual warm-up periods will be similar to those at steady state, although the engine temperature will be lower. Then the crevice volume will be larger, the solubility of the gasoline vapor in oil will be higher, and the extent of in-cylinder and exhaust HC oxidation will be reduced. All these factors contribute to a higher level of engine-out HC emissions (see Chapter 3).

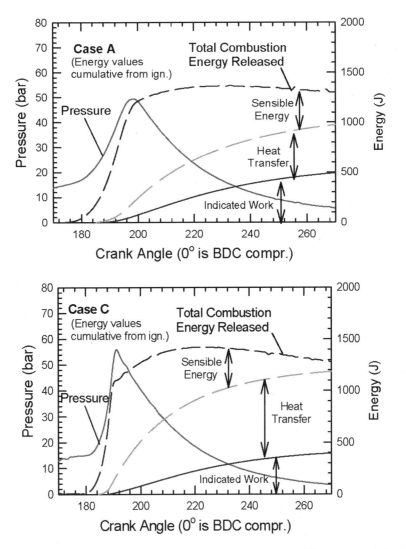

Figure 1.11 Heat release analysis of cycles in the first round of firing. Case A (top): The engine speed accelerated from 680 rpm at ignition to 790 rpm at EVO. Case C (bottom): The engine speed accelerated from 220 to 550 rpm. The in-cylinder equivalence ratio of charge was approximately the same for both cases, with the spark at 10° BTDC for both [Castaing et al. (2000)].

There is, however, a significant amount of HC emissions associated with the cranking and run-up processes (see top trace of Figure 1.3). The first peak in that figure corresponds to the pumping out of the residual fuel from the previous shutdown. The second peak is due to the fueling of the current startup process. Because of the difficulty in evaporating fuel at the cold-start temperature, a large amount of fuel must be injected to achieve a sufficiently combustible mixture (see Figure 1.7). A significant amount of liquid fuel enters the combustion chamber (see Figure 1.8). Part of the fuel that does not burn will escape the combustion chamber as HC emissions. (The other part will go into the oil sump.) To put values in perspective, for the particular data of Figure 1.3, the total HC emissions value of 71 mg associated with the first (16 mg) and second (55 mg) peaks is a large fraction of the allotted 110 mg of total FTP HC emissions for the super ultra-low emissions vehicles (SULEV II) regulation. Thus, there is substantial interest in reducing the HC emissions in the cranking and run-up process. The strategies to achieve such are discussed in Chapters 2, 5 through 9, and 14.

1.4 The Engine Startup Process for Direct Injection Spark Ignition Engines

In PFI engines, a large part of the fuel goes into the liquid film on the port walls before it is transported into the cylinder. This indirect process causes fuel delay, and special compensations must be made in the engine calibration to obtain good transient and cold-start engine behavior. The port fuel film is eliminated in direct injection spark ignition (DISI) engines. Thus, such engines could have significantly better cold-start performance. A comprehensive review of DISI engines was provided by Zhao *et al.* (1997).

Potential improvement in cold-start performance of the DISI over PFI engines is shown in Figure 1.12 [Takagi *et al.* (1998)]. Calibrating for the same engine stability, substantially less metered enrichment was required, and significantly less engine-out HC emissions were obtained. However, note that the results were obtained with an externally driven high-pressure fuel pump. For satisfactory operation of direct injection, the fuel pressure should be 5–12 MPa so that droplets of the order of 20 μm can be produced. In practice, such high pressure cannot be provided by the pump in the cranking process. Nominal fuel pressure at startup is of the order of 0.4 MPa. As such, large droplets (~100 μm) are produced, and there is significant wall wetting on the combustion chamber surfaces.

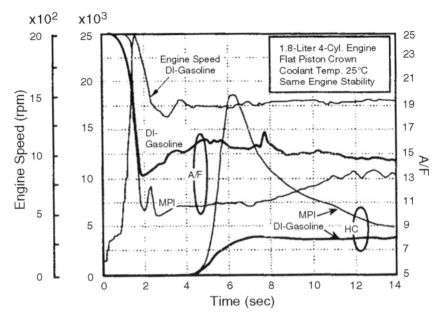

Figure 1.12 Reduction of cold-start HCs in a DISI engine, compared to a PFI engine (Takagi et al. (1998)). Note that the result was obtained with an externally driven high-pressure fuel pump so that there was no fuel pressure lag in the startup process.

Koga et al. (2001) did a careful accounting of the fuel pathways for the DISI engine in cold start. The method was to trap the cylinder content (using hydraulically activated valves) to determine the remaining fuel in the cylinder after the exhaust stroke. The study compared the operations of multipoint injection (MPI), normal DISI (with a 0.4 MPa fuel pressure at start), and high-starting-fuel-pressure DISI (with a minimum of 5 MPa fuel pressure at start) engines. Compared to the MPI engine, less enrichment was required for achieving the same engine stability in startup for the DISI engines (for both the normal one and the high-starting-fuel-pressure one). However, the direct injection (DI) configurations produced more in-cylinder wall wetting and, as a result, more engine-out HCs. Figure 1.13 shows the fuel accounting for the first cycle.

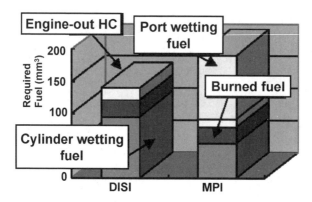

Figure 1.13 First cycle fuel pathway for normal DISI and MPI engines. The required fuel was the minimum amount needed for stable combustion [Koga et al. (2001)].

To reduce in-cylinder wetting and to lower engine-out emissions, Koga *et al.* (2001) used the following strategies for direct injection:

- **Late intake valve opening**—To increase the cylinder vacuum level at the time of injection, to promote flash boiling. Also, the manifold/cylinder pressure difference enhances the intake flow and promotes the in-cylinder mixture preparation process.

- **Late injection in the compression stroke**—To reduce the amount of cylinder wall wetting.

- **Heated fuel injection**—To promote vaporization and flash boiling.

Figure 1.14 shows the effectiveness of these strategies on engine-out HC emissions. There were incremental reductions in emissions with combinations of the strategies. Note that the conventional DISI (i.e., with the startup injection pressure of 0.4 MPa) had significantly higher HC emissions than the MPI engine. Even with a startup injection pressure of 5 MPa, the DISI engine was inferior. (That this result is contradictory to the results depicted in Figure 1.9 is probably because of the improvement of the PFI equipment in the few years between when the data were taken.) A better DISI engine cold-start HC emissions performance than the MPI engine was achieved only when a heated injector was used. Thus, because of the significant in-cylinder wall wetting produced

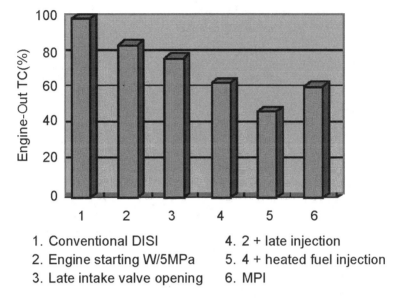

Figure 1.14 Comparison of engine-out HC emissions for different fuel injection strategies. The y-axis is the cumulative HC emissions for the first 10 seconds in the cold start, normalized by strategy 1 (Koga et al. (2001)].

by direct injection, especially at low startup fuel pressures, it is not clear at this point that DISI engines offer an HC emissions advantage over MPI engines in the startup transient.

The behavior of the engine shutdown process with DISI engines has not been reported in the literature. If the engine is shut down by turning off the injection first with the ignition still running, there will be no misfired cycle. Then, without the presence of the port fuel film, it is expected that there will be little residual fuel in the cylinder. Thus, there should be little contribution of the shutdown process to the next startup emissions.

1.5 Summary

The engine startup and shutdown transients are important processes that affect vehicle total emissions. With the advancement of fast light-off catalysts, emissions in the startup process become the major part of the total trip emissions. The startup calibration must provide good driveability, low engine-out emissions, and the means to enable fast catalyst light-off. Precisely controlling the engine operation in startup is difficult because there are uncertainties about the state of the engine (i.e., the amount of residual fuel from the last stop, the starting cam position), the mixture preparation process, and the fuel properties. The calibration margin has been diminishing steadily because of the increasingly more stringent emissions regulations. An important area for development is in improving the effectiveness of fuel delivery to form the combustible charge. Because of the difficulty in fuel evaporation at cold start, only a small amount of the injected fuel is vaporized and burned in the engine cycle. This fact is true for both PFI and DISI engines. The large amount of residual fuel constitutes a major source of HC emissions. Various means are under development to better the mixture preparation process. Innovations include injector improvements (e.g., injector with heated tips), directed intake air motion, and late IVO with a VVT system. These devices will become mainstream when their cost effectiveness is demonstrated and when the emissions regulations are tightened.

Another important area that requires attention is in the development of an effective mixture preparation model to be used for controlling the cranking and startup processes. Future engine calibration for low emissions will require sequential control of every injection event in the startup sequence and advanced strategies such as split injection. Because of the highly dynamic nature of the process, model-based fuel management is needed to minimize the calibration effort and to maximize the robustness of the fuel control procedure.

1.6 References

1. Acke, F. and Marsh, P. (2001), "LEV II Applications Based on a Lean Start Calibration," SAE Paper No. 2001-01-1311, Society of Automotive Engineers, Warrendale, PA.

2. Alkidas, A.C. and Drews, R.J. (1996), "Effects of Mixture Preparation on HC Emissions of a S.I. Engine Operating Under Steady-State Cold Conditions," SAE Paper No. 961958, Society of Automotive Engineers, Warrendale, PA.

3. Barker, D.A., Gibbs, L.M., and Steinke, E.D. (1988), "The Development of Proposed Implementation of the ASTM Driveability Index for Motor Gasoline," SAE Paper No. 881668, Society of Automotive Engineers, Warrendale, PA.

4. Castaing, B.M., Cowart, J.S., and Cheng, W.K. (2000), "Fuel Metering Effects on Hydrocarbon Emissions and Engine Stability During Cranking and Startup in a Port Fuel Injected Spark Ignition Engine," SAE Paper No. 2000-01-2836, Society of Automotive Engineers, Warrendale, PA.

5. Cowart, J. and Cheng, W. (1999), "Intake Valve Thermal Behavior During Steady-State and Transient Engine Operation," SAE Paper No. 1999-01-3643, Society of Automotive Engineers, Warrendale, PA.

6. Graham, J.P., Evans, B., Reuter, R.M., and Steury, J.H. (1991), "Effect of Volatility on Intermediate-Temperature Driveabilty with Hydrocarbon-Only and Oxygenated Gasolines," SAE Paper No. 912432, Society of Automotive Engineers, Warrendale, PA.

7. Henein, N.A., Tagomori, M.K., Yassine, M.K., Asmus, T.W., Thomas, C.P., and Hartman, P.G. (1995), "Cycle-by-Cycle Analysis of HC Emissions During Cold Start of Gasoline Engines," SAE Paper No. 952402, Society of Automotive Engineers, Warrendale, PA.

8. Kapus, P. and Poetscher, P. (2000), "ULEV and Fuel Economy—A Contradiction?" SAE Paper No. 2000-01-1209, Society of Automotive Engineers, Warrendale, PA.

9. Kidokoro, T., Hoshi, K., Hiraku, K., Satoya, K., Watanabe, T., and Fujiwara, T. (2003), "Development of PZEV Exhaust Emission Control System," SAE Paper No. 2003-01-0817, Society of Automotive Engineers, Warrendale, PA.

10. Klein, D. and Cheng, W.K. (2002), "Spark Ignition Engine Hydrocarbon Emissions Behaviors in Stopping and Restarting," SAE Paper No. 2002-01-2804, Society of Automotive Engineers, Warrendale, PA.

11. Koga, N., Miyashita, S., Takeda, K., and Imatake, N. (2001), "An Experimental Study on Fuel Behavior During the Cold Start Period of a Direct Injection Spark Ignition Engine," SAE Paper No. 2001-01-0969, Society of Automotive Engineers, Warrendale, PA.

12. Lang, K. and Cheng, W.K. (2004), "Effect of Intake Cam Phasing on First Cycle Fuel Delivery and HC Emissions in an SI Engine," SAE Paper No. 2004-01-1852, Society of Automotive Engineers, Warrendale, PA.

13. Meyer, R., Yilmaz, E., and Heywood, J. (1998), "Liquid Fuel Flow in the Vicinity of the Intake Valve of a Port-Injected SI Engine," SAE Paper No. 982471, Society of Automotive Engineers, Warrendale, PA.

14. Nishizawa, K., Momoshima, S., and Koga, M. (2000), "Nissan's Gasoline SULEV Technology," SAE Paper No. 2000-01-1583, Society of Automotive Engineers, Warrendale, PA.

15. Roberts, C.E. and Stanglmaier, R.H. (1999), "Investigation of Intake Timing Effects on the Cold Start Behavior of a Spark Ignition Engine," SAE Paper No. 1999-01-3622, Society of Automotive Engineers, Warrendale, PA.

16. Russ, S., Lavoie, G., and Dai, W. (1999a), "SI Engine Operation with Retarded Ignition: Part 1—Cyclic Variations," SAE Paper No. 1999-01-3506, Society of Automotive Engineers, Warrendale, PA.

17. Russ, S., Thiel, M., and Lavoie, G. (1999b), "SI Engine Operation with Retarded Ignition: Part 2—HC Emissions and Oxidation," SAE Paper No. 1999-01-3507, Society of Automotive Engineers, Warrendale, PA.

18. Samenfink, W., Albrodt, H., Frank, M., Gesk, M., Melsheimer, A., Thurso, J., and Matt, M. (2003), "Strategies to Reduce HC Emissions During the Cold Starting of a Port Fuel Injected Gasoline Engine," SAE Paper No. 2003-01-627, Society of Automotive Engineers, Warrendale, PA.

19. Santoso, H. and Cheng, W.K. (2002), "Mixture Preparation and Hydrocarbon Emissions Behaviors in the First Cycle of SI Engine Cranking," SAE Paper No. 2002-01-2805, Society of Automotive Engineers, Warrendale, PA.

20. Shayler, P.J., Darnton, N.J., and Ma, T. (1997), "Factors Influencing Drive Cycle Emissions and Fuel Consumption," SAE Paper No. 971603, Society of Automotive Engineers, Warrendale, PA.

21. Shin, Y., Min, K., and Cheng, W.K. (1995), "Visualization of Mixture Preparation in a Port-Fuel Injection Engine During Engine Warm-Up," SAE Paper No. 952481, Society of Automotive Engineers, Warrendale, PA.

22. Stanglmaier, R., Hall, M., and Matthews, R. (1997), "In-Cylinder Fuel Transport During the First Cranking Cycles in a Port Injected 4-Valve Engine," SAE Paper No. 970043, Society of Automotive Engineers, Warrendale, PA.

23. Takagi, Y., Itoh, T., Muranaka, S., Iiyama, A., Iwakiri, Y., Urushihara, T., and Naitoh, K. (1998), "Simultaneous Attainment of Low Fuel Consumption, High Output Power and Low Exhaust Emissions in Direct-Injection SI Engines," SAE Paper No. 980149, Society of Automotive Engineers, Warrendale, PA.

24. Takeda, K., Yaegashi, T., Sekiguchi, K., Saito, K., and Imatake, N. (1995), "Mixture Preparation and HC Emissions of a 4-Valve Engine with Port Fuel Injection During Cold Starting and Warm-Up," SAE Paper No. 950074, Society of Automotive Engineers, Warrendale, PA.

25. Zhao, F.Q., Lai, M.C., and Harrington, D.L. (1997), "A Review of Mixture Preparation and Combustion Control Strategies for Spark-Ignited Direct-Injection Gasoline Engines," SAE Paper No. 970627, Society of Automotive Engineers, Warrendale, PA.

26. Zimmerman, F., Bright, J., Ren, W., and Imoehl, B. (1999), "An Internally Heated Tip Injector to Reduce HC Emissions During Cold Start," SAE Paper No. 1999-01-0792, Society of Automotive Engineers, Warrendale, PA.

CHAPTER 2

Mixture Formation Processes

Ron Matthews and Matt Hall
The University of Texas

2.1 Introduction

Virtually every gasoline-powered car or light truck currently for sale in the United States has a port fuel injected (PFI) gasoline engine. For such engines, 60 to 95% of tailpipe hydrocarbon (HC) emissions result from the first 30 to 120 seconds of cold start and warm-up [e.g., Kim and Foster (1985); Crane *et al.* (1997); Witze and Green (1997); Stovell *et al.* (1999)], with the percentage increasing and the time period decreasing as the engine is designed and calibrated to meet more stringent emissions standards, such as the partial zero emissions vehicle (PZEV) standard [e.g., Kidoro *et al.* (2003)]. Here, cold start occurs when the coolant and ambient air temperatures are the same, even if the ambient temperature is 30°C (86°F), which is the maximum starting temperature for the Federal Test Procedure (FTP). The dominant effect of the initial tens of seconds of operation on tailpipe HC emissions results from two factors. First, engine-out HCs are high during cold start and warm-up for several reasons. The most important of these is that the engine must be "over-fueled" to achieve rapid cold start and acceptable idle during warm-up. Second, the exhaust catalyst is inefficient at oxidizing engine-out HCs until it reaches light-off temperature (typically 260°C, or 500°F). Both of these factors are influenced by mixture preparation, which is the subject of this chapter.

The sources of HC emissions are reviewed in Chapter 3. Any factors that affect the unburned fuel loading for the oil film, deposit, and crevice sources, such as the equivalence ratio and surface temperatures, will affect the engine-out

HC emissions. That is, rich cold start and warm-up will increase engine-out HCs due to crevice, deposit, and oil film effects, and due to the exhaust valve leakage source. Cold in-cylinder surfaces also will increase the flame quench source. However, the source that may be most important during cold start and warm-up, but essentially disappears after the engine is fully warmed up, is the liquid fuel source. This source is a major element of mixture formation during starting and warm-up and is the subject of Section 2.2 of this chapter.

Mixture preparation is the process by which the fuel vapor is mixed with fresh inducted air, with the residual burned gases that remain in the cylinder from the previous cycle, and with any recirculated exhaust gases (EGR). It also encompasses aspects such as cylinder-to-cylinder differences in fuel and air delivery.

For most PFI engines on the market today, the goal of mixture preparation is to achieve the most homogeneous mixture possible by the time of ignition. Although an engine operating at an overall lean air/fuel (A/F) ratio may benefit from a well-formed stratified charge, in most engines, mixture inhomogeneities lead to increased emissions and can have a negative effect on performance, such as idle quality. For starting and warm-up specifically, the goals are to minimize engine-out HCs and catalyst light-off time. As will be discussed in Section 2.5 later in this chapter, these two goals may conflict, leading to a variety of strategies for minimizing tailpipe HCs during starting and warm-up.

Many factors influence the air/fuel/residual mixing process. Fuel injection system hardware and controls are important, including items such as the type of injector used, how it is targeted, and the timing of the injection. These topics are the subject of Section 2.3 of this chapter. Mixture preparation also is influenced by the flow in the intake port and the bulk motion and turbulence in the cylinder, as will be discussed in Section 2.4.

Strategies for decreasing HC emissions during cold start and warm-up are discussed in Section 2.5. Finally, Section 2.6 summarizes this chapter and draws conclusions.

2.2 Liquid Fuel as a Source of Hydrocarbon Emissions

As already noted, 60 to 95% of tailpipe HC emissions over the FTP cycle are emitted during the cold start and warm-up period. A dominant source of these emissions is the need for over-fueling. Typically, the injector is aimed at the top of the intake valve because this is the hottest surface in the intake system.

However, this surface is at ambient temperature during a cold start. Chen *et al.* (1996) found that only 20% of gasoline will evaporate under this condition, in reasonable agreement with equilibrium calculations that 10 to 20% of the fuel vaporizes during the first few cycles of a cold start [Boyle *et al.* (1993); Santoso and Cheng (2002)]. Thus, the engine must be over-fueled by 5X to provide sufficient fuel vapor to attain ignition and initial starting. However, the engine calibrator must be concerned about the "worst-case gasoline." That is, the engine control module (ECM), the engine control unit (ECU), or the programmable control module (PCM) must be calibrated to assure rapid start and stable idle for a high-driveability-index (high-DI) gasoline, which is not as volatile as the typical gasoline. Over-fueling by 10X or more may be required to assure rapid start and stable idle for low-volatility gasolines. However, because there is no volatility sensor in the fueling system, the low-volatility calibration is used for all gasolines, independent of their volatility. Typically, 8 to 15 times the stoichiometric amount of gasoline is injected during the first several cycles of the cold start and warm-up transient [Heywood (1998)]. Due to the increased difficulty in evaporating high-DI gasolines, the in-cylinder fuel vapor/air ratio tends to shift lean, increasing cyclic variability. Kishi *et al.* (1998) report Honda's use of crankshaft speed fluctuations to enrich the mixture to automatically compensate for high-DI gasolines. Although this allows for more optimum mixture strength during warm-up, cranking enrichment must still account for high-DI gasolines. The DI and other fuel effects on cold-start HC emissions are discussed in detail in Chapter 7.

Because little gasoline evaporates during cold start and warm-up, a fuel film is formed in the intake port and on top of the intake valves. Between intake strokes, some of this film will evaporate. This fuel vapor will then be drawn into the combustion chamber, along with any vapor from the injection for the current cycle plus any vapor from in-cylinder fuel films, during the ensuing intake stroke.

Most, if not all, PFI engine control algorithms incorporate "fuel compensation" to account for the vapor from the films that evaporates between cycles. This is accomplished using a port wall film model, such as that posed by Hires and Overington (1981). The goal is to decrease the pulse width for the current cycle (possibly to zero) such that the vapor from the current injection plus the vapor from the film remaining from prior cycles yields the target in-cylinder vapor/air mixture. These models can account for the fact that the rate of film evaporation from the various puddles (e.g., upstream port, downstream port, valve, head, liner) is limited by the surface temperatures, which increase throughout warm-up [Batteh and Curtis (2003)].

Cowart and Cheng (1999) examined the temperature on the port side of the intake valve during starting and warm-up. Alkidas (2001) performed a similar analysis but also examined the temperature of the intake port wall. As illustrated in Figure 2.1, during starting and warm-up, the intake valve temperature history is divided into two stages. The intake valve temperature during the first stage has an exponential behavior with a duration of approximately 1 minute. Following this initial stage, the valve temperature increases quasi-linearly. In contrast, the port warms much more slowly, with a linear increase that closely follows the coolant outlet temperature. The energy balance that determines the transient intake valve temperature is dominated by heat input from the

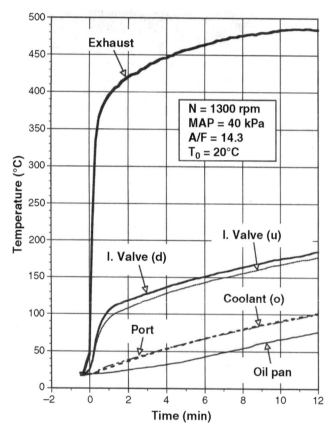

Figure 2.1 Transient temperatures during starting and warm-up at 1300 rpm. Intake valve temperatures are shown for both the downstream (d) and upstream (u) surfaces of the valve [Alkidas (2001)].

combustion gases, the effective thermal inertia of the valve (for which the effective mass is that of the valve body plus one-third of the stem), and heat losses. The heat losses are dominated by conduction through the valve seat and evaporation of liquid fuel on the port side of the valve. Because the far side of the valve is shaded by the valve stem, it heats faster due to little or no fuel to evaporate, as illustrated in Figure 2.1.

The question arises, "If fuel compensation achieves its goal, why are engine-out HC emissions still higher during starting and warm-up?" There are three primary reasons. First is the increased strength of the flame quench, crevice, deposit, and oil film sources when the combustion chamber walls are cold. Second, the target mixture is generally rich during warm-up to provide adequate idle stability. Third, no fuel compensation is possible for the first cycle, and a film remains and builds for subsequent cycles, even when fuel compensation is used. This problem is compounded by the need to calibrate at least the cranking enrichment for a high-DI gasoline, which has fewer light fractions and thus does not evaporate as readily. Shear between the flowing gas and the liquid forces the port wall film to propagate toward the combustion chamber when the intake valves are open [e.g., Servati and Yuen (1984)]. Also, the film on the intake valve is rapidly swept into the combustion chamber. The result is that liquid fuel gets into the cylinder.

The dynamics of the liquid fuel behavior in the port and how it flows into the cylinder during intake affect emissions. Using phase Doppler anemometry (PDA), Meyer *et al.* (1998) were able to discern four mechanisms for the flow of liquid fuel into the combustion chamber. The four mechanisms include the following:

1. Fuel stripping from the intake valve during intake valve opening (IVO)

2. The direct flow of liquid fuel into the combustion chamber from the injector with open-valve injection (OVI)

3. Transport during the middle of the intake stroke driven by the high-speed flow of intake air

4. Droplets formed when the fuel in the film on the face of the valve and the valve seat is squeezed during intake valve closing

In a follow-on study, Meyer and Heywood (1999) found that the type of transport affected the in-cylinder droplet size. Closed-valve injection (CVI) and OVI where the spray impinged on the port and valve yielded larger droplets than OVI where the spray entered the combustion chamber unimpeded. Visualization studies [Stanglemeier et al. (1997); Kelly-Zion et al. (1997)] conducted in motored engines showed that large fuel droplets can be stripped off the intake valve(s) entering the cylinder during intake. They found that OVI yields direct impingement of small droplets of liquid fuel onto the liner below the exhaust valves. Surface tension effects result in the formation of a fuel film from these droplets, impeding vaporization. This film persists and builds up over successive cycles when the liner is cold. Some of the liner film is scraped into the crankcase as "lost fuel." Some of the liner film is scraped up closer to the exhaust valves, where it can more easily contribute to engine-out HCs. With CVI, larger droplets, but fewer in number, impinge on the piston top. Witze and Green (1997) showed that films also form on the combustion chamber side of the head. Closed-valve injection produces head films mainly on the intake side, whereas OVI produces films that cover a larger area, including between the exhaust valves of a pentroof four-valve head. The head films from OVI begin to form on earlier cycles than for CVI. Witze and Green (1997) also imaged pool fires that result from the diffusion of evaporated fuel from the in-cylinder films and subsequent ignition. However, flame passage does not completely evaporate in-cylinder fuel films; the rate of evaporation is dictated by the surface temperature. The effects of injection timing are discussed in more detail in Section 2.3.2. The fate of the fuel that impacts in-cylinder surfaces is the focus of the remainder of this section.

The impingement of liquid fuel on in-cylinder surfaces has been shown to significantly increase HC emissions. The emissions are affected by the amount of liquid fuel hitting the surface, where the wetting occurs, and the volatility of the fuel. Stanglmaier et al. (1999) developed a spark-plug-mounted directional injection probe to examine the fate of liquid fuel that impinges on different surfaces of the combustion chamber. The probe deposited a controlled amount of liquid fuel on a given location in the combustion chamber at a desired crank angle while the engine was operated predominantly (~90%) on premixed liquefied petroleum gas (LPG). This allowed the HC emissions due to wall wetting to be studied independently from other sources. Although this series of injector probe studies was performed to examine wall wetting in direct injection spark ignition (DISI) engines, the results are relevant to PFI engines due to the operating conditions. Two operating conditions were studied: (1) 1000-rpm idle, and (2) 1500-rpm, 262-kPa brake mean effective pressure (BMEP), with ignition at minimum spark advance for best torque (MBT). The overall mixture strength (propane plus liquid fuel) was slightly lean (an equivalence ratio $\phi = 0.9$, which corresponds to an excess air ratio $\lambda = 1.1$) to minimize the sensitivity of the HC emissions to minor variations in stoichiometry. In general, most of the experiments were conducted with a coolant temperature of 45°C. As will be noted in Section 2.3.1 later in this chapter, the cold coolant is insufficient to completely simulate cold-start conditions, but cold start was not the focus of this series of experiments. The predicted piston temperature, calculated using Ford's GESIM code, was 422 K for the mid-load condition.

The measurements by Stanglmaier *et al.* (1999) showed that the location of the wetting affected HC emissions. The largest increases in HC emissions occurred when the wetting was on the cylinder liner just below the exhaust valves. Wetting on the piston top gave the next highest emissions, and the smallest increase occurred for wetting the liner just below the intake valves, as illustrated in Figure 2.2. These observations suggest that an increase in residence time associated with the distance that evaporated fuel must travel to the exhaust port increases the extent of fuel oxidation and/or increases the probability of retention within the cylinder, thereby lowering HC emissions.

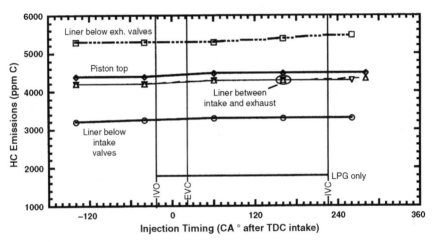

Figure 2.2 *The effects of in-cylinder surface wetting location and time available for evaporation on engine-out HC emissions for steady-state operation with cold coolant [Stanglmaier* et al. *(1999)].*

The HC emissions levels were also found to be independent of the coolant temperature and the injection timing. This suggests that the evaporation of impinged fuel is a relatively slow process. In follow-up studies, Li *et al.* (1999, 2000, and 2001) used a Fast-Spec [Mizaikoff *et al.* (1998)] and a wide-range lambda sensor to measure how long it took for the increase in HC emissions associated with liquid fuel impingement to drop after the injection probe was turned off. They found that with 1.5 mg of California Phase II reformulated gasoline (CP2-RFG) injected, it took from three to four cycles for the HC emissions to fall 64% (the $\frac{1}{e}$ time constant) from their initial levels, as shown in Figure 2.3.

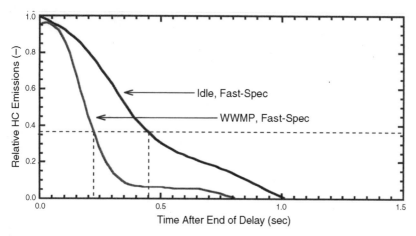

Figure 2.3 *Illustration of slow evaporation of 1.5 mg of gasoline off the piston once the source of liquid fuel is turned off [adapted from Li et al. (2001)].*

The results from further studies [Alger *et al.* (2001); Huang *et al.* (2001a and 2001b); Li *et al.* (2001); Matthews *et al.* (2001)] suggest that heat transfer considerations, including the Leidenfrost effect, explain that surface temperatures that are not much higher or lower than the boiling point of the fuel can significantly decrease the evaporation rate, even for multicomponent fuel blends. Although this may be important at higher loads when the surface temperatures are relatively high, it probably is not important during cold start and warm-up. However, they also found that the increase in HC emissions due to wall wetting was proportional to the amount of liquid fuel impacting the in-cylinder surface. Depositing 10% of the fuel as a liquid CP2-RFG on the top of the piston resulted in approximately 30 to 70% increases in HC emissions for the idle and part-load conditions, respectively. This is illustrated in Figure 2.4, which shows the results of experiments that were designed to demonstrate that the large increase in HC emissions for CP2-RFG compared to LPG were not the effects of fuel structure. For experiments conducted at both 1000-rpm

Mixture Formation Processes

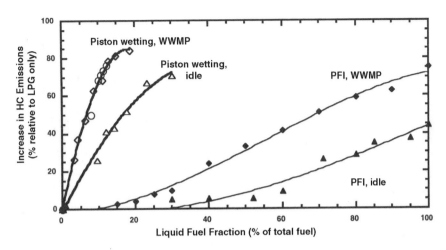

Figure 2.4 Effects of liquid fuel impingement on the piston compared to PFI of the same mass of liquid fuel, with the remainder of the fuel supplied as LPG vapor [Li et al. (2001)].

idle with cold coolant and at 1500-rpm, 2.62-bar BMEP, and MBT timing (the Ford World Wide Mapping Point, or WWMP) with cold coolant, the effect of fuel structure is apparent by comparing PFI with 100% liquid fuel versus 0% liquid fuel (100% LPG). However, the effect of piston wetting is much more pronounced. At idle with cold coolant, wetting the piston with 30% liquid fuel increases the engine-out HC emissions by approximately 70%, whereas introducing the same mass of gasoline via the port fuel injector has little effect on the HC emissions. At idle, the piston wetting results are approximately linear. The stronger effect at the mid-speed, mid-load condition is of interest due to the acceleration to cruise that begins 20 seconds into the FTP. This almost linear dependence of the HC emissions on the wetting mass also might be expected for wetting the head surface, but possibly not for the liner because some of the mass that impacts the liner gets scraped into the crankcase.

Landsberg *et al.* (2001) used a similar injection probe but mounted it in the intake port to examine the effects of the location of intake port wetting. They operated the engine on prevaporized Indolene (certification gasoline) and injected a small amount of liquid fuel onto the port wall near the intake valve seat. They examined port wetting with the intake valve closed and with it open. For these experiments, the engine was fully warmed up and operating at 1500 rpm, 0.5-bar intake manifold absolute pressure (MAP). Their results are discussed in more detail in Section 2.3.2 with respect to injection timing and in Section 2.3.3 with respect to injector targeting. In all cases, the engine-out HCs increased roughly linearly with increasing port wetting mass due to a corresponding increase in in-cylinder wall wetting. Additionally, using their data-calibrated model for the fully warmed-up production engine with CVI at 1500 rpm and 0.5-bar MAP, they estimate that 5% of the injected fuel enters the cylinder as a liquid. This is responsible for the 11% increase in engine-out HC emissions for stock injection compared to prevaporized fuel delivery at this fully warmed-up operating condition.

Nitschke (1993) performed a simulated cold start by motoring for less than 20 seconds to set the speed at 1300 rpm, then firing from initially cold conditions (25°C air, oil, and coolant). The fuel used was CP2-RFG. With stoichiometric fueling, which produced a high misfire frequency, the specific reactivity of the engine-out exhaust was 25% lower over the first 30 seconds than for warm steady-state operation. Approximately 75% of the exhaust HCs were attributed to unreacted fuel during this first 30 seconds, compared to only 53% at steady state. A rich starting pulse eliminated misfires and reduced nonmethane organic gas (NMOG) emissions and ozone-forming potential (OFP) compared to the stoichiometric fueling schedule. Although Nitschke's work does not directly address the liquid fuel source, it supports the results of Kaiser *et al.* (1994), which indicate that retention of the heavier fuel components—as liquid and/or in the exhaust system—is responsible for many of Nitschke's findings.

Kaiser *et al.* (1994) performed similar experiments but for actual starts and using federal certification gasoline. They examined two starting strategies. The first was similar to production: a rich pulse for the first cycle (approximately twice the amount required to achieve $\lambda \sim 0.7$) to achieve firing on the second cycle, followed by rich operation tapering off to stoichiometric at approximately 50 seconds after start. The spark timing was retarded 5 to 10°CA from MBT. The second strategy involved higher idle speed, lean operation ($\lambda = 1.05$ for the first 60 seconds), and more aggressive spark retard (20 to 25°CA retarded from MBT). They found that the HC emissions initially were enriched in light (high-vapor-pressure) alkanes and depleted in heavy (low-vapor-pressure) aromatics, compared to the composition of the fuel. Over the first 10 seconds, the contribution by the light alkanes fell rapidly, while that by the heavy aromatics increased rapidly for both starting strategies. Iso-octane and toluene, which have similar boiling points, behaved similarly (increasing for the first ~10 seconds and then falling) for the base strategy, but differently for the aggressive strategy (with iso-octane emissions falling over the first 7.5 seconds and then stabilizing). They note that their results indicate retention of the heavy ends in the intake manifold, intake port, and exhaust system (they sampled from the tailpipe of a complete exhaust system but with a bare brick "catalyst") until the engine warms, and that the light alkanes in the fuel vaporize easily and provide the fuel vapor required to start the engine. They conclude that techniques for decreasing total HC emissions during cold start may affect the exhaust reactivity in a manner that could either increase or decrease the apparent effectiveness of the HC emissions reduction. This is illustrated in Figure 2.5. They also conclude that, due to the importance of the distillation properties, the composition of the fuel can affect the exhaust reactivity.

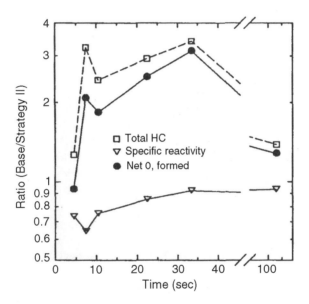

Figure 2.5 *Effects of starting strategy on transient HC emissions, specific reactivity, and OFP for a cold start and warm-up [Kaiser* et al. *(1994)].*

Kaiser *et al.* (1996) performed additional cold-start experiments that were reported two years later. They used the same engine as previously—a 4.6-L two-valve V8. They examined two fueling systems (conventional PFI and prevaporized central fuel injection [CFI]) and two starting loads (2.57-bar gross indicated mean effective pressure [IMEP] [0.34-bar MAP] and 5.15-bar gross IMEP [0.58-bar MAP]). Here, recall that the engine becomes lightly loaded 15 seconds into the FTP when the transmission is shifted into drive, and the load increases starting at 20 seconds for the first acceleration and cruise. They found little difference in the total HC emissions as a function of time for the light-load condition. They concluded that "the HC emissions are insensitive to the fuel preparation for this engine under light load conditions." However, the PFI system required a rich pulse for approximately the first 2 seconds ($\lambda \sim 0.2$), whereas much less initial over-fueling was required for the prevaporized system ($\lambda \sim 0.6$). (The need for any startup over-fueling with the pre-vaporized system is surprising. Fox *et al.* (1992) found that no over-fueling was needed for their prevaporized starting experiments. Also, as will be discussed in Section 2.5 later in this chapter, the use of a highly volatile starting fuel allows starting with a stoichiometric mixture.) Additionally, the PFI system operated somewhat leaner (intake ϕ) than the prevaporized system from approximately 2 to 12 seconds. For this light-load cold start, the prevaporized system had much less lost fuel (into the crankcase) than for PFI operation. Of perhaps more importance, for cold start to the mid-load condition, they found that the prevaporized system produced 80% lower engine-out HC emissions during the initial 5 to 10 seconds. They note that there is less backflow of residual into the intake port at the higher load, thus decreasing the efficiency of evaporation. Additionally, the PFI exhaust contained a higher percentage of heavy fuel components, but the difference compared to the prevaporized system decreased with time. They note that this suggests that liquid fuel enters the combustion chamber during the early stages of a cold start; the heavy ends of the in-cylinder wall films evaporate more slowly, such that they might evaporate during the late expansion stroke or during the exhaust stroke. Several experiments and models [e.g., Takeda *et al.* (1995); Skippon *et al.* (1996)] conclude that the heavy ends of gasoline are the major contributors to the port wall film. This is true of in-cylinder wall films as well [e.g., Lee and Morley (1995), Zhu *et al.* (2001)].

Swindal *et al.* (1995) note that multicomponent fuel evaporation can be complicated. They propose that droplet evaporation is a batch distillation process for cold operation and during the early part of the intake stroke. In batch distillation, the lighter components evaporate first, leaving smaller droplets that are enriched in heavier fuel components. For warm operation and late in the compression stroke, they propose that droplet evaporation is probably a combination of batch and is diffusion limited. If the thermal diffusivity is much greater than the mass diffusivity, all components evaporate at similar rates, and the fuel droplet has essentially constant composition. They performed experiments in an optical access engine to examine the fate of fuel droplets that are entrained in the airstream but do not impact a surface. They used tracers with various boiling points and used planar laser-induced fluorescence (PLIF) to image a plane within the combustion chamber. They found that the low-boiling-point tracers (<100°C) produced an essentially homogeneous charge during the compression stroke, with the lowest-boiling-point tracer yielding a uniform distribution earlier. However, the higher-boiling-point tracers produced not only charge inhomogeneity but also surviving droplets at the latest crank angle observed. This was true not only for cold operation but also for warm operation.

Weaver *et al.* (2003) performed similar experiments, including steady-state firing with cold coolant. As can be deduced from Figure 2.1, this type of cold start simulation does not fully encompass all of the surface temperatures that are critical to fuel evaporation, especially the intake valve temperature. However, this work showed that there are liquid films within the cylinder, even for steady firing with cold coolant, and related research [Russ *et al.* (1999b)] showed that some of this liquid escapes into the exhaust port and manifold. Weaver *et al.* (2003) used only a mid-boiling-point tracer but also could detect liquid droplets within the laser sheet. Similar to Swindal *et al.* (1995), they found that the mid-boiling-point tracer revealed a homogeneous mixture, with few droplets remaining, by 20° before top dead center (BTDC) of compression. In fact, the equivalence ratio distribution in the illuminated plane, and the coefficient of variation (COV) of this distribution, were indistinguishable from those for prevaporized fueling. However, this does not mean that all fuel in the in-cylinder films had vaporized before ignition. After combustion and during exhaust, their images revealed evidence of either large films or vapor clouds in the vicinity of the intake and exhaust valves.

McGee *et al.* (2000) compared CVI with a conventional port injector to prevaporized fueling. They found that PFI produced 40 to 50% greater inhomogeneity than vapor fueling, regardless of the temperature or load. They concluded that the in-cylinder liquid is a greater factor in HC emissions than mixture stratification.

That is, difficulty in evaporating the heavier components of the gasoline results in in-cylinder wall wetting and may lead to charge inhomogeneity of the heavy ends late in compression. In turn, the charge inhomogeneity could increase HC emissions via rich pockets, not only for cold operation but also for warm conditions. The evidence indicates that in-cylinder fuel films are a more important source of HC emissions than charge inhomogeneity.

2.3 Fuel Injection Hardware and Controls

In a PFI engine, individual electronically controlled fuel injectors deliver metered amounts of gasoline to the individual engine cylinders with which they are paired. Each injector is located in the intake manifold or port and generally directs its spray of gasoline toward the intake valve(s) to promote evaporation because this is the hottest surface in the intake system. The amount of gasoline delivered is determined by the ECM control strategy. The delivered equivalence ratio is the most important factor governing performance and emissions. Under most driving conditions, the amount of fuel delivered is proportional to the airflow into the engine and is regulated to maintain a nearly stoichiometric A/F ratio. A stoichiometric A/F ratio is necessary to allow the three-way catalyst to simultaneously reduce nitrogen oxides (NOx) and oxidize carbon monoxide (CO) and unburned HCs at peak efficiency after light-off. Under certain transient conditions, however, such as cold start and warm-up and hard accelerations, additional fuel is introduced, pushing the A/F mixture richer than stoichiometric, with corresponding consequences for emissions production. The effects on HC emissions depend on the injector type, injection timing, injector targeting, and other factors. These hardware and control factors are the subjects of the following subsections.

2.3.1 Injector Types

Yang *et al.* (1993) studied the effects of injector type and injection timing for steady-state operation at 3.78-bar IMEP and 1500 rpm with both cold and warm coolant. Although this was not a starting and warm-up experiment, the results show general agreement with similar starting and warm-up experiments. However, note that cold coolant is not the same as cold start because the steady-state surface temperatures are warmer, and the crevice volumes are smaller for steady-state operation, even with cold coolant. The single-cylinder, two-valve engine had an AVL swirl ratio of 3.0. (AVL is an Austrian company, AVL GmbH.) Yang *et al.* examined three types of injectors:

1. A production pintle injector that produced droplets with a Sauter mean diameter (SMD) of 300 μm

2. An air-assist injector (AAI) that produced an SMD of 40 μm

3. An air-forced injector (AFI) that produced an SMD of 14 μm

The AFI was tested with two different cone angles, but there was little difference between the results, possibly because the small droplets allowed almost complete vaporization (i.e., little port wetting). The injectors were aimed at the base of the valve stem. As illustrated in Figure 2.6, they found little effect of injection timing on HC emissions with CVI, except that in the cold-coolant case with the larger droplets, the HC emissions began to increase for the start of injection ~80°CA before IVO. However, for the larger droplet sizes with OVI, there was a large increase in engine-out HC emissions, especially with cold coolant. They also found an increase in fuel consumption and lost fuel for the same conditions that produced increased HC emissions. Also, the contribution to the HC emissions by the heavier aromatics increased, while that from the lighter fuel components and partial oxidation products decreased.

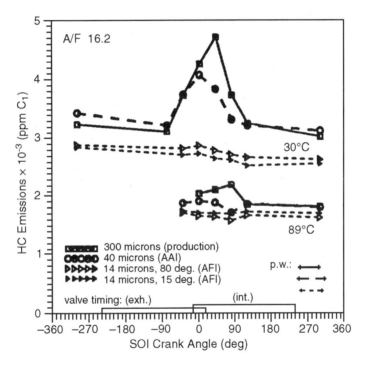

Figure 2.6 *Effects of injector type (droplet size) and injection timing on engine-out HC emissions for steady-state operation [Yang et al. (1993)].*

They attributed their results predominantly to the effects of droplet size and injection timing on liner wetting.

Felton et al. (1995) also compared AFIs (10 µm SMD) to pintle-type injectors (~150 µm SMD). They imaged films in the intake port for simulated cold starts using Unocal RF-A gasoline. The pintle injector produced thicker films than the AFI. The port films from both injectors persisted more than 10 minutes at 1000 rpm, 2.6-bar BMEP. The early spike in the engine-out HCs corresponded to early growth of the port films, and the long-term decrease in the port film was accompanied by a gradual decrease in HC emissions.

Meyer and Heywood (1999) used PDA to compare AAIs and pencil stream injectors. In spite of the large difference in spray droplet sizes (49 µm SMD for the AAI versus 176 µm for the pencil stream), there was little difference in the in-cylinder droplet size distribution for either OVI or CVI. However, starting with the pencil stream injector was superior to that for the AAI due to less fuel impingement on the port walls. (As discussed previously, the port wall heats much more slowly than the intake valve.)

Quader et al. (1991) performed experiments that represent an extreme of small droplets. They compared production port injection with premixed fully vaporized fuel delivery. Completely prevaporizing the fuel reduced engine-out HC emissions by 57% and decreased the percentage of HCs above T_{50} (species with a boiling point above the temperature at which 50% of the fuel has evaporated) in the exhaust HCs. Also, fuels with lower T_{90} produced lower HC emissions. For experiments performed at –7°C, the HC emissions from PFI operation increased substantially compared to the results at 24°C, while the premixed prevaporized case showed a much smaller increase. They noted the importance of vaporization of the heavier ends of the fuel, as was discussed in more detail in Section 2.2 of this chapter.

Weaver et al. (2003) also compared prevaporized fueling with PFI. As discussed previously, for measurements near the time of ignition, they found little difference in the homogeneity of the mixture distribution in-cylinder using a PLIF tracer that mimicked T_{50}. However, as also noted previously, in similar experiments Swindal et al. (1995) found inhomogeneities late in compression using a T_{90} tracer. When coupled with the findings of Quader et al. (1991) that prevaporizing the fuel results in decreased engine-out emissions of species above T_{50}, it must be concluded that the rich pockets of vapors above T_{50} are important, or that the abundance of species above T_{50} in the fuel films is important, or both. The bulk of the evidence reviewed in this chapter indicates that the in-cylinder films dominate.

Ireton and Kaiser (2001) examined the effects of fuel preparation on engine-out HCs for Phase I of the FTP (cold start and warm-up) compared to Phase III (hot start). Each of the three accelerations during the first 100 seconds of the FTP produced large increases in the HC emissions that were not observed for the same accelerations during Phase III. They conclude that the reduced efficiency of fuel evaporation in the intake port during warm-up plays an important role during the accelerations and high-speed operating conditions. Use of dual cone injectors decreased the HCs emitted during these accelerations by a factor of 3, resulting in a 25% reduction in tailpipe HCs over the FTP.

Improved atomization increases the surface-to-volume ratio of the droplets and thus increases the probability of vaporization of the injected spray before impacting the port or valves. Zimmerman *et al.* (1999) estimate that droplets smaller than 10 μm will remain in the airflow and will avoid wall impaction. Samenfink *et al.* (2003) postulate that a 25-μm SMD is required to meet super ultra-low emissions vehicle (SULEV) standards. They show that an AAI is not required to obtain droplets of such small size; a more conventional multihole injector operating at 1.6 to 2.0 MPa can achieve this target. Thus, there is also a trend toward PFI injectors having more and smaller-diameter holes. In some of its engines, Ford recently changed from a four-hole to a twelve-hole injector, citing better atomization and evaporation of the fuel, leading to lower HC emissions [Carney (2003)]. Kidoro *et al.* (2003) report that improvement in the twelve-hole injector used for Toyota's 2002 Model Year ultra-low emissions vehicle (ULEV) allows operation approximately one A/F ratio leaner, as will be illustrated in Figure 2.10 of Section 2.4, with respect to airflow effects). Kim *et al.* (2003a) examined injectors with four to eighteen holes, including two twelve-hole injectors, one producing a single stream and the other a dual stream. They used a color imaging capturing technique (CICT) that they developed to quantify the films on the liner and the piston top. The four-hole and eighteen-hole injectors produced the most in-cylinder wetting, while the ten-hole and dual-stream twelve-hole injectors produced the least.

2.3.2 Injection Timing

During normal operation, the injectors are fired sequentially—injection begins at the same crank angle relative to top dead center (TDC) of compression for each injector/cylinder pair. Typically, the start of injection is timed to occur right after intake valve closing (IVC) to provide the maximum amount of time for fuel evaporation. However, sequential injection can occur only after the

ECU has synchronized the cam and crank sensor signals to determine TDC for all cylinders. That is, starting the engine poses special problems. The options are either to batch fire the injectors or to delay injection until the signals are synchronized. Until recently, batch firing was used to achieve rapid start. However, batch firing results in some cylinders having OVI and others CVI (but probably not with the optimum CVI timing). The effects of injection timing were briefly introduced in Section 2.3.1 with respect to injector types. A more detailed discussion is provided in this section.

In a series of experiments, Alkidas and coworkers [Alkidas (1994); Alkidas and Drews (1996); Stache and Alkidas (1997)] examined the effects of injection timing, coolant temperature, and fuel volatility for steady-state operation. A portion of the results from the Stache and Alkidas (1997) study is provided in Figure 2.7. In agreement with Yang et al. (1993), they found that the HC emissions are almost independent of injection timing for CVI unless the start of injection is ~80°CA or later before IVO. With OVI, and especially with cold coolant, a portion of the liquid fuel enters the combustion chamber and impacts in-cylinder surfaces. It also appears that if the CVI start of injection is ~80°CA or later before IVO, some liquid enters the combustion chamber.

It is also of interest to compare the results for the three fuels. Iso-octane has a boiling point that is near T_{40} compared to the certification gasoline (Indolene), while iso-pentane has a boiling point near the initial boiling point of the Indolene. Iso-pentane produced the lowest HC emissions, while iso-octane produced the highest for all but a small range of OVI timings with the cold coolant. At first, it appears that this disagrees with the results of the injector probe experiments discussed in Section 2.2. The injector probe experiments showed that n-pentane (which is approximately the same volatility as iso-pentane) produces higher HC emissions than iso-octane *if the same mass* of both impacts an in-cylinder surface. Thus, the reason that iso-pentane produced the lowest HC emissions in the experiments by Alkidas and coworkers (1997) is that more of it vaporized before impacting an in-cylinder surface.

However, it is also of interest that iso-pentane produced increased HC emissions with OVI. The results from Swindal et al. (1995) indicate that volatile fuels should vaporize early and become well mixed. Thus, surviving droplets or rich pockets of iso-pentane are not expected. It appears that the reason for the increase in HC emissions for iso-pentane with OVI is that a relatively small mass impacts an in-cylinder surface.

Stache and Alkidas (1997) used a fast-response flame ionization detector (FFID) to compare the time-resolved HC emissions for iso-octane and Indolene with

Mixture Formation Processes

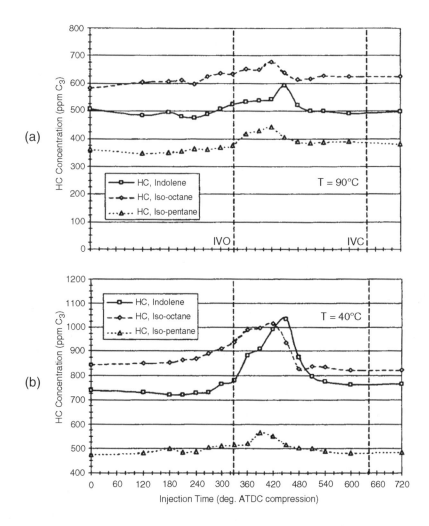

Figure 2.7 *Effects of injection timing, coolant temperature, and fuel volatility on engine-out HC emissions for steady-state operation: (a) coolant T = 90°C, (b) coolant T = 40°C [Stache and Alkidas (1997)].*

each using the OVI timing that produced the maximum emissions. For both coolant temperatures, they found that the increase in emissions for OVI compared to CVI was dominated by liner wetting for Indolene, but by liquid and vapor fuel in the piston ring pack for iso-octane. The heavy ends of the gasoline allow the liner film to survive. This finding was confirmed by a recent model, comparing a multicomponent gasoline to iso-octane [Zhu *et al.* (2001)]. For iso-octane, the cooler conditions in the ring pack allow the liquid fuel to survive longer and/or result in a high vapor concentration in the ring pack.

Landsberg et al. (2001) used an injection probe to wet the intake port near the valve seat with the valve open and with it closed. The engine was fully warmed up and operated with prevaporized certification gasoline at 1500 rpm and 0.5-bar MAP. Delivery of liquid fuel to the port wall with the valves closed produced an increase in HC emissions that was about half that observed with open-valve port wetting for all probe locations. They note that this suggests that the hot backflow produces the same amount of port film evaporation for all three probe locations: ~40%, according to the combination of their data with a model. The higher emissions with open-valve wetting were attributed to more liquid fuel entering the cylinder.

Using a unique experimental system, Imatake et al. (1997) were able to quantify the amount of port and cylinder wall wetting during the starting transient and for steady-state operation. During the early cycles of warm-up with OVI, they found that the liquid fuel in the intake port was 9 times the injected amount, and the in-cylinder liquid was 2.5 times the injected amount. Closed-valve injection produced 7% more port wall wetting and 16% less in-cylinder wall wetting. Both the port wall wetting and the in-cylinder wall wetting increased with decreasing volatility of the gasoline (increasing T_{10}, T_{50}, T_{90}, and the end boiling point, which all varied together).

As noted in Section 2.2, it has been found that OVI yields more liner wetting under the exhaust valves and more extensive head wetting, whereas CVI primarily yields piston wetting with large droplets when the valves and port are cold. Kim et al. (2003a) used their new CICT to quantify liner and piston wetting for OVI versus CVI and five different multihole injectors. They found that the large drops impacting the piston with CVI could be almost eliminated with some injector designs. However, they also found that all five injectors produced a fine mist on the piston with OVI. When quantified, OVI produced more piston wetting than CVI for four of the five injectors. As expected, all five injectors produced more liner wetting with OVI than CVI. Closed-valve injection produced almost no liner wetting under the exhaust but extensive liner wetting under the intake. Open-valve injection produced extensive liner wetting under the exhaust for all five injectors, but also some liner wetting under the intake.

Rottenkolber et al. (1999) found cold-start HC emissions 25 to 80% higher for OVI versus CVI, depending on injector targeting. Castaing et al. (2000)

measured both the in-cylinder λ, using an FFID, and the exhaust HC emissions for the first fired cycle during cranking in one cylinder of a four-valve, four-cylinder engine. They varied the injector pulse width during startup and found that when the in-cylinder λ was near stoichiometric, engine-out HCs were only slightly higher for OVI versus CVI. However, for in-cylinder $\lambda < 0.95$, HC emissions with OVI were approximately three times those for CVI. Open-valve injection led to charge inhomogeneity, which, in combination with the direct wall wetting, increased cold-start HC emissions. Injection timing is discussed in Section 2.3.3 with respect to the manifold fuel film.

McGee *et al.* (2000) also compared OVI and CVI, using a conventional port injector. For throttle ramp transients with cold coolant and valve targeting, OVI produced smaller A/F ratio excursions and a film mass in the intake system that was half that with CVI. However, CVI produced lower engine-out HC emissions, allowed more spark retard, and improved mixture homogeneity for all loads and temperatures.

2.3.3 Other Injection Parameters

This section discusses the effects of injector targeting, in-cylinder λ, manifold air pressure, revolutions per minute (rpm), ambient temperature, and fuel heating.

Injector targeting affects engine-out HC emissions via its direct and indirect impacts on in-cylinder wall wetting. Landsberg *et al.* (2001) used an injector probe to wet three locations of the intake port wall near the valve seat. Other than this small mass of liquid, the engine operated on prevaporized Indolene at 1500 rpm, 0.5-bar MAP, fully warm. The location and amount of liquid delivered to the port wall affected the amount of liquid entering the cylinder and thus the HC emissions. Wetting the intake port wall closest to the exhaust valve side of the engine with the intake valve open produced the highest HC emissions: seven times higher than delivery of the same mass of fuel as vapor. Wetting the port wall farthest from the exhaust side produced the lowest increase in HCs, but still three times higher than delivery of the same mass of fuel as vapor. Wetting the port wall at an intermediate location near the seat produced a 5X increase in HCs. Wetting the port with a closed intake valve produced similar trends but with approximately half the increase in HCs because of fuel evaporation during the backflow process.

Russ et al. (1998) investigated the effect of injector targeting and injection timing on the A/F ratio for throttle ramp transients during warm-up. The exhaust A/F ratio was measured for the transient, and the change in the liquid stored in the port and cylinder was estimated. Air/fuel ratio fluctuations were found that were due to changes in the amount of fuel stored in liquid films in the intake port. In this study, the injector was targeted to wet either the backs of the two intake valves or the floor of the port below the valves. Russ et al. (1998) found that during the transient, approximately 20% less fuel mass was stored in the films with the intake valve targeting, and that there was approximately 30% less storage in port films with OVI versus CVI.

As discussed previously, McGee et al. (2000) compared OVI with CVI and examined the effects of targeting. The two targets they investigated were the top of the valve and the port floor. For throttle ramp transients with cold coolant, OVI with valve targeting produced the smallest A/F ratio excursions, while CVI with port targeting produced the worst. Port targeting produced a larger intake film mass for both OVI and CVI and for both cold and warm coolant (although the difference in film mass between valve and port targeting was small with cold coolant). Surprisingly, port targeting produced improved homogeneity for both OVI and CVI for both loads examined and for both cold and warm coolant. Port targeting also produced lower HC emissions for OVI for both speeds and temperatures examined. However, port targeting increased HC emissions with CVI for all cases.

Kim et al. (2003b) showed that a 5° change in injector targeting produced a 45% change in liner wetting for their test conditions. Perhaps the most important reason that injector targeting affects in-cylinder wall wetting is that the port is much cooler than the port side of the intake valve [Alkidas (2001)], as was shown in Figure 2.1. As also noted in Section 2.2, the port side of the intake valve does not heat evenly. A dominant source of heat loss from the valve body is via fuel evaporation [Cowart and Cheng (1999); Alkidas (2001)]. The downstream portion of the valve is shielded from the fuel spray by the valve stem and thus heats faster. The portion of the valve that is wetted by the fuel will heat even more slowly by fuel enrichment during starting and warm-up. Thus, the reason for targeting the intake valve(s)—and avoiding port wall wetting—is that this is the hottest area in the port, promoting vaporization [Cowart and Cheng (1999); Rottenkolber et al. (1999); Alkidas (2001)].

As noted previously, the most significant factor contributing to cold-start HC emissions is the need for over-fueling (which, in turn, leads to in-cylinder surface wetting) to achieve a rapid start. To achieve ignition during cold start, a sufficient quantity of the injected liquid fuel must vaporize and find its way to the spark plug by the time of spark. It can take several engine cycles for this to occur. Combustion usually begins in the first cycle in which the fuel vapor/air ratio exceeds the lean flammability limit of the mixture [Quader and Majkowski (1999)]. A fuel vapor/air in-cylinder λ of approximately 0.6 is needed for ignition. Fox *et al.* (1992) found that the excess air ratio must reach approximately 0.8 before stable engine firing is achieved. They sampled in-cylinder gases during cold start, measuring fuel concentrations with an FFID. The time for stable operation is thought to depend on the thermal inertia of the intake port and valve as they are heated by the reverse flow of burned gas during the gas exchange process. Koenig *et al.* (1997) measured time-resolved in-cylinder fuel concentrations during simulated cold starts in a motored four-valve PFI engine via infrared (IR) absorption using a fiber-optic spark plug. Both studies found that the parameters most important to building a combustible mixture in the cylinder are the inlet manifold air pressure and the amount of fuel injected. Over-fueling provides more light ends of the fuel for vaporization, and a low manifold pressure strongly promotes fuel vaporization. Both studies also found little difference between CVI and OVI on cold-start performance. Fox *et al.* (1992) noted that OVI at part load had no noticeable effect on the starting and warm-up processes. Koenig *et al.* (1997) observed that OVI and CVI yielded the same fuel vapor concentration near the spark gap at TDC of compression, and in both cases, the TDC fuel concentration was approximately twice as large with a MAP of 0.5 versus 1 atm. Both studies also found engine speed and injection timing to have lesser effects on the mixture formation and cold-start characteristics. Fox *et al.* (1992) also found that a significant amount of fuel appeared lost to the crankcase during cold start.

The amount of over-fueling that is required depends on the ambient temperature. Quader and Majkowski (1999) spectroscopically measured the fraction of injected gasoline vaporized to achieve first combustion at two different temperatures in a Waukesha CFR engine. They found that at 22°C, 57% of the injected fuel had vaporized in the first cycle in which combustion took place, versus only 30% at a temperature of 0°C. Thus, the required enrichment was almost twice as great at the lower temperature. They found that the measured fuel vapor concentrations never reached a level as great as the calculated equilibrium-based fuel vapor-air equivalence ratios. They also found that after the first cycle with combustion, dilution with residual gas could contribute to misfires in subsequent cycles, requiring that the residual-free fuel vapor/air equivalence ratio be greater than the lean flammability limit in succeeding cycles (typically, $\lambda \sim 0.6$).

Santoso and Cheng (2002) estimate that the cranking/starting process may be responsible for more than 60% of the allowable HC emissions for a SULEV vehicle. Therefore, they examined the first two cycles of engine starting. They motored the engine on a dyno at three different speeds, with a corresponding MAP from actual engine starting experiments, and injected only once. Figure 2.8 shows a portion of their results for the first cranking cycle. The in-cylinder equivalence ratio $\left(\phi = \dfrac{1}{\lambda}\right)$ increases with increased injected amount, as expected, and with increased cranking speed. The high

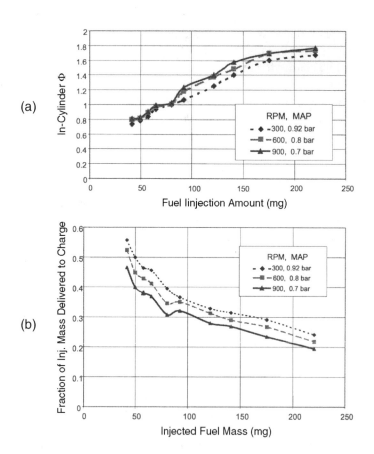

Figure 2.8 Effect of mass of fuel injected and cranking speed (and corresponding MAP) on in-cylinder mixture strength and fraction of fuel delivered to charge: (a) in-cylinder equivalence ratio, and (b) fraction of injected mass delivered to charge [Santoso and Cheng (2002)].

MAP for the first cycle inhibits vaporization. The decreased MAP achieved with the higher cranking speed allows more fuel vaporization. Although increasing the injected amount increases the in-cylinder mixture strength, more mass also remains in the intake port to contribute to in-cylinder wall wetting on subsequent cycles. Although the in-cylinder equivalence ratio is higher at the higher speed, the engine-out HCs decreased with increasing speed. They explained this by examining heat release rates for the various cranking speeds with the same in-cylinder ϕ. On a crank angle basis, combustion was faster at the lower speeds, yielding lower temperatures during the later portion of the expansion stroke. In turn, this yields less in-cylinder and port oxidation at the lower speeds. They also found that the fuel vapor/air equivalence ratio on the first cycle had to be 0.7 to 0.9 ($1.1 < \lambda < 1.4$) to obtain a robust first fire. The minimum in engine-out HCs (with 40°C coolant) occurred for $\phi \sim 0.9$. Additionally, they found that no fuel injection was required to obtain ignition on the second cycle if sufficient fuel was delivered on the first cycle to yield an in-cylinder ϕ during the second cycle of approximately 0.8.

Kim *et al.* (2003a) also used their new CICT system, which was discussed previously, to examine the effects of cranking speed on liner wetting for one of the five injectors in their study. They found that the wetted area and color intensity (an indication of film mass) both increased as the cranking speed increased from 150 to 250 rpm, but then decreased at 300 rpm. They hypothesize that secondary atomization, from fuel droplets bouncing off the liner, may be responsible for the change in behavior at the highest speed.

Fuel rail and fuel injector heating have been investigated as means to improve atomization, and thereby vaporization, from port fuel injectors. Sunwoo *et al.* (1999) heated the fuel rail of a PFI injection system. They examined a two-hole, a six-hole, and a pintle-type injector. They found that the SMD of the droplets decreased between 3 and 25% when the fuel was heated from 25 to 40°C. Zimmerman *et al.* (1999) examined a prototype heated tip injector for its effect on droplet size and HC emissions. The injector was able to heat the fuel residing in the injector from 20 to 65°C in 5 seconds. The objective was to induce flash boiling to enhance vaporization. They found that the heated injector reduced the SMD of the droplets from approximately 100 to approximately 20 µm with almost no droplets bigger than 50 µm. They also found a benefit for cold-start HC emissions.

2.4 Flow Field Effects

Airflow velocities into an engine, bulk flow velocities in-cylinder, and turbulence intensity in-cylinder all tend to scale approximately linearly with engine speed. Increased flow velocities enhance the rates of mixing between the fuel vapor and the air. At lower engine speeds, such as cranking and idling during the warm-up period, the flow velocities and turbulence are low. Many manufacturers install charge motion control valves (CMCVs) in the intake runners, which can be actuated to partially or completely block the flow in a runner at low engine speeds. In some configurations, one of the two runners in a dual-intake valve engine is blocked to both increase intake velocities and impose a swirling bulk flow. Some engines have CMCVs in both runners that will block the lower portion of each runner to enhance tumble at low speeds [e.g., Carney (2003); Kidoro et al. (2003)].

Several automakers recently have found new reasons to employ CMCVs. The traditional role of CMCVs, as a means of enhancing burn rate at low speed, was to provide better combustion stability leading to enhanced tolerance to EGR and lean operation. Their new role is for enhanced cold-start performance. The better mixture preparation and enhanced burn rates during cold start allow extreme spark retard, leading to higher exhaust gas temperatures for faster catalyst light-off [Nishizawa et al. (2000); Kishi et al. (1999); Takahashi et al. (1998)]. Charge motion control valves can improve mixture preparation through enhanced secondary atomization of fuel, increased rates of fuel evaporation, better wall-film atomization and vaporization, and improved mixing of fuel-air and EGR or residual/mixture. These factors may lead to a reduced cold-start enrichment requirement. Also, CMCVs provide an opportunity to control mixture distribution through charge stratification.

Charge motion control using CMCVs can be implemented in several ways. One method is through intake masking where a masking plate is located upstream of the injector, partially blocking the entire flow to that cylinder. A second means is through port deactivation, where one of the two intake runners downstream of the injector is blocked by a rotating plate positioned just upstream of one of the two intake valves. A third method is valve deactivation implemented by reducing the lift of one of the intake valves using the variable valve timing and lift (VVT/L) system of the engine.

One study [Kapus and Poetscher (2000)] noted that for a given minimum fuel consumption, port deactivation exhibited lower HC emissions when compared with valve deactivation running on a reduced valve lift. It was believed that with valve deactivation, the fuel spray injected into the port is deflected off the valve running on low valve lift. This may result in larger droplets as fuel is stripped from the valve and/or more surface impingement of fuel within the combustion chamber. In comparison, with port deactivation, both valves are open; thus, the fuel flowing into the cylinder through the two intake valves is comparable to open-port operation for which these flows were properly matched in the design phase.

Castaing *et al.* (2000) found that the most important effect of the piston starting position was on the instantaneous revolutions per minute of the first firing cycle, ranging from approximately 200 to 600 rpm. As already noted, this affects bulk flow velocities and turbulence. The faster speeds were achieved when the piston started near TDC compression, and the slower speeds when starting from the middle of the exhaust or intake strokes. They found that at the slower speeds, the combustion process was fast relative to the engine speed, resulting in advanced combustion phasing such that a typical spark timing of 10° BTDC was too advanced. For the cold starts, as the in-cylinder vapor/air mixture strength was increased from stoichiometric to rich, the cranking engine-out HC values rose from approximately 3000 to 6000 ppm C_1. When the mixture was very rich ($\lambda < 0.65$, $\phi > 1.54$) the HC emissions rose sharply due to incomplete combustion. The effect of cranking speed was small and may have been masked by other factors.

Weaver *et al.* (2003) examined the effects of enhanced swirl, using a CMCV, on mixture homogeneity. Using a conventional port injector, a PLIF tracer that boils near T_{50}, and steady-state firing with cold coolant, they found no effect of charge motion on the mixture homogeneity or COV of the equivalence ratio in the imaging plane. As discussed earlier, there may have been an effect on the homogeneity of the heavier fuel species. However, with the CMCV closed (one port partially blocked to enhance swirl), they found a large increase in droplets between the intake valves at 90° ATDC of intake compared to the images with the CMCV open. In both cases, those droplets that did not impact a wall vaporized before the time of ignition.

Imatake *et al.* (1997) found that, for OVI, in-cylinder wall wetting was approximately 30% smaller with swirl than without during the early cycles of a transient from low load to wide open throttle with 30°C coolant. Moriyoshi *et al.* (1998) showed that swirl inhibits radial diffusion of the fuel. That is, swirl tends to constrain droplets within the center of the chamber, inhibiting droplet impact on the liner.

Oguma *et al.* (2003) report the effects of swirl during cold-start cranking and idle. Their swirl control valve is fully closed during cranking and idle to promote improved swirl at low speeds. They also incorporated injectors that produced 71-μm droplets. The result was less liner wetting, a decreased requirement for over-fueling, decreased engine-out HC emissions, and faster catalyst light-off, as illustrated in Figure 2.9.

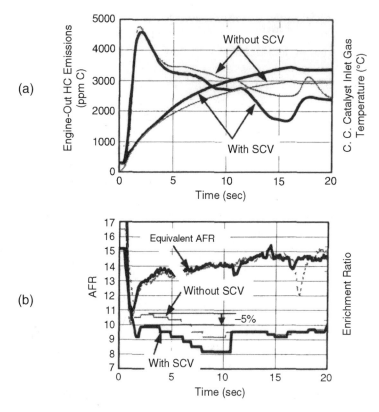

Figure 2.9 *Effects of enhanced swirl on required enrichment on (a) engine-out HC emissions and catalyst inlet temperature, and (b) air/fuel ratio and enrichment ratio [Oguma et al. (2003)].*

Kidoro *et al.* (2003) report the effects of enhanced tumble at idle. They used an intake air control valve (IACV) with twelve-hole injectors that produced an SMD of 70 μm. The IACV system forces the flow through the top of the intake ports to promote tumble. As shown in Figure 2.10, the IACV system allows leaner operation.

Mixture Formation Processes

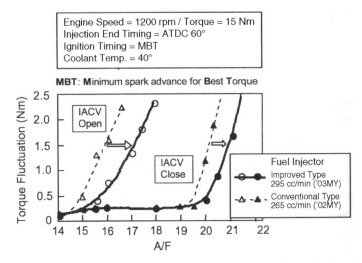

Figure 2.10 *Effects of enhanced tumble on dilute operation for a high-speed idle with cold coolant [Kidoro et al. (2003)].*

Although the goal is to have a homogeneous mixture throughout the combustion chamber at the time of ignition, several studies suggest that some degree of heterogeneity may not inhibit the combustion process and may even increase the rate of combustion. Cho and Santavicca (1993) measured flame kernel growth rates after spark ignition for incompletely mixed propane-air mixtures. The measurements were made in a flow tunnel at ambient conditions of 1 atm and 300 K with mixtures having an overall stoichiometry of $\lambda = 1$. Five cases were studied: a premixed case, and four cases having RMS mixture fluctuations in fuel/air equivalence ratio of 6, 13, 22, and 33%. Their results showed that fluctuations in the local mixture strength due to incomplete mixing increased flame wrinkling and that the wrinkling increased as the mixture inhomogeneity increased. Interestingly, the rates of flame kernel growth, however, were statistically the same for the first four cases and slower for the most inhomogeneous mixture. Zhou *et al.* (1998) measured flame propagation speeds in a combustion bomb where the degree of inhomogeneity of a propane/air mixture was varied. They measured flame propagation speeds for a range of equivalence ratios and at different levels of turbulence intensity. They found that with a variation of the mixture from a homogeneous distribution to different degrees of heterogeneity, the flame propagation speed first increased, reaching a maximum value for a particular degree of heterogeneity, and then decreased as the mixture became more heterogeneous. Using Schlieren imaging, they found that the degree of flame wrinkling increased as the mixture became more heterogeneous, in agreement with the observations of Cho and Santavicca (1993).

2.5 Strategies to Improve Fuel Preparation and Reduce Liquid Fuel Effects During Cold Start and Warm-Up

Various engine management strategies have been adopted by automakers in recent years to minimize the tailpipe emissions of HCs during cold start and warm-up. Some are focused on minimizing engine-out emissions; others aim at decreasing the catalyst light-off time. Most of these strategies are discussed in Chapters 3 through 9. Only those aspects related to mixture preparation and in-cylinder wall wetting are discussed in this section.

Several techniques can be used to decrease the catalyst light-off time. Catalysts, including methods for attaining fast light-off, are the subject of Chapters 5 through 8 inclusive. However, two factors that affect light-off are related to mixture preparation or at least mixture stoichiometry. Thus, these factors that affect catalyst light-off are discussed below.

The effects of spark retard are the subject of Chapter 5. Retarding the spark timing from MBT results in decreased crevice packing (i.e., decreased engine-out HCs), increased exhaust gas temperatures, and increased HC oxidation in the exhaust system [e.g., Russ *et al.* (1999a)]. It also results in decreased indicated torque and thus decreased idle speed [e.g., Choi *et al.* (2000); Ashford *et al.* (2003)]. If the airflow is increased (e.g., via the idle air bypass system), the increased exhaust enthalpy yields decreased catalyst light-off time. However, many gasoline engines are intolerant of significant spark retard during the warm-up period. Russ *et al.* (1999a and 1999b) examined the effects of spark retard, primarily at 1200 rpm, 2.5-bar IMEP, $\lambda = 1.06$, and 20°C fluids with a 97 RON gasoline. For these conditions, cyclic variations in the IMEP were dominated by variability in the crank angle location of 50% mass fraction burned (MFB), with the late burn having little effect as this fraction burns near exhaust valve opening (EVO) with retarded timing and contributes little to the IMEP. They found that increasing swirl and tumble decreased cyclic variability, thereby allowing additional spark retard. They also found that the HC concentration in the exhaust port was much lower when the fluids were at normal operating temperatures, illustrating the importance of fuel evaporation. McGee *et al.* (2000) found that vapor fueling allowed more spark retard than liquid fueling.

At least one method for decreasing the catalyst light-off time requires high engine-out HC emissions. Borland and Zhao (2002) of DaimlerChrysler point out that decreasing catalyst-in HCs is of more importance than decreasing engine-out HCs prior to light-off. Their system uses the normal rich starting and warm-up strategy plus secondary air injection (air injection into the exhaust upstream of the catalyst) to promote thermal oxidation in the exhaust system prior to the catalyst. This system is the subject of Chapter 6.

Other strategies are focused on decreasing engine-out HC emissions during starting and warm-up. These strategies are discussed below.

Fuel compensation, as discussed in Section 2.2, is now routinely used and will not be discussed further. Roberts and Stanglmaier (1999) showed that retarding the intake cam phasing improved the IMEP and cyclic variability during cold start and warm-up. They note several techniques that could be employed to take advantage of these factors to reduce engine-out HC emissions for vehicles equipped with variable valve timing (VVT) or cam phasing systems. Batteh and Curtis (2003) showed that cam phasing can have a strong influence on A/F ratio excursions during the cold-start transient.

A strategy reported by Honda [Kishi *et al.* (1999)] to minimize the need for over-fueling makes use of AAIs to sequentially inject the fuel when the intake valves are open. Most previous studies that have examined the effect of injection timing on HC emissions have found that engine-out emissions are greater with open valve injection due to the direct impingement of liquid fuel on the cylinder walls. The Honda system works because, by injecting with an open valve, more fuel enters the cylinder where it can vaporize as opposed to wetting the closed intake valves, significantly reducing the liquid fuel deposition in the port. They then use in-cylinder bulk flow and the improved atomization associated with the AAIs to keep the injected fuel from being deposited on the cylinder walls. (Note that Imatake *et al.* [1997] found that swirl decreased liner wetting with OVI but did not eliminate it. However, they did not use an AAI.) The result is that a greater fraction of the injected fuel mass is vaporized.

Kidoro *et al.* (2003) report on Toyota's development of a 2003 Model Year PZEV and compare it with its 2002 Model Year ULEV counterpart. To generate the PZEV, they used improved twelve-hole injectors, modified the control and base timing of their VVT system, added an IACV, retarded the spark timing, increased the idle speed for cold start and warm-up, and made various changes to the catalyst system. The IACV system forces the flow to the top of the intake ports to promote tumble. They state that the IACV-induced tumble allows both leaner operation and retarded ignition timing, with the combination of these three factors producing a decrease in nonmethane hydrocarbon (NMHC) emissions of 40 to 50% over the FTP. The improved twelve-hole injector produces an SMD of 70 µm. The new injector plus tumble decrease port and liner wetting and allow leaner operation. The VVT system used for their 2002 Model Year ULEV was not used during cold start and warm-up but was modified to operate during this period for 2003. However, VVT cannot be incorporated until the oil pressure is sufficiently high after starting to allow valve actuation. Instead, they changed the overlap to 6°CA when the VVT is in the neutral position (inactive). When the intake valve opens, reverse flow of the hot residual from the prior cycle promotes evaporation of the port wall film and atomization and evaporation of fuel droplets. They also note that it promotes reburning of the HCs from the prior cycle. The change in overlap during the cold start and early warm-up process allowed the use of less enrichment, as shown in Figure 2.11. They also developed a VVT position feedback sensor to allow VVT operation while the oil viscosity is still high, allowing VVT operation earlier in the warm-up process. Use of this neutral VVT timing, plus VVT during the warm-up and less starting enrichment, produced a 20% decrease in NMHC emissions.

Oguma *et al.* (2003) reported on Nissan's development of a third generation of PZEV technology. The changes they made that are related to mixture preparation are a change in the fuel injector design, the use of a continuously variable swirl control valve (CV-SCV), an increase in cranking speed, and optimized ignition timing. The new injector design decreased the droplet size (presumably SMD) from 129 to 71 µm. This improves atomization and vaporization, thus decreasing port and in-cylinder fuel films. This allows approximately a 3% reduction in over-fueling during the period of approximately 2 to 7 seconds after start. Overall, the engine-out HCs decreased approximately 5% "right after engine start." A CV-SCV was used on the Nissan first-generation (2000 Model Year) PZEV. The swirl control valve (SCV) is fully closed during cranking and idling to increase swirl [Nishizawa *et al.* (2001)]. The increased swirl and smaller droplets yield less liner wetting. This allows a decrease in over-fueling by approximately 5% from the cranking pulse up to approximately 11 seconds after starting. Together with the retarded ignition timing, the engine-out HCs decreased approximately 12%, and the catalyst inlet temperature increased by ~30°C.

At least two systems undergoing development directly address the need for cold start and warm-up over-fueling and the resulting wall wetting problems

Mixture Formation Processes

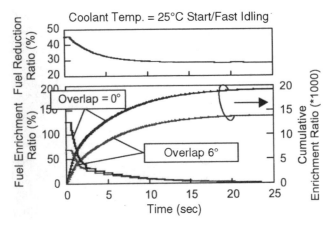

Figure 2.11 *Decreased fuel enrichment requirement due to the change in overlap for a cold start to a high idle [Kidoro et al. (2003)].*

by either prevaporizing the fuel or generating a very volatile starting fuel from gasoline. These systems are discussed below.

Delphi is developing an onboard partial oxidation reformer (POx) system to overcome the need for over-fueling during starting and warm-up [Kirwan et al. (1999 and 2002); Quader et al. (2003)]. This system is discussed in detail in Chapter 7 and will not be discussed further here.

Another promising new technology to reduce the need for over-fueling is the onboard distillation system (OBDS) under development at the University of Texas [Ashford et al. (2003)]. Heat from the engine coolant system is used to selectively vaporize the volatile components of the gasoline, which are then recondensed and stored in a separate tank. The use of this volatile fuel (mostly C_5s) for cold starting and warm-up eliminates the need for enrichment, allowing stoichiometric or slightly lean starting and lean warm-up ($\lambda \sim 1.3$). Additionally, this volatile fuel is much more tolerant of spark retard during cold starts and warm-up than is gasoline, allowing approximately 40°CA of retard from stock, compared to only ~20°CA for gasoline. An OBDS was installed on a 2001 Lincoln Navigator, with ~10,000 miles on the catalyst, to explore the emissions reductions possible on a large vehicle with a large-displacement (5.4-L V8) engine. The fuel and spark calibration of the PCM were modified to exploit the benefits of the OBDS startup fuel. Lean starting and warm-up alone resulted in a 43% decrease in total hydrocarbon (THC) over the FTP, partially due to a 23% decrease in the light-off time. When the additional benefit of aggressive spark retard was added to the lean warm-up, the light-off time was decreased by 55% (compared to stock), and the reactivity adjusted NMOG emissions fell by 82% to 0.02 g/mi, half the ULEV II 50,000-mile standard (Ashford et al. (2005)). The engine-out and tailpipe HC emissions over the first 90 seconds of the FTP are provided in Figure 2.12.

2.6 Summary

Cold start and warm-up pose special problems for HC emissions control over the FTP cycle. When the surface temperatures of the engine are cold, the sources of engine-out HCs are high, resulting in high tailpipe HCs until the catalyst reaches light-off temperature. Because of their temperature dependency, engine-out HCs due to the flame quench, oil film, deposit, and crevice sources are inherently higher during cold start and warm-up than after the surfaces have reached normal operating temperatures. However, the most important source during this period results from the need to over-fuel the engine due to the low efficiency of fuel evaporation off cold surfaces. This generally results in a rich fuel vapor/air mixture for at least the first several seconds of the FTP. In turn, this increases engine-out HCs from the exhaust valve leakage source. More importantly, the liquid fuel source is extremely high during cold start and warm-up. It has been shown that the liquid fuel source can persist well after catalyst light-off, yielding high tailpipe HCs during the first three accelerations of Phase I of the FTP.

If batch firing is used while awaiting synchronization, some cylinders will have OVI while others will have CVI, but possibly within ~80°CA of IVO when HCs begin to increase for steady-state operation. In general, OVI yields liner wetting under the exhaust valves and head wetting around the intake valves and between the exhaust valves. Closed-valve injection produces more port and valve wetting than OVI, plus some piston wetting and head wetting between the intake valves. In-cylinder fuel films persist and build up over successive cycles when the surfaces are cold. Some of the liner film is scraped into the crankcase as "lost fuel," while some of the liner film is scraped up closer to the exhaust valves. The in-cylinder wetting location is important because the closer that location is to the exhaust port, the less likely evaporated fuel from that location will undergo in-cylinder oxidation or be retained within the cylinder. That is, wetting locations closer to the exhaust port contribute more to engine-out HCs. Also, the head films from OVI begin to form on earlier cycles than for CVI. Even if injection is delayed until after synchronization such that CVI can be

Mixture Formation Processes

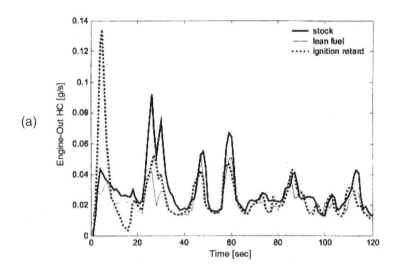

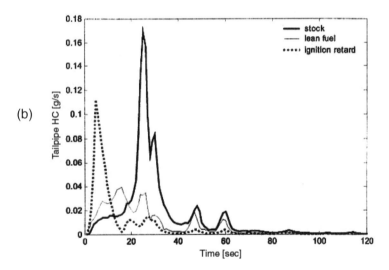

Figure 2.12 *Engine-out (top) and tailpipe (bottom) HC emissions for the first 120 seconds of the FTP, comparing the stock system to use of the OBDS system: (a) engine-out HC, and (b) tailpipe HC [Ashford et al. (2003)].*

used exclusively, and even if the start of injection for CVI is right after IVC to provide the maximum time for fuel evaporation, little fuel will evaporate because the port and intake valve are cold. The reverse flow right after IVO is advantageous because it redistributes the fuel in the port and aids evaporation. However, port films can persist for more than 10 minutes, even with the very small droplets obtained via AFI. Port wetting is important because it leads to in-cylinder wall wetting.

The four mechanisms that produce in-cylinder wall wetting are as follows:

1. Fuel stripping from the intake valve during intake valve opening

2. The direct flow of liquid fuel into the combustion chamber from injectors with OVI

3. Transport during the middle of the intake stroke driven by the high-speed flow of intake air

4. Droplets formed when the fuel in the film on the face of the valve and the valve seat is squeezed during intake valve closing

Injector targeting is important because the rate of evaporation is dictated by the surface temperature. The intake valve heats much faster than the port wall, for which the temperature is very close to that of the coolant outlet throughout warm-up. Fuel on the port wall nearest the exhaust valve(s) produces higher HC emissions than port films farthest from the exhaust valve(s). The location of the port films affects the location of in-cylinder films that are produced by the port films. Flame passage does not completely evaporate in-cylinder fuel films because the in-cylinder surface temperatures increase relatively slowly. In-cylinder fuel films, especially in the vicinity of the exhaust, contribute to exhaust HCs because evaporation from cold surfaces is slow. Some of these in-cylinder films will evaporate late in the expansion stroke, when in-cylinder gas temperatures are too low for complete oxidation, or during the exhaust stroke. Also, in-cylinder fuel films can lead to pool fires [Witze and Green (1997)], which can lead to emissions of both HCs and particulates.

Difficulty in evaporating the heavier components of the gasoline result in in-cylinder wall wetting. The lighter fractions of the fuel, up through at least T_{50}, become well mixed with the air by the time of ignition. Some rich pockets of the heavier ends may remain until the time of ignition. Although these inhomogeneities affect combustion stability, the evidence indicates that inhomogeneities do not affect cold startability and probably have only a small

Mixture Formation Processes

effect on cold-start HC emissions. The heavier ends in the fuel films, especially in-cylinder films, dominate cold-start HC emissions.

The beginning of the FTP, and the corresponding calibration, can be divided into stages. The first stage is cranking/starting. No fuel compensation can be used for the first injection because there is no liquid fuel left from the prior cycle. Although there may still be fuel in some cylinders from the prior shutdown [Klein and Cheng (2002)], this is difficult to account for in a cranking calibration. The engine starts on the lighter (more volatile) ends of the gasoline, leaving the heavier ends behind in fuel films. Also, the crank pulse must be calibrated to compensate for high-DI gasolines, which are less abundant in the light ends. Additionally, there is no backflow of hot residual to aid evaporation for the first firing cycle. Thus, eight to fifteen times the stoichiometric amount of gasoline typically is injected during the first several cycles of the FTP cold start (20 to 30°C). An in-cylinder fuel vapor/air ϕ of 0.7 to 0.9 ($1.1 < \lambda < 1.4$) is required on the first cycle to obtain a robust first fire. The minimum in engine-out HCs occurs for an in-cylinder fuel vapor/air $\phi \sim 0.9$. Even if a richer in-cylinder vapor/air mixture is used for the first fire, there is obviously a lot of liquid fuel remaining in the films, and these films are enriched in the heavier, more difficult to evaporate components. If sufficient fuel is injected on the first cycle, enough vapor may be generated between cycles that the extreme of fuel compensation—no injection—can be used for the second cycle. However, a higher mass injected results in thicker port and valve films, and consequently thicker in-cylinder films, which evaporate more slowly.

The second stage is warm-up at idle in neutral (from the end of crank/flare to 15 seconds of the FTP). The fuel films from cranking/starting persist and build. Because it takes approximately one minute for the intake valves to complete the first (rapid) stage of warm-up, fuel evaporation remains a problem. Also, the port and in-cylinder surface temperatures have changed little from the temperature before cranking. Most of the valve film and much of the downstream port film are swept into the cylinder, where some of this fuel is deposited on the relatively cold in-cylinder surfaces. Fuel compensation can be used during this and subsequent stages. However, the consumer expects a stable idle, and this normally requires a rich vapor/air mixture with little spark retard. It has been shown that crankshaft speed fluctuations can be used to compensate for high-DI gasolines during this period. Once the speed fluctuation signals are detected and processed, this allows the minimum fueling required (as a function of gasoline volatility) to maintain a stable idle. That is, it allows the warm-up calibration enrichment to be based on certification gasoline but dynamically compensates for high-DI gasolines to maintain idle quality during the warm-up transient.

The third stage is warm-up at idle in drive (15 to 20 seconds of the FTP). The increase in MAP impedes evaporation of fuel films both in the intake and in-cylinder. The intake valves are still in the first stage of heating, and the port and in-cylinder surface temperatures have changed little. Evaporation from these films remains a problem.

The next stage is the first acceleration and cruise (20 to 114 seconds of the FTP). The intake valves complete the rapid stage of heating approximately 40 to 50 seconds into the FTP (less than 60 seconds due to the increased load) but do not reach normal operating temperature. The port and liner continue to heat slowly. The increased MAP impedes vaporization. The first acceleration generates high engine-out HCs due to transfer of fuel film from the intake to in-cylinder surfaces as a result of the increased flow velocities past the valves. Typically, the catalyst will light off during this stage.

The key to minimizing engine-out HCs during cold start and warm-up is to minimize in-cylinder surface wetting, especially at locations near the exhaust valves. Both tumble and swirl, in conjunction with small droplets, decrease in-cylinder wetting, allow warm-up with mixtures that are not as rich (perhaps even lean), and allow more spark retard. Charge motion control valves have been shown to be effective in increasing both tumble and swirl, not only for idle speeds but also during cranking. Improved atomization can significantly decrease port wetting but does not eliminate the problem of poor vaporization during cold start and warm-up. Air-forced injectors produce the smallest droplets, followed by AAIs and then by multihole injectors. In fact, multihole injectors with high rail pressure can produce droplets nearly as small as those produced via AFIs. However, several PZEVs in production use droplets that have an SMD of the order of 70 μm, which is larger than those from AFIs (~10 μm) or AAIs (~25 μm). Open-valve injection minimizes port wetting (with proper targeting) but must be used in conjunction with small droplets and a strong bulk flow that minimizes impaction on in-cylinder surfaces, especially those near the exhaust valves.

Variable valve timing, or cam phasing, can be used to enhance the backflow to improve vaporization from port films during the starting and warm-up process.

Two technologies are being developed that either eliminate or greatly minimize fuel wetting in the intake and, thereby, in-cylinder. These reformer and onboard distillation technologies merit continued development.

2.7 References

1. Alger, T., Huang, Y., Hall, M.J., and Matthews, R.D. (2001), "Liquid Film Evaporation Off the Piston of a Direct Injection Gasoline Engine," SAE Paper No. 2001-01-1204, Society of Automotive Engineers, Warrendale, PA; also in *Journal of Engines*, Vol. 110, pp. 1295–1306.

2. Alkidas, A.C. (1994), "The Effect of Fuel Preparation on Hydrocarbon Emissions of a S.I. Engine Operating Under Steady-State Conditions," SAE Paper No. 941959, Society of Automotive Engineers, Warrendale, PA.

3. Alkidas, A.C. (2001), "Intake-Valve Temperature Histories During S.I. Engine Warm-Up," SAE Paper No. 2001-01-1704, Society of Automotive Engineers, Warrendale, PA.

4. Alkidas, A.C. and Drews, R.J. (1996), "Effects of Mixture Preparation on HC Emissions of an SI Engine Operating Under Steady-State Cold Conditions," SAE Paper No. 961958, Society of Automotive Engineers, Warrendale, PA.

5. Ashford, M., Matthews, R., Hall, M., Kiehne, T., Dai, W., Curtis, E., and Davis, G. (2003), "An On-Board Distillation System to Reduce Cold Start Hydrocarbon Emissions," SAE Paper No. 2003-01-3239, Society of Automotive Engineers, Warrendale, PA.

6. Ashford, M.D. and Matthews, R.D. (2005), "Further Development of an On-Board Distillation System for Generating a Highly Volatile Cold-Start Fuel," SAE Paper No. 2005-01-0233, Society of Automotive Engineers, Warrendale, PA; also in *Journal of Fuels and Lubricants*, Vol. 4, pp. 131–137.

7. Batteh, J.J. and Curtis, E.W. (2003), "Modeling Transient Fuel Effects with Variable Cam Timing," SAE Paper No. 2003-01-3126, Society of Automotive Engineers, Warrendale, PA.

8. Borland, M. and Zhao, F. (2002), "Application of Secondary Air Injection for Simultaneously Reducing Converter-in Emissions and Improving Catalyst Light-Off Performance," SAE Paper No. 2002-01-2803, Society of Automotive Engineers, Warrendale, PA.

9. Boyle, R.J., Doam, D.J., and Finlay, I.C. (1993), "Cold-Start Performance of an Automotive Engine Using Prevaporized Gasoline," SAE Paper No. 930710, Society of Automotive Engineers, Warrendale, PA.

10. Carney, D. (2003), "Global I4 Goes PZEV," *Automotive Engineering*, Vol. 111, No. 7, pp. 76–78.

11. Castaing, B.M., Cowart, J.S., and Cheng, W.K. (2000), "Fuel Metering Effects on Hydrocarbon Emissions and Engine Stability During Cranking and Startup in a Port-Fuel Injected Spark Ignition Engine," SAE Paper No. 2000-01-2836, Society of Automotive Engineers, Warrendale, PA.

12. Chen, K.C., Cheng, W.K., Van Doren, J.M., Murphy, J.P. III, Hargus, M.D., and McSweeney, S.A. (1996), "Time-Resolved, Speciated Emissions from an SI Engine During Starting and Warm-Up," SAE Paper No. 961955, Society of Automotive Engineers, Warrendale, PA.

13. Cho, Y.-S. and Santavicca, D.A. (1993), "The Effect of Incomplete Fuel-Air Mixing on Spark-Ignited Flame Kernel Growth," SAE Paper No. 932715, Society of Automotive Engineers, Warrendale, PA.

14. Choi, M.-S., Sun, H.-Y., Lee, C.-H., Myung, C.-L., Kim, W.-T., and Choi, J.-K. (2000), "The Study of HC Emission Characteristics and Combustion Stability with Spark Timing Retard at Cold Start in Gasoline Engine Vehicle," SAE Paper No. 2000-01-1082, Society of Automotive Engineers, Warrendale, PA.

15. Cowart, J. and Cheng, W. (1999), "Intake Valve Thermal Behavior During Steady-State and Transient Engine Operation," SAE Paper No. 1999-01-3643, Society of Automotive Engineers, Warrendale, PA.

16. Crane, M.E., Thring, R.H., Podnar, D.J., and Dodge, L.G. (1997), "Reduced Cold-Start Emissions Using Rapid Exhaust Port Oxidation (REPO) in a Spark-Ignition Engine," SAE Paper No. 970264, Society of Automotive Engineers, Warrendale, PA.

17. Felton, P.G., Kyritsis, D.C., and Fulcher, S.K. (1995), "LIF Visualization of Liquid Fuel in the Intake Manifold During Cold Start," SAE Paper No. 952464, Society of Automotive Engineers, Warrendale, PA.

18. Fox, J.W., Min, K.D., Cheng, W.K., and Heywood, J.B. (1992), "Mixture Preparation in a SI Engine with Port Fuel Injection During Starting and Warm-Up," SAE Paper No. 922170, Society of Automotive Engineers, Warrendale, PA.

19. Heywood, J.B. (1998), "Motor Vehicle Emissions Control: Past Achievement, Future Prospects," in *Handbook of Air Pollution from Internal Combustion Engines*, Sher, E. (ed.), Academic Press, New York, NY.

20. Hires, S.D. and Overington, M.T. (1981), "Transient Mixture Strength Excursions—An Investigation of Their Causes and the Development of a Constant Mixture Strength Fueling Strategy," SAE Paper No. 810495, Society of Automotive Engineers, Warrendale, PA.

21. Huang, Y., Alger, T., Matthews, R.D., and Ellzey, J.E. (2001a), "The Effects of Fuel Volatility and Structure in HC Emissions from Piston Wetting in DISI Engines," SAE Paper No. 2001-01-1205, Society of Automotive Engineers, Warrendale, PA; also in *Journal of Fuels Lubricants*, Vol. 110, pp. 912–929.

22. Huang, Y., Matthews, R.D., Ellzey, J.E., and Dai, W. (2001b), "The Effects of Fuel Volatility, Load, and Speed on HC Emissions Due to Piston Wetting," SAE Paper No. 2001-01-2024, Society of Automotive Engineers, Warrendale, PA; also in *Journal of Engines*, Vol. 110, pp. 1878–1889.

23. Imatake, N., Saito, K., Morishima, S., Kudo, S., and Ohhata, A. (1997), "Quantitative Analysis of Fuel Behavior in Port-Injected Gasoline Engines," SAE Paper No. 971639, Society of Automotive Engineers, Warrendale, PA.

24. Ireton, J.R. and Kaiser, E.W. (2001), "Can Fuel Preparation Affect Engine-Out Hydrocarbon Emissions During an FTP (75CVS) Cycle Test?," SAE Paper No. 2001-01-1312, Society of Automotive Engineers, Warrendale, PA.

25. Kaiser, E.W., Siegl, W.O., Baidas, L.M., Lawson, G.P., Cramer, C.F., Dobbins, K.L., Roth, P.W., and Smokovitz, M. (1994), "Time-Resolved

Measurement of Speciated Hydrocarbon Emissions During Cold Start of a Spark-Ignited Engine," SAE Paper No. 940963, Society of Automotive Engineers, Warrendale, PA.

26. Kaiser, E.W., Siegl, W.O., Lawson, G.P., Connolly, F.T., Cramer, C.F., Dobbins, K.L., Roth, P.W., and Smokovitz, M. (1996), "Effect of Fuel Preparation on Cold-Start Hydrocarbon Emissions from a Spark-Ignited Engine," SAE Paper No. 961957, Society of Automotive Engineers, Warrendale, PA.

27. Kapus, P. and Poetscher, P. (2000), "ULEV and Fuel Economy—A Contradiction?" SAE Paper No. 2000-01-1209, Society of Automotive Engineers, Warrendale, PA.

28. Kelly-Zion, P.L., Styron, J.P., Lee, C-F., Lucht, R.P., Peters, J.E., and White, R.A. (1997), "In-Cylinder Measurements of Liquid Fuel During the Intake Stroke of a Port Injected Spark Ignition Engine," SAE Paper No. 972945, Society of Automotive Engineers, Warrendale, PA.

29. Kidoro, T., Hoshi, K., Hiraku, K., Satoya, K., Watanabe, T., Fujiwara, T., and Suzuki, H. (2003), "Development of PZEV Exhaust Emission Control System," SAE Paper No. 2003-01-0817, Society of Automotive Engineers, Warrendale, PA.

30. Kim, C. and Foster, D.E. (1985), "Aldehyde and Unburned Fuel Emissions Measurements from a Methanol-Fueled Texaco Stratified-Charge Engine," SAE Paper No. 852120, Society of Automotive Engineers, Warrendale, PA.

31. Kim, H., Yoon, S., Lai, M-C., Quelhas, S., Boyd, R., Kumar, N., and Yoo, J-H. (2003a), "Correlating Port Fuel Injection to Wetted Fuel Footprints on Combustion Chamber Walls and UBHC in Engine Start Processes," SAE Paper No. 2003-01-3240, Society of Automotive Engineers, Warrendale, PA.

32. Kim, M., Cho, H., Cho, Y., and Min, K. (2003b), "Computational and Optical Investigation of Liquid Fuel Film on the Cylinder Wall of an SI Engine," SAE Paper No. 2003-01-1113, Society of Automotive Engineers, Warrendale, PA.

33. Kirwan, J., Quader, A., and Grieve, M. (2002), "Fast Start-Up On-Board Gasoline Reformer for Near Zero Emissions in Spark Ignition Engines,"

SAE Paper No. 2002-01-1011, Society of Automotive Engineers, Warrendale, PA.

34. Kirwan, J.E., Quader, A.A., and Grieve, M.J. (1999), "Advanced Engine Management Using On-Board Partial Oxidation Reforming for Meeting Super-ULEV (SULEV) Emissions Standards," SAE Paper No. 1999-01-2927, Society of Automotive Engineers, Warrendale, PA.

35. Kishi, N., Kikuchi, S., Seki, Y., Kato, A., and Fujimori, K. (1998), "Development of the High Performance L4 Engine ULEV System," SAE Paper No. 980415, Society of Automotive Engineers, Warrendale, PA.

36. Kishi, N., Kikuchi, S., Suzuki, N., and Hayashi, T. (1999), "Technology for Reducing Exhaust Gas Emissions on Zero Level Emission Vehicles (ZLEV)," SAE Paper No. 1999-01-0772, Society of Automotive Engineers, Warrendale, PA.

37. Klein, D. and Cheng, W. (2002), "Spark Ignition Engine Hydrocarbon Emissions Behaviors in Stopping and Restarting," SAE Paper No. 2002-01-2804, Society of Automotive Engineers, Warrendale, PA.

38. Koenig, M.J., Stanglmaier, R.H., Hall, M.J., and Matthews, R.D. (1997), "Mixture Preparation During Cranking in a Port-Injected 4-Valve SI Engine," SAE Paper No. 972982, Society of Automotive Engineers, Warrendale, PA.

39. Landsberg, G.B., Heywood, J.B., and Cheng, W.K. (2001), "Contribution of Liquid Fuel to Hydrocarbon Emissions in Spark Ignition Engines," SAE Paper No. 2001-01-3587, Society of Automotive Engineers, Warrendale, PA.

40. Lee, G.R. and Morley, C. (1995), "Fuel-Wall Impaction as a Mechanism for Increased Hydrocarbon Emissions from Fuel Heavy Ends," SAE Paper No. 952523, Society of Automotive Engineers, Warrendale, PA.

41. Li, J.W., Huang, Y., Alger, T.F., Matthews, R.D., Hall, M.J., Stanglmaier, R.H., Roberts, C.E., Dai, W., and Anderson, R.W. (2000), "Liquid Fuel Impingement on In-Cylinder Surfaces as a Source of Hydrocarbon Emissions from Direct Injection Gasoline Engines," ASME Paper No. 2000-ICE-270, in *Fuel Injection, Combustion, and Engine Emissions,* ICE Vol. 34-2, ASME, New York, NY, pp. 17–26.

42. Li, J.W., Huang, Y., Alger, T.F., Matthews, R.D., Hall, M.J., Stanglmaier, R.H., Roberts, C.E., Dai, W., and Anderson, R.W. (2001), "Liquid Fuel Impingement on In-Cylinder Surfaces as a Source of Hydrocarbon Emissions from Direct Injection Gasoline Engines," *Journal of Gas Turbines and Power*, Vol. 123, pp. 659–668.

43. Li, J.W., Matthews, R.D., Stanglmaier, R.H., Roberts, C.E., and Anderson, R.W. (1999), "Further Experiments on In-Cylinder Wall Wetting in Direct Injected Gasoline Engines," SAE Paper No. 1999-01-3661, Society of Automotive Engineers, Warrendale, PA; also in *Journal of Fuels and Lubricants*, Vol. 108, pp. 2213–2224.

44. Matthews, R., Huang, Y., Alger, T., Hall, M., Ellzey, J., Stanglmaier, R., Roberts, C., Dai, W., and Anderson, R. (2001), "The Piston Wetting Source of HC Emissions from Direct-Injected Spark Ignition Engines," *Direkteinspritzung im Ottomotor III*, Spicher, U. (ed.), Expert Verlag, Renningen-Malmsheim, Germany, pp. 291–313.

45. McGee, J., Curtis, E., Russ, S., and Lavoie, G. (2000), "The Effects of Port Fuel Injection Timing and Targeting on Fuel Preparation Relative to a Pre-Vaporized System," SAE Paper No. 2000-01-2834, Society of Automotive Engineers, Warrendale, PA.

46. Meyer, R. and Heywood, J.B. (1999), "Effect of Engine and Fuel Variables on Liquid Fuel Transport into the Cylinder in Port Injected SI Engines," SAE Paper No. 1999-01-0563, Society of Automotive Engineers, Warrendale, PA.

47. Meyer, R., Yilmaz, E., and Heywood, J.B. (1998), "Liquid Fuel Flow in the Vicinity of the Intake Valve in a Port Injected SI Engine," SAE Paper No. 982471, Society of Automotive Engineers, Warrendale, PA.

48. Mizaikoff, B., Fuss, P., and Hall, M.J. (1998), "Fast-Spec: An Infrared Spectroscopic Diagnostic to Measure Time-Resolved Exhaust Hydrocarbon Emissions from SI Engines," *Twenty-Seventh Symposium (International) on Combustion*, The Combustion Institute, Pittsburgh, PA.

49. Moriyoshi, Y., Nomura, H., and Saisyu, Y. (1998), "Evaluation of a Concept for DI Gasoline Combustion Using Enhanced Gas Motion," SAE Paper No. 980152, Society of Automotive Engineers, Warrendale, PA.

50. Nishizawa, K., Mitsuishi, S., Mori, K., and Yamamoto, S. (2001), "Development of Second Generation of Gasoline P-ZEV Technology," SAE Paper No. 2001-01-1310, Society of Automotive Engineers, Warrendale, PA.

51. Nishizawa, K., Momoshima, S., Koga, M., Tsuchida, H., and Yamamoto, S. (2000), "Development of New Technologies Targeting Zero Emissions for Gasoline Engines," SAE Paper No. 2000-01-0890, Society of Automotive Engineers, Warrendale, PA.

52. Nitschke, R.G. (1993), "Reactivity of SI Engine Exhaust Under Steady-State and Simulated Cold-Start Operating Conditions," SAE Paper No. 932704, Society of Automotive Engineers, Warrendale, PA.

53. Oguma, H., Koga, M., Momoshima, S., Nishizawa, K., and Yamamoto, S. (2003), "Development of Third Generation of Gasoline P-ZEV Technology," SAE Paper No. 2003-01-0816, Society of Automotive Engineers, Warrendale, PA.

54. Quader, A.A., Kirwan, J.E., and Grieve, M.J. (2003), "Engine Performance and Emissions Near the Dilute Limit with Hydrogen Enrichment Using an On-Board Reforming Strategy," SAE Paper No. 2003-01-1356, Society of Automotive Engineers, Warrendale, PA.

55. Quader, A.A. and Majkowski, R.F. (1999), "Cycle-by-Cycle Mixture Strength and Residual-Gas Measurements During Cold Starting," SAE Paper No. 1999-01-1107, Society of Automotive Engineers, Warrendale, PA.

56. Quader, A.A., Sloane, T.M., Sinkevitch, R.M., and Olsen, K.L. (1991), "Why Gasoline 90% Distillation Temperature Affects Emissions with Port Fuel Injection and Premixed Charge," SAE Paper No. 912430, Society of Automotive Engineers, Warrendale, PA.

57. Roberts, C.E. and Stanglmaier, R.H. (1999), "Investigation of Intake Timing Effects on the Cold Start Behavior of a Spark Ignition Engine," SAE Paper No. 1999-01-3622, Society of Automotive Engineers, Warrendale, PA.

58. Rottenkolber, G., Dullenkopf, K., Wittig, S., Kolmel, A., Feng, B., and Spicher, U. (1999), "Influence of Mixture Preparation on Combustion and Emissions Inside an SI Engine by Means of Visualization, PIV and IR Thermometry During Cold Operating Conditions," SAE Paper No. 1999-01-3644, Society of Automotive Engineers, Warrendale, PA.

59. Russ, S., Stevens, J., Aquino, C., Curtis, E., and Fry, J. (1998), "The Effects of Injector Targeting and Fuel Volatility on Fuel Dynamics in a PFI Engine During Engine Warm-Up: Part 1—Experimental Results," SAE Paper No. 982518, Society of Automotive Engineers, Warrendale, PA.

60. Russ, S., Lavoie, G., and Dai, W. (1999a), "SI Engine Operation with Retarded Ignition: Part 1—Cyclic Variations," SAE Paper No. 1999-01-3506, Society of Automotive Engineers, Warrendale, PA.

61. Russ, S., Thiel, M., and Lavoie, G. (1999b), "SI Engine Operation with Retarded Ignition: Part 2—HC Emissions and Oxidation," SAE Paper No. 1999-01-3507, Society of Automotive Engineers, Warrendale, PA.

62. Samenfink, W., Albrodt, H., Frank, M., Gesk, M., Melsheimer, A., Thurso, J., and Matt, M. (2003), "Strategies to Reduce HC-Emissions During the Cold Starting of a Port Fuel Injected Gasoline Engine," SAE Paper No. 2003-01-0627, Society of Automotive Engineers, Warrendale, PA.

63. Santoso, H. and Cheng, W. (2002), "Mixture Preparation and Hydrocarbon Emissions Behaviors in the First Cycle of SI Engine Cranking," SAE Paper No. 2002-01-2805, Society of Automotive Engineers, Warrendale, PA.

64. Servati, H.B. and Yuen, W.W. (1984), "Deposition of Fuel Droplets in Horizontal Intake Manifolds and the Behavior of Film Flow on Its Walls," SAE Paper No. 840239, Society of Automotive Engineers, Warrendale, PA.

65. Skippon, S.M., Nattrass, S.R., Kitching, J.S., Hardiman, L., and Miller, H. (1996), "Effects of Fuel Composition on In-Cylinder Air/Fuel Ratio During Fueling Transients in an SI Engine, Measured Using Differential Infra-Red Absorption," SAE Paper No. 961204, Society of Automotive Engineers, Warrendale, PA.

66. Stache, I. and Alkidas, A.C. (1997), "The Influence of Mixture Preparation on the HC Concentration Histories from an SI Engine Running Under Steady-State Conditions," SAE Paper No. 972981, Society of Automotive Engineers, Warrendale, PA.

67. Stanglmaier, R.H., Hall, M.J., and Matthews, R.D. (1997), "In-Cylinder Fuel Transport During the First Cranking Cycles in a Port Injected 4-Valve SI Engine," SAE Paper No. 970043, Society of Automotive Engineers, Warrendale, PA; also in *Journal of Engines* Vol. 106, pp. 56–72.

68. Stanglmaier, R.H., Li, J.W., and Matthews, R.D. (1999), "The Effect of In-Cylinder Wall Wetting on HC Emissions from SI Engines," SAE Paper No. 1999-01-0502, Society of Automotive Engineers, Warrendale, PA; also in *Journal of Engines*, Vol. 3, pp. 533–542.

69. Stovell, C., Matthews, R.D., Johnson, B.E., Ng, H., and Larsen, B. (1999), "Emissions and Fuel Economy of a 1998 Toyota with a Direct Injection Spark Ignition Engine," SAE Paper No. 1999-01-1527, Society of Automotive Engineers, Warrendale, PA.

70. Sunwoo, M., Yoon, P., Park, S., Eo, Y., and Cheon, D. (1999), "An Experimental Study of Influences of Fuel-Rail Heating on Fuel Atomization," SAE Paper No. 1999-01-0793, Society of Automotive Engineers, Warrendale, PA.

71. Swindal, J.C., Dragonetti, D.P., Hahn, R.T., Furman, P.A., and Acker, W.P. (1995), "In-Cylinder Charge Homogeneity During Cold-Start Studied with Fluorescent Tracers Simulating Different Fuel Distillation Temperatures," SAE Paper No. 950106, Society of Automotive Engineers, Warrendale, PA.

72. Takahashi, H., Momoshima, S., Ishizuka, Y., Tomita, M., and Nishizawa, K. (1998), "Engine-Out and Tail-Pipe Emission Reduction Technologies of V-6 LEVs," SAE Paper No. 980674, Society of Automotive Engineers, Warrendale, PA.

73. Takeda, K., Yaegashi, T., Saito, K., Sekiguti, K., and Imatake, N. (1995), "Mixture Preparation and HC Emissions of a 4-Valve Engine with Port Injection During Cold Starting and Warm-Up," SAE Paper No. 950074, Society of Automotive Engineers, Warrendale, PA.

74. Weaver, C., Wooldridge, S., Johnson, S., Sick, V., and Lavoie, G. (2003), "PLIF Measurements of Fuel Distribution in a PFI Engine Under Cold Start Conditions," SAE Paper No. 2003-01-3236, Society of Automotive Engineers, Warrendale, PA.

75. Witze, P.O. and Green, R.M. (1997), "LIF and Flame-Emission Imaging of Liquid Fuel Films and Pool Fires in an SI Engine During a Simulated Cold Start," SAE Paper No. 970866, Society of Automotive Engineers, Warrendale, PA.

76. Yang, J., Kaiser, E.W., Siegl, W.O., and Anderson, R.W. (1993), "Effects of Port-Injection Timing and Fuel Droplet Size on Total and Speciated Exhaust Hydrocarbon Emissions," SAE Paper No. 930711, Society of Automotive Engineers, Warrendale, PA.

77. Zhou, J., Nishida, K., Yoshizaki, T., and Hiroyasu, H. (1998), "Flame Propagation Characteristics in a Heterogeneous Concentration Distribution of a Fuel-Air Mixture," SAE Paper No. 982563, Society of Automotive Engineers, Warrendale, PA.

78. Zhu, G.S., Reitz, R.D., Kim, J., and Takabayashi, T. (2001), "Characteristics of Vaporizing Continuous Multi-Component Fuel Sprays in a Port Fuel Injection Gasoline Engine," SAE Paper No. 2001-01-1231, Society of Automotive Engineers, Warrendale, PA.

79. Zimmerman, F., Bright, J., Ren, W., and Imoehl, B. (1999), "An Internally Heated Tip Injector to Reduce HC Emissions During Cold Start," SAE Paper No. 1999-01-0792, Society of Automotive Engineers, Warrendale, PA.

CHAPTER 3

Cold-Start Hydrocarbon Emissions Mechanisms

James A. Eng
General Motors Research

3.1 Introduction

With the lowering of the hydrocarbon (HC) emissions standard, a proportionally larger fraction of the total emissions from the Federal Test Procedure (FTP) is emitted during the cold-start portion of the first cycle. At the ultra-low emissions vehicle (ULEV) standard, 80 to 90% of the tailpipe HC emissions are emitted during the first test cycle of the FTP [Takeda *et al.* (1995)]. The proportion of HC emissions emitted during the cold start is expected to increase further at the super ultra-low emissions vehicle (SULEV) standard. Tailpipe HC emissions during the cold start are high because the catalyst is not at its operating temperature to efficiently oxidize the HCs. Thus, tailpipe HC emissions during the cold start are essentially equal to engine-out HC emissions until the catalyst reaches its light-off temperature (typically around 300°C). To shorten the time to obtain catalyst light-off, advanced engine control strategies are used, such as retarding the spark to increase the exhaust gas temperature. However, because the rate of catalyst heating is limited by the thermal stresses produced in the catalyst [Gulati (1999)], it will not be possible to obtain the required HC emissions reduction through fast catalyst light-off and control strategies alone. Engine-out HC emissions also must be reduced to meet the SULEV emissions standard. To reduce engine-out HC emissions during the cold start, it is essential to understand the HC emissions mechanisms from spark ignition (SI) engines.

Engine-out HC emissions from SI engines are the result of two processes. The first is the storage of raw fuel within the cylinder that escapes combustion. This is the initial source of HCs. The second is the oxidation of the raw HCs. After the primary combustion process has consumed the major portion of the combustible (i.e., vaporized) fuel within the cylinder, the unburned HCs either can be completely consumed into CO, CO_2, and H_2O, or can be partially oxidized into lower-molecular-weight HC species that exit the engine and contribute to engine-out HC emissions. The rate of mixing between the unburned HCs and the high-temperature burned gas during the expansion stroke is an important process for post-flame HC consumption.

The sources of HC emissions from engines have been studied extensively for more than 40 years. Most of the studies have been performed on engine dynamometers at steady-state conditions. As a result, we currently have a good understanding of the important HC emissions mechanisms under steady-state conditions. As the significance of cold-start HC emissions has become more important, a large number of experiments have been performed at cold-engine structure and fluid temperatures to determine the sources of the increased HC emissions from cold-engine operation. In addition, during the past decade, an increasing number of studies have been performed to investigate the effects of engine speed and load transients on HC emissions. These tests have been performed both on simulated startup transients on dynamometers and on cold-start tests of production engines. From these experiments, we are starting to increase our knowledge base about the effects of cold, transient engine operation on HC emissions.

This chapter reviews all known HC emissions mechanisms from port fuel injected (PFI) SI engines, starting with fully warmed-up, steady-state conditions and then systematically discussing the effects of reduced coolant temperature and transient engine operation on HC emissions. The important physical processes for each mechanism at steady-state conditions are discussed, and the known physics is used to estimate changes in the storage mechanism for cold, transient engine operation.

3.2 Global Engine Behavior During a Cold Start

To obtain a proper perspective of engine behavior and HC emissions during a cold start, it is helpful to examine the global aspects of engine operation during starting. During the first cycle of the FTP, the engine is started and idled for 15 seconds before the transmission is shifted from neutral into drive. The first acceleration begins at 20 seconds after the start of the test. The engine speed and intake manifold absolute pressure (MAP) during a representative cold start are displayed in Figure 3.1 for the first 80 engine cycles. Both the engine speed and MAP change within an engine cycle as the engine responds to the torque demand and valve events from each individual cylinder. Thus, the values displayed in the figure are average values over a given engine cycle.

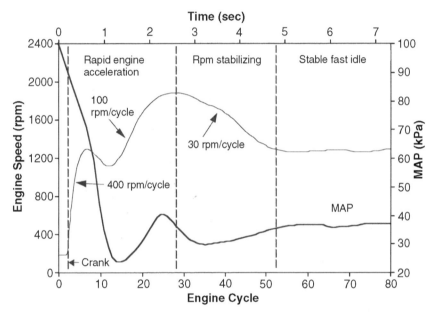

Figure 3.1 Engine speed and intake manifold pressure during a representative cold start.

The cold-start transient can be divided roughly into the following four phases [Sampson and Heywood (1995)]:

1. Crank
2. Rapid engine acceleration
3. Speed stabilization
4. Stable fast idle

During cranking, the engine speed is 200 rpm, and the MAP is at atmospheric pressure. After cranking, there is a rapid increase in engine speed as the engine transitions into a run condition. In the example shown in Figure 3.1 there are two different engine accelerations as the engine speeds up. Initially, the engine speed changes at a rate of 400 rpm per cycle, and then the acceleration decreases to 100 rpm per cycle until a peak engine speed of 1900 rpm is reached at engine cycle 28. The MAP responds to changes in both the engine speed and the commanded airflow during starting. In the example considered here, a minimum MAP of 22 kPa is obtained during engine cycle 14. After engine cycle 28, the engine speed decreases relatively slowly, and the MAP increases slightly until a stable fast idle condition is reached. The engine speed and MAP are stable during the remainder of the cold start until the engine is shifted into drive.

The engine speed and cylinder pressure from cylinder 1 are displayed in more detail for the first 10 engine cycles shown in Figure 3.2. The engine typically is cranked for a number of revolutions to enable the engine controller to synchronize the fuel injection pulses with the crankshaft position. In this example, the engine was cranked for two revolutions before cylinder 1 fired during engine cycle 3. After the first firing, there is a rapid increase in engine speed. The primary effect of the rapid engine acceleration is to reduce the overall time available to complete combustion before the exhaust valve opens. The higher engine speeds bring with them larger intake mixture motion

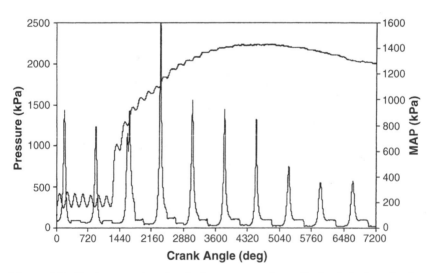

Figure 3.2 Engine speed and cylinder pressure for cylinder 1 during the first 10 engine cycles of a cold start.

levels so that the in-cylinder turbulence levels are increased, thus hastening combustion. A partial burn occurs when the turbulence levels are not high enough to increase the bulk burn rate so that combustion is completed before the exhaust valve opening (EVO). The decrease in the MAP leads to an increase in the in-cylinder residual levels that can have a negative impact on flame propagation by diluting the charge. A beneficial effect of the reduced MAP during a cold start is an increased amount of hot residual gases blowing back into the intake port, which helps fuel vaporization and improves fuel preparation [Kidokoro *et al.* (2003)].

3.2.1 Required Fueling Levels

The amount of fuel required during a cold start is significantly higher than that needed for warmed-up operation. At cold conditions, only approximately 20% of the fuel is vaporized during the first few engine cycles [Cheng and Santoso (2002)]. As a result, the engine must be over-fueled significantly to create a combustible mixture. This over-fueling results in large amounts of liquid fuel entering the cylinder, which can be a major source of engine-out HC emissions. Figure 3.3 is a plot of the minimum fuel required per cylinder for stable combustion without misfire during a cold start of a four-cylinder PFI engine [Takeda *et al.* (1995)]. The amount of fuel required for the first cycle is significantly more than required for later cycles. The required fuel after 300 cycles is 26 mg/cylinder/cycle. Note that the steady fueling level of 26 mg is equal to 17% of the fuel injected during the first cycle, which agrees well with the 20% fuel vaporization estimate by Cheng and Santoso (2002).

The fuel used in the experiments shown in Figure 3.3 had a temperature for 50% fuel evaporation (T_{50}) equal to 105°C. The volatility of the fuel obviously will have a large impact on the fueling levels required to cold start an engine. For example, Takeda *et al.* (1995) measured a 50% increase in the minimum fueling level for the first engine cycle when the T_{50} temperature was increased from 91 to 120°C. This minimum fueling level increase also was required for the second and third engine cycles during the cold start. The T_{50} distillation temperature is an important fuel property for reducing HC emissions during the cold start and engine warm-up. The effect of fuels on engine-out and tailpipe HC emissions is discussed in Chapter 7.

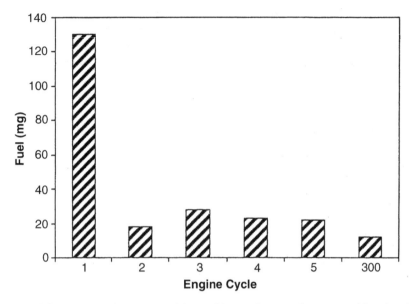

Figure 3.3 *Minimum fuel required for stable combustion during a cold start of a four-cylinder engine [Takeda* et al. *(1995)].*

3.2.2 Fuel Accounting

Comparisons between the total amount of fuel that is injected into the engine with the fuel mass that was burned and the cumulative engine-out HC and CO emissions indicate that a substantial fraction of the fuel is unaccounted for during the transient. Estimates of the unaccounted fuel during a cold start vary widely and are dependent on the engine and the cold-start control strategy. The cumulative amount of unaccounted fuel during the first 10 engine cycles of a cold start was measured by Sampson and Heywood (1995) to be 150 to 250 mg/cylinder. This amount of fuel was equal to roughly 35% of the total injected fuel during the first 10 engine cycles (approximately 2 seconds duration). Shayler *et al.* (1999) estimated that during engine warm-up tests, 500 to 1000 mg/cylinder was unaccounted for during the first 100 seconds. This unaccounted fuel is stored within the fuel films in the intake port and manifold, in-cylinder liquid fuel films, and liquid fuel carried with blowby gas from the cylinder into the crankcase with the lubrication oil. The fuel entering the crankcase is desorbed over a period of minutes as the oil temperature increases, and the vaporized fuel is returned to the engine via the positive crankcase ventilation system. The amount of unaccounted fuel depends on the engine temperature and the volatility of the fuel. At sub-zero temperatures, the amount of unaccounted fuel during engine warm-up tests increases to 2000 to 3000 mg/cylinder/cycle [Shayler et al. (1999)].

The amount of unaccounted fuel can be estimated by comparing the measured air/fuel (A/F) ratio delivered to the engine with that measured in the exhaust gas by a wide-range air/fuel (WRAF) sensor. The difference between these two A/F measurements is typically on the order of 2 A/F ratios following a cold start [Shayler et al. (1997)]. These differences in A/F ratio can exist for up to 100 seconds following engine startup. After the engine has reached a stable operating condition, even with forced cold coolant and intake air temperatures, the A/F measurements agree closely, and the unaccounted fuel amounts to less than 2% of the total fuel flow. This indicates that the effective "loss" of fuel through the engine is a result of the transient process.

The distribution of the fuel mass between the fuel films in the intake port and cylinder was measured directly by Takeda et al. (1995) in an innovative series of experiments. In these experiments, the valvetrain of a production engine was replaced with a hydraulic valve actuation system that was able to shut the valves at any specified time during a start. The fuel stored within the intake port and cylinder was measured during each cycle of the cold start by closing the valves and rapidly stopping the engine to trap the liquid fuel within the engine. The fuel was then vaporized by purging the cylinder and intake port with air heated to 200°C, and the vaporized fuel in the purge air was measured with an HC flame ionization detector (FID) and converted to a mass basis. Figure 3.4 shows the results for the first three engine cycles of a cold start. After the first cycle, 70% of the injected fuel was contained within the port and cylinder fuel film layers and was carried into the next cycle. A significant amount of fuel is carried between cycles, which makes fuel control during a cold start challenging. Interestingly, the data show that the trapped fuel per cycle (the injected fuel plus the residual fuel from the previous cycle) is rather constant at 130 mg/cycle. The exhaust HC emissions for the first three engine cycles are equal to 7 to 15% of the trapped fuel.

The total weighted tailpipe HC emissions allowed at the SULEV emissions standard over the FTP test cycle is 111 mg. Using the weighting factor of 0.43 for the first bag (0 to 505 seconds of the FTP test), this translates into roughly 250 mg from the first portion of the test. This 250 mg serves as a useful benchmark for cold-start HC emissions. Depending on the engine and control strategy, the unaccounted fuel during the cold start can range from 500 to

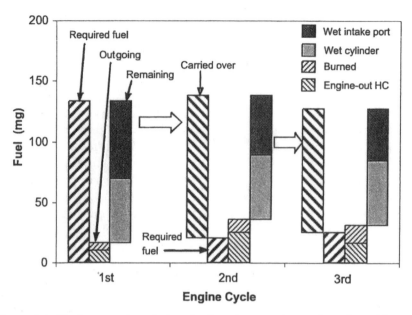

Figure 3.4 Fuel accounting during the first three engine cycles of a cold start [Takeda et al. (1995)].

2000 mg/cylinder. For a four-cylinder engine, this translates into an amount of fuel storage within the engine that is roughly 10 to 30 times larger than the total allowed engine-out HC emissions. In other words, engine-out HC emissions are the small difference between the injected fuel and the fuel storage within the engine. Again, the large amount of unaccounted fuel within the engine makes determining the sources of emissions and their relative importance during the cold start challenging. One of the key factors for developing low-emissions engines will be to minimize the amount of excess fuel injected into the engine during the cold start.

3.3 Hydrocarbon Storage Mechanisms

The mechanisms by which fuel can be stored in the cylinder and can escape combustion are as follows:

1. Storage in combustion chamber crevices
2. Absorption into oil layers and deposits

3. Liquid fuel within the cylinder that is too rich to burn
4. Quenching on combustion chamber surfaces
5. Partial burns

Figure 3.5 schematically shows these mechanisms. Although exhaust valve leakage can be a source of HC emissions for poorly maintained engines, this mechanism is largely an issue of engine maintenance and is not a fundamental source of HC emissions from engines. During a cold start, the engine is operated fuel-rich to avoid misfire, and the rich A/F ratio leads to increased HC emissions. Also shown in Figure 3.5 are the liquid fuel films, which are known to exist within the intake port and cylinder. No meaningful discussion of cold-start HC emissions can be made without considering the effects of fuel storage within these areas during the transient. Each of these mechanisms will be systematically discussed in the following sections.

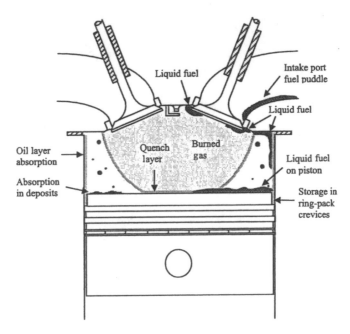

Figure 3.5 *Hydrocarbon storage mechanisms for cold start.*

3.3.1 Storage in Crevices

During compression, unburned fuel enters crevices into which the flame cannot propagate and escapes combustion during the main combustion process. The main combustion process is typically taken to be the period during which 90% of the energy is released, as determined from heat release analysis. Crevices within the combustion chamber into which the fuel can be stored are in the exposed threads around the spark plug, in the crevices around the circumference of the intake and exhaust valves, in the cylinder head gasket, and in the piston ring-pack crevices. By far the largest crevice volume in the engine is the piston ring-pack. In modern engines, the ring-pack can account for up to 80% of the total crevice volume. Of the ring-pack volume, the largest volumes are those above and behind the top compression ring.

Wentworth (1968) was the first to identify the large effect of combustion chamber crevices and cylinder blowby levels on HC emissions. Figure 3.6 shows the effect of cylinder blowby flow rate on exhaust HC emissions at fully

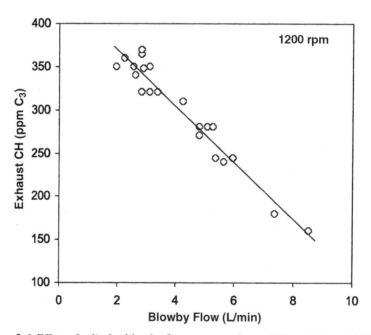

Figure 3.6 Effect of cylinder blowby flow rate on exhaust HC emissions; 1200 rpm engine speed, 54 kPa intake manifold pressure [Wentworth (1968)].

warmed-up, steady-state conditions. There is a linear decrease in HC emissions with an increase in the blowby flow rate. The mechanism for the reduction in HC emissions is simple: unburned fuel that enters the rings flows into the crankcase and is not returned to the cylinder as a source of unburned HCs during the expansion stroke.

The important effect of the cylinder blowby flow rate on crevice HC emissions should be kept in mind when interpreting experiments that are designed to determine the relative influence of any given storage mechanism on HC emissions. Any modification to the engine that affects the cylinder blowby will have a direct effect on the baseline crevice HC emissions, and the results of the experiment will be confounded with changes in the crevice contribution to HC emissions. This problem has been particularly true of experiments designed to investigate the effects of oil layer absorption on HC emissions by comparing emissions from engines operated without lubrication to those operated with lubrication [Ishizawa *et al.* (1987)]. Without an oil layer to seal the compression rings, the blowby levels are increased significantly, which results in a decrease in crevice HC emissions. When the engine is operated with oil, the increase due to oil layer absorption is confounded with the increase in crevice HC emissions due to lower blowby levels. Because the important physical parameters for the crevice storage mechanism are the cylinder pressure and the engine speed, the crevice HC mechanism applies equally well to transient engine operation during a cold start.

It is generally agreed that piston ring-pack crevices are the largest source of engine-out HC emissions at steady-state, fully warmed-up operation. Estimates for the contribution of ring-pack crevices to HC emissions at these conditions range from 30% [Cheng *et al.* (1993)] to as much as 70 to 90% [Eng *et al.* (1997); Alkidas (1999)]. Given this wide variation in the importance of crevice HC emissions, it is perhaps not too surprising to find that the literature abounds with conflicting results. For example, Min *et al.* (1995) observed only a relatively small effect of top-land crevice volume on HC emissions, and other experiments have indicated that the piston top-land volume has little effect on HC emissions during engine warm-up [Sterlepper *et al.* (1993)]. Experiments performed by Alkidas *et al.* (1995) demonstrated that a complex relationship exists between the piston side clearance and flame propagation into the crevice. Much of the conflicting results in the literature are undoubtedly a result of undesired (and unmeasured) changes in cylinder blowby flow rate with modified piston designs. It also is known that subtle changes in piston design have been shown to increase oil consumption significantly [Yoshida *et al.* (1996)]. Because oil consumption rates have been correlated with cylinder blowby flow rates, it is implied that changes also will take place in crevice HC emissions with small changes in piston ring-pack design.

Typical top-land side clearances in modern engines range from 0.25 to 0.4 mm. The heat transfer rates are high when the gas flows into the top-land crevice. As a result, it is commonly assumed that the gas temperature in the crevice is equal to the cylinder wall temperature [Namazian and Heywood (1982)]. Accordingly, as the coolant temperature is reduced, the contribution of the crevices becomes larger due to the increase in gas density within the crevice. In addition, the piston top-land side clearance increases due to thermal contraction of the piston. Russ et al. (1995) measured a 4% increase in HC emissions per 10°C change in coolant temperature in the cylinder block while the cylinder head temperature was maintained at a constant temperature. Roughly the same changes in HC emissions were obtained from experiments performed with isopentane (which vaporizes at 28°C at atmospheric pressure) and gasoline. These results indicate that there would be an expected 25% increase in steady-state HC emissions as the coolant temperature is reduced from 90 to 20°C.

3.3.2 Absorption in Oil Layers and Deposits

In the oil layer absorption mechanism, a portion of the fuel escapes oxidation during combustion by being absorbed into the oil layers and deposits during the intake and compression strokes. The concentration of the fuel vapor at the interface between the oil and the bulk gas is typically assumed to be given by Henry's law for a dilute system in equilibrium

$$X_F = \frac{p_F}{H}$$

where

X_F = mole fraction of fuel vapor in the oil

p_F = partial pressure of the fuel in the bulk gas

H = Henry's constant

Thus, fuel vapor continues to enter the oil layer as the cylinder pressure rises throughout the compression process and as combustion takes place. The fuel desorbs from the oil as the pressure decreases during the expansion stroke and can be a potential source of unburned HCs within the cylinder.

As mentioned in the previous section, experiments performed without cylinder lubrication and/or with different oil formulations are complicated by changes in cylinder blowby levels that change the baseline crevice HC emissions level. Thus, these experiments cannot be used to determine the relative influence of oil layers on HC emissions. The most unambiguous means to demonstrate the relative effect of oil layers on HC emissions is to compare HC emissions from fuels with different levels of solubility in the oil. Experiments performed with a fully blended gasoline and iso-pentane at different engine coolant temperatures exhibited the same decrease in emissions as the coolant temperature was increased from 70 to 100°C [Russ et al. (1995)]. Because iso-pentane has a much lower solubility in oil than gasoline, these results indicate that the oil solubility does not contribute significantly to HC emissions. A variation of this approach was developed by Kaiser et al. (1995) to further investigate the effects of fuel solubility. Experiments were performed with a full-boiling-range fuel doped with different HCs with significantly different solubilities in oil. The experiments were performed over a range of coolant temperatures from 60 to 120°C with a fully formulated mineral-based 5W-30 oil. The exhaust gas was speciated to determine if there were differences among the emissions from the various components added to the gasoline. If fuel oil absorption were important, then it would be expected that components with a high solubility should have relatively higher concentrations within the exhaust gas. However, the results did not show any significant differences between the various dopants, which again demonstrated that the effect of fuel oil absorption on HC emissions is small at these temperatures. Based on the experimental results, Kaiser et al. (1995) estimated that the contribution of oil layer absorption to HC emissions at operating temperatures higher than 60°C is less than 5%. Additional evidence for the small contribution of oil layer absorption to HC emissions was provided by Linna et al. (1997) in a set of experiments performed with a range of lubricant formulations and fuels with different oil solubilities. These experiments showed little effect of oil layer absorption on HC emissions at steady-state conditions.

Results from a one-dimensional cyclic absorption–desorption model indicate that oil layer diffusion is not a rate-limiting process and that the oil layer is nearly saturated with fuel [Dent and Lakshminarayanan (1983)]. Accordingly, the oil layer contribution to HC emissions is expected to vary linearly with the thickness of the oil layer within the combustion chamber. During cold engine operation, it is expected that the oil layer thickness will be larger than during warm operation due to the increased viscosity of the oil. There also may be more oil present within the combustion chamber, and the solubility of the fuel in the oil is increased at the lower temperatures. The solubility for a given fuel component is estimated to increase by a factor of 15 to 20 as the temperature is decreased from 380 to 300 K.

Making estimates for the effect of oil layer absorption along the lines of Kaiser *et al.* (1995), the contribution of oil layer absorption to HC emissions at 300 K is estimated to be less than 5 to 10%. However, note that these estimates are developed from steady-state data, and during engine warm-up, the effect may be different. In particular, the change in the oil layer thickness within the chamber during cold start is not known. While oil consumption rates during steady-state operation are controlled to low levels to avoid poisoning the catalyst, some experimental results indicate that oil consumption increases significantly during transient operation [Yilmaz *et al.* (2001)].

3.3.3 Liquid Fuel

The effect of liquid fuel on HC emissions is the easiest of the HC emissions mechanisms to understand. Liquid fuel enters the cylinder and impacts a cold surface where it does not vaporize, and it can exit the cylinder to become a source of HC emissions. The in-cylinder fuel films are very thin (on the order of 50 to 300 μm), such that the fuel is at the temperature of the surface [Shin *et al.* (1994)]. Because the higher-boiling-point components in gasoline vaporize over a range of temperatures from 100 to 200°C, much of the fuel remains unburned within the cylinder for a significant time following a cold start [Shayler *et al.* (1999)].

The effects of liquid fuel on HC emissions can easily be observed at steady-state conditions by performing an injection timing sweep. Figure 3.7 shows the variation in HC emissions with fuel injection timing at steady-state conditions [Yang *et al.* (1993)]. The experiments were performed with three different injection systems giving Sauter mean diameters (SMDs) of 300, 40, and 14 microns. An air-assist injection (AAI) system was used to generate the 14-micron spray. The tests were performed at 90 and 30°C coolant temperatures, as well as at rich and lean A/F ratios. The general effects of injection timing were the same at both A/F ratios; therefore, only data from the rich A/F ratios are shown in the figure. The valve timings also are displayed in the figure for reference.

With 90°C coolant temperature, the intake port wall is hot enough to vaporize a large fraction of the fuel, and the residual gas blowback into the intake port during the valve overlap period effectively atomizes and vaporizes the fuel contained in the wall film. Thus, the effect of injection timing on HC emissions

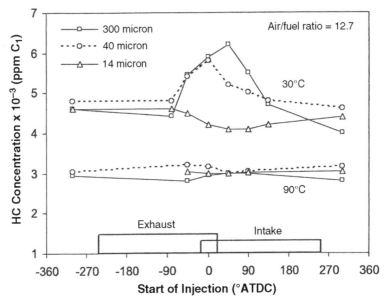

Figure 3.7 *Effect of injection timing on engine-out HC emissions at steady-state conditions; 1500 rpm engine speed, 378 kPa indicated mean effective pressure (IMEP) [Yang et al. (1993)].*

is small. However, at the lower coolant temperature, the injection timing has a pronounced effect on HC emissions. When fuel is injected during an open intake valve, there is a significant increase in HC emissions. With the 300-micron spray and open-valve injection (OVI), there is a 50% increase in HC emissions. By comparison, the 14-micron injector did not exhibit an increase in emissions during the open-valve period. With better fuel atomization, the fuel remains airborne and does not rely on heat transfer from the port wall and intake valve to vaporize. Speciation of the exhaust gas HCs showed variations in the heavier components in the exhaust with OVI at cold temperatures. With OVI, the percent of xylenes and methyl-benzenes in the exhaust gas increased. Interestingly, the contribution of these components to emissions actually decreased for the 14-micron injector. These trends are consistent with the physical model that the increase in HC emissions is due to wall impaction of liquid fuel that is then too cold to vaporize within the cylinder. After combustion, the heavier ends of the fuel are vaporized but are not oxidized by the high-temperature burned gas and contribute to HC emissions.

The effects of the location of the liquid fuel within the cylinder on steady-state HC emissions were investigated by Stanglmaier et al. (1999). The engine was fueled with gaseous propane, and a specially designed fuel injector was used to deposit small amounts of liquid fuel at discrete points within the cylinder. The experiments were performed with both 90 and 36°C coolant temperatures. The liquid fuel accounted for roughly 15% of the total fuel flow rate to the engine. By comparing changes in the exhaust gas A/F ratio with and without liquid fuel injection, it was estimated that for warm operation, nearly all of the fuel deposited on the piston top was vaporized within a given engine cycle. By comparison, only 80% of the fuel deposited on the cylinder wall evaporated. Fuel injected near the exhaust valves had the largest impact on HC emissions, while fuel injected near the intake valves had the least impact. The reason for the increase in emissions when the fuel was injected near the exhaust valve is that liquid fuel was able to flow directly out of the engine during exhaust blowdown. This behavior also has been observed by Fry et al. (1995) in visualization experiments during engine warm-up tests.

It is known that large amounts of liquid fuel enter the cylinder during engine warm-up. Models for fuel dynamics during a warm-up transient have indicated that the mass of the cylinder fuel film can rapidly increase to as much as 120 to 150 mg [Curtis et al. (1998)]. The model results indicate that the in-cylinder fuel film decreases to 20 mg over a period of 100 seconds. As stated in the introduction to this chapter, there is more than enough fuel within the cylinder during the cold start to account for the HC emissions from the engine, and it is generally believed that liquid fuel is a significant, if not the largest, contributor to cold-start HC emissions. While the influence of liquid fuel on HC emissions has not been completely ascertained, our understanding of the transport of the fuel has increased significantly during the last 10 years. Numerous experiments have been performed to directly visualize liquid fuel within the cylinder. The reader is referred to the paper by Witze (1999) for a comprehensive review of experimental techniques that can be applied to measure liquid fuel effects. Some of the phenomena that have been observed are pool fires located on the top of the piston late in the expansion stroke and the burning of liquid fuel layers around the intake valves [Campbell et al. (1996)].

Shin et al. (1994) observed liquid fuel entering the engine in a PFI optical engine. The different mechanisms of fuel interaction with the intake valve as it closed were identified. The liquid fuel in the cylinder was distributed among the fuel near the intake valves that flowed from the valve, liquid droplets that impinged on the combustion chamber, and in other isolated puddles. The largest amount of fuel was located near the intake valve. While the liquid fuel in other portions of the chamber was observed to disappear soon after the start of combustion, the fuel film near the intake valve existed for up to 60 seconds. Fry *et al.* (1995) observed liquid fuel in the cylinder during a warm-up transient that would collect near the intake valve. During exhaust blowdown, a rivulet of fuel was observed to flow from the cylinder directly into the exhaust. Because the exhaust port walls are cold during the initial cycles of a cold start, it would be expected that much of this fuel would reside on the exhaust port walls until the temperature is high enough to vaporize the fuel.

One of the difficulties in determining the effects of either fuel preparation or fuel volatility on cold-start performance is that significant differences in the amount of vaporized fuel can result from changes in either fuel injection or fuel volatility. This results in a change in the A/F vapor ratio that must be taken into consideration to separate the effect of liquid fuel from the large effect of A/F ratio on HC emissions. The effects of mixture preparation on cold-start engine performance and emissions were investigated by Kaiser *et al.* (1996). Simulated cold-start tests were performed using a prevaporized, central fuel injection (CFI) system and a conventional PFI system. The tests were performed at 23°C coolant temperature using a fully blended gasoline. The engine control unit was adjusted to minimize over-fueling during the test. Figure 3.8 shows the exhaust gas A/F equivalence ratios measured with a WRAF sensor for the first 100 seconds of the test. Although differences exist between the two fueling systems, for the first 30 seconds, the difference is less than 1.5 A/F ratios between the two tests.

Figure 3.9 displays the engine-out HC emissions resulting from the two fuel systems. There are significant differences in HC levels early during the test. Hydrocarbon emissions from the PFI system are nearly 80% higher than those from the CFI fueling system. It is significant to note that the WRAF sensor measurements indicate that the PFI system was operating closer to stoichiometry than the CFI system. This lower A/F ratio should have resulted in a decrease in HC emissions rather than an increase. After 40 seconds, the differences in HC emissions have been reduced to roughly 15%. The elevated HC emissions from the PFI system persist throughout the entire duration of the test. These experimental results suggest that a larger in-cylinder fuel film is produced by the PFI fuel system. The liquid fuel layer is also the cause of the lean shift in the exhaust A/F ratio early during the test. The increase in HC emissions during the first 40 seconds is a result of the liquid fuel layer slowly vaporizing as the engine warms up. Computational studies of the vaporization of a thin liquid fuel layer show results that are in general agreement with these conclusions [Oliveira and Hochgreb (2000)].

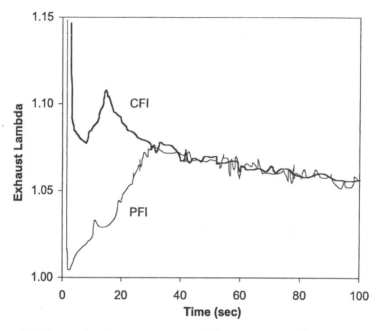

Figure 3.8 *Comparison between measured exhaust gas equivalence ratio during a simulated cold-start test using PFI and prevaporized CFI fuel systems; 1200 rpm, 58 kPa intake manifold pressure [Kaiser* et al. *(1996)].*

3.3.4 Quench Layers

When the flame extinguishes at the combustion chamber wall, a layer of unreacted fuel remains next to the wall. It was initially believed that the major source of HC emissions resulted from quench layers within the combustion chamber [Daniels (1957)]. However, later experimentation performed with in-cylinder sampling valves to measure the HC concentrations near the wall indicated that, although a layer of unreacted fuel remained when the flame extinguished on the wall, the concentration quickly dropped after flame arrival [LoRusso *et al.* (1983)].

Computational studies of laminar head-on flame quenching using both one-step chemistry [Adamczyk and Lavoie (1978)] and detailed chemistry [Westbrook *et al.* (1981)] indicated that the HCs are quickly consumed in the high-temperature burned gas near the wall. After the flame quenches at the wall, the unburned fuel diffuses into the high-temperature burned gas and

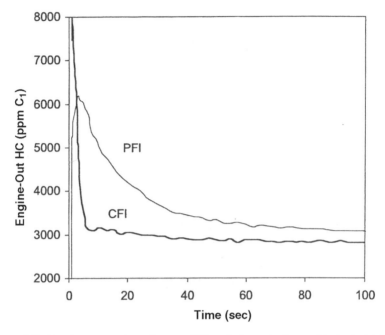

Figure 3.9 Total measured engine-out HC emissions during cold start obtained from PFI and CFI fueling systems; 1200 rpm, 58 kPa intake manifold pressure [Kaiser et al. (1996)].

is consumed on a time scale of roughly 1 msec. The post-flame oxidation rate is limited by the diffusion rate of the unburned fuel into the high-temperature burned gas. The fuels used in the studies by Westbrook *et al.* (1981) were methane and methanol. Although computational studies using more complex fuels indicated that intermediate HC species oxidation rates were slower [Kiehne *et al.* (1986)], the results indicated that even the intermediate HCs were consumed during a time scale of 1 to 2 msec. For fully warmed-up, steady-state operation, the quench layer contribution to HC emissions is estimated to be less than 5% [Cheng *et al.* (1993)].

Recognizing that the post-quench HC consumption rate is diffusion controlled means that estimates can be made for the relative increase in quench-layer HC emissions as the temperature is reduced. A characteristic time for diffusion is $\tau \sim \frac{\delta^2}{D}$, where δ is the thermal boundary layer thickness and D is the molecular diffusion coefficient for the fuel. The molecular diffusion coefficients for HC fuels at atmospheric pressure and 400 K range from 0.13 to 0.17 cm^2/sec. Using a thermal boundary layer thickness of $\delta \sim 0.1$ mm just after quenching, a characteristic diffusion time for the unburned fuel into the high-temperature burned gas is 0.6 to 0.8 msec. This is on the same order as the consumption time calculated from the computational models. Molecular diffusion rates scale with temperature approximately as $D \sim T^{\frac{3}{2}}$. Thus, as the temperature of the wall is decreased, the rate of diffusion will decrease, and the consumption rate will correspondingly decrease. For the same quenching distance, a characteristic diffusion time for a wall temperature of 300 K will be roughly 1.6 times larger than that for a wall temperature of 400 K. Using the lower bounds for the molecular diffusivity, the diffusion time at 300 K is roughly 1.25 to 1.5 msec. Note that this time is still significantly shorter than the total time available during the expansion stroke.

The quenching distance for a flame scales inversely with the flame temperature [Glassman (1997)]; therefore, the quenching distance varies with equivalence ratio in direct response to changes in the adiabatic flame temperature. Westbrook et al. (1981) calculated that at a pressure of 10 atmospheres, the quenching distance for methane/air flames increased from 0.12 to 0.2 mm as the fuel/air equivalence ratio was changed from $\Phi = 1.0$ to $\Phi = 1.4$. Thus, under cold-start conditions where the engine is operated fuel-rich, it is expected that the quench layer thickness will be increased relative to that for stoichiometric operation. Assuming a 0.2-mm quench layer thickness and a diffusivity of 0.07 cm^2/sec at 300 K, the characteristic diffusion time is less than 6 msec. At an engine speed of 1200 rpm (typical of a cold start), the time for one stroke is 25 msec, which provides more than enough time for the quench-layer HCs to be consumed before the exhaust valve opens. Therefore, while the contribution of the quench-layer HCs may increase under cold-start conditions, the increase is expected to be small relative to the other HC sources. In-cylinder sampling of the HCs in the quench layer could be performed to experimentally determine the fate of the quench-layer HCs during a cold start.

3.3.5 Partial Burns

The cylinder pressure and temperature decrease rapidly during exhaust blowdown, and the bulk gas reactions are rapidly quenched following the exhaust blowdown. A partial burn takes place when the fuel in the cylinder is not completely consumed before the exhaust valve opens. The fuel that remains when the flame extinguishes will exit the engine and be a significant source of HC emissions. Identifying a partial burn under steady-state conditions is straightforward because the amount of trapped fuel in the cylinder is known, and heat release analysis can be used to estimate the amount of fuel burned on any given cycle. Identifying a partial burn under cold-start conditions is much more complicated because the amount of combustible (i.e., vaporized) fuel in the cylinder on any given cycle is not well known. In addition, the rapid engine speed and MAP changes make determining the fresh air mass delivered per cycle difficult. As a practical matter, the IMEP often can be used to determine when a misfire or partial burn occurs. A complete misfire is easy to identify because the IMEP is negative. When the engine speed is not changing rapidly, a good metric that can be used to identify partial burns is when the IMEP of a cycle is more than 20% lower than the previous and successive cycle.

Flame propagation in engines is controlled by the in-cylinder turbulence levels. During cranking and low engine speeds, the bulk mixture motion generated by the intake flow is low, and as a result, the turbulence during combustion is low. Turbulence levels near top dead center (TDC) scale with the mean piston speed as $U' \sim 0.5\ S_p$ [Bracco et al. (1985)], where U' is the turbulence intensity and S_p is the mean piston speed. When the engine speed increases after starting, there is roughly a linear increase in the in-cylinder turbulence levels. However, the increase in engine speed also results in a shorter amount of time available for completing flame propagation before the exhaust valve opens. There is competition between the increase in flame speeds from the higher turbulence and the shorter overall residence times within which the charge must be burned. As the intake manifold pressure decreases during the initial engine speed increase, the residual gas increases from zero for the first cycle to near 25%. The residual gas dilutes the charge, which reduces the maximum flame temperatures and decreases the laminar flame speeds early during the cold start. A positive effect of the increased residuals is that they retain the HC within the cylinder so that it can be burned on succeeding cycles.

At the low SULEV emissions level, the vehicle will fail to meet the emissions standard if the engine has a misfire during the cold start. At the lower emissions levels, it is extremely important that the engine and control strategy are developed to make the engine robust enough that it does not misfire or have significant partial burns taking place. The cold-start calibration is developed with the minimum required fueling levels and spark timings to ensure that the engine does not misfire. Thus, the effect of partial burns on cold-start emissions is small because, as a rule, a great amount of time is spent on developing the cold-start calibration to specifically avoid them. However, partial burns may become important at low ambient temperatures or with low-volatility fuels.

3.3.6 Rich Air/Fuel Operation

The effect of A/F ratio on HC emissions from SI engines is well known. Hydrocarbon emissions increase more or less linearly on either side of the stoichiometric A/F ratio. At steady-state conditions, the minimum HC emissions occur near A/F equivalence ratios of $\Phi = 0.90 \sim 0.95$. The variation in HC emissions with A/F ratio is strictly one of chemistry. On the rich side, there is not sufficient oxygen to complete the oxidation of all the fuel. On the lean side, sufficient oxygen is present, but the maximum temperatures are reduced due to the increased dilution, and the amount of post-flame oxidation is reduced. The primary reason the engine is over-fueled during the cold start is due to poor vaporization of the fuel. As discussed in Chapter 7, only 10 to 20% of the fuel vaporizes during the first few cycles of the cold start. Additionally, a large percentage of the fuel injected early is stored in fuel films in the intake port and cylinder and does not participate in combustion. The engine must be over-fueled to account for both of these effects, so that the fuel that vaporizes and enters the cylinder will form a combustible mixture. Significant reductions in HC emissions could be achieved if the engine could be operated closer to stoichiometry during the cold start. Many manufacturers have incorporated mixture control devices specifically to improve the mixture motion within the cylinder at low engine speeds and to allow the engine to be operated closer to stoichiometry [Takeda *et al.* (1995); Kidokoro *et al.* (2003)].

3.4 Hydrocarbon Transport Mechanisms

A common feature of all HC emissions mechanisms is that the unburned liquid fuel is distributed within the combustion chamber on surfaces. Very little unburned fuel is expected to exist within the burned gas because it will be rapidly oxidized to CO on water (H_2O). During the expansion and exhaust strokes, the unburned gaseous HCs diffuse from the boundary layer into the burned gas. Both molecular and turbulent diffusion are expected to play a role in the mixing rate between the boundary layer HCs and the high-temperature burned gas. The interaction of the piston with the boundary layer on the cylinder liner as the piston moves upward during the exhaust stroke also is expected to be important. The transport mechanisms of the unburned HCs are discussed in this section. The processes taking place under warm engine operation will be discussed first, so that they can be contrasted with the mechanisms occurring during cold engine operation.

3.4.1 Transport Mechanisms at Warm Conditions

At warm operating conditions and with closed-valve injection (CVI) timing, the largest source of HC emissions is from the piston ring-pack crevices. As the cylinder pressure increases due to the compression stroke and combustion, unburned fuel and air are forced into the crevices until the cylinder pressure reaches a maximum. Immediately following peak pressure, these gases begin to outgas from the piston ring-pack. The unburned fuel and air flow out of the crevice and are deposited within the thermal boundary layer along the cylinder wall as the piston moves downward during the expansion stroke. The unburned fuel and air in the boundary layer are mixed with the high-temperature combustion products by both molecular diffusion and turbulence in the bulk gases and either are completely consumed or are partially oxidized to CO and intermediate HCs. These flow processes have been observed in transparent engines [Namazian and Heywood (1983)] and are shown schematically in Figure 3.10. The fuel that is desorbed from oil layers and deposits early during the expansion stroke is rapidly consumed as it diffuses into the high-temperature burned gas. The cylinder pressure and temperature decrease rapidly following the blowdown, and the reactions are quenched when the temperature is below 1500 K. As

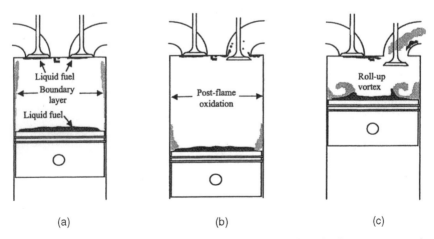

Figure 3.10 Hydrocarbon transport mechanisms within the cylinder at warm conditions. (a) Expansion stroke. (b) Exhaust blowdown. (c) Mid-exhaust blowdown.

the piston moves upward during the exhaust stroke, the unburned HCs in the boundary layer are scraped from the cylinder wall into a roll-up vortex on the top of the piston. During the valve overlap period when the intake valve opens, there is a backflow into the intake, and a portion of the gas in the exhaust port reenters the cylinder and is retained within the residual gas.

Indirect evidence of these HC transport mechanisms within the cylinder is provided by examining fast-response flame ionization detector (FFID) HC measurements in the exhaust port. Figure 3.11 is a plot of typical FFID characteristics observed in the exhaust port of an SI engine. The locations of the EVO and exhaust valve closing (EVC) are denoted in the figure. The initial small increase in HCs after EVO is believed to be caused by exhaust valve leakage and HC storage in crevices near the exhaust valve. The HC concentration quickly drops during the main exhaust flow period, indicating that virtually all the fuel in the main volume of the combustion chamber has been completely oxidized. The HC concentration rises near the end of the exhaust stroke as a portion of the crevice gas contained in the roll-up vortex escapes through the exhaust valve. The high HC concentration in the exhaust port at EVC exists throughout the closed-valve period because there is no mass flow.

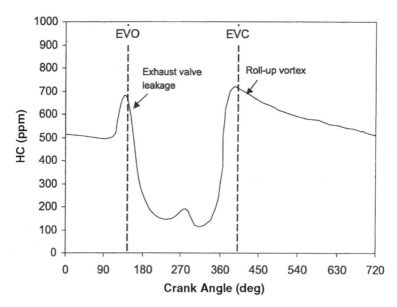

Figure 3.11 Exhaust HC concentration measured with FFID during warm engine operation [Alkidas (1997)].

As shown by these FFID measurements, the last mass to exit the cylinder has a high HC concentration. Thus, a large fraction of the HCs remaining at the end of the exhaust process are retained within the cylinder in the residual gas. Daniels *et al.* (1962) measured HC concentrations in the residual gas of a single-cylinder CFR engine by rapidly deactivating the intake and exhaust valves and dumping the cylinder contents into an evacuated chamber. Gas chromatography (GC) analysis of the sample showed HC concentrations 10 times higher than the HCs measured in the exhaust. It was estimated that 30% of the unburned HCs are retained within the cylinder. The amount of HC retention within the cylinder depends strongly on the valve timing. In general, an increase in valve overlap area increases the residuals and decreases engine-out HC emissions because more of the unburned HCs contained in the roll-up vortex are contained within the combustion chamber. However, if the increase in internal residuals results in combustion instability, the HC emissions will again increase.

3.4.2 Transport Mechanisms at Cold Conditions

Fast-response flame ionization detector measurements during cold engine operation and during cold-start transients show significantly different behavior than that observed for warm engine operation. Figure 3.12 shows FFID measurements for one engine cycle obtained from an engine cold start [Roberts and Stanglmaier (1999)]. The data in this figure are characteristic of an engine cycle after the engine speed has stabilized and thus are approximately 50 to 60 cycles after the cold start.

Before examining the details of Figure 3.12, note that the HC levels for this cold engine operation are significantly higher than the HC emissions for the warm engine operation shown in Figure 3.11. In Figure 3.12, immediately following EVO, the HC emissions increase significantly. The spark timing was retarded for this cold-start condition so that proportionally more of the mass leaves the cylinder during the blowdown pulse. It is known that as the engine operating temperature is reduced, a proportionally larger amount of liquid fuel enters the cylinder. Thus, the large increase in HC emissions following EVO is most likely due to liquid fuel located near the exhaust and intake valves exiting the cylinder with the initial high-blowdown pulse. After the peak in HC concentration, the HC emissions begin to decrease significantly to a level that is slightly lower than the concentration that was in the port before the exhaust valve opened. This significant decrease reflects the lower HC concentration in the bulk burned gas. Toward the end of the exhaust stroke, the HC emissions begin to increase again. Note that the magnitude of the increase is not as large

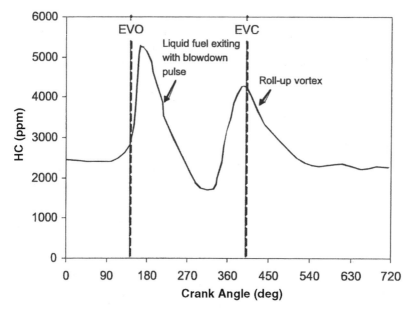

Figure 3.12 Fast-response flame ionization detector HC measurements in the exhaust port of an engine during cold start after the engine revolutions per minute are stabilized [Roberts and Stanglmaier (1999)].

as that observed following exhaust blowdown. The increase in HC emissions near the end of the exhaust stroke is due to the piston roll-up vortex exiting the cylinder.

3.5 Hydrocarbon Oxidation

After the fuel escapes the primary combustion process, it also must avoid being oxidized both within the cylinder and the exhaust port in order to contribute to engine-out HC emissions. At fully warmed-up, steady-state conditions, a significant amount of post-flame oxidation takes place within the cylinder. Both computational [Hochgreb and Oliveira (1999)] and experimental [Eng et al. (1997)] results indicate that 50 to 90% of the unburned HCs are consumed within the cylinder before the exhaust valve opens. Given the large amount of oxidation taking place, it is important to have at least a general understanding of HC reaction mechanisms and the effect of in-cylinder conditions on post-flame HC oxidation rates. This section will include a brief review of HC oxidation mechanisms, followed by a discussion of the pertinent studies on post-flame HC oxidation in internal combustion engines.

3.5.1 *Hydrocarbon Oxidation Mechanisms*

There are three fundamental regimes of HC oxidation chemistry, although not all fuels exhibit each regime. In the low-temperature regime, the reaction rate increases with increasing temperature, and the dominant reactions are oxygen addition to form alkylperoxy radicals (RO_2) and olefins, followed by radical isomerization and decomposition. As the temperature is increased further, the alkylperoxy radicals decompose back into initial reactants, the production of olefins and hydroperoxyl radicals (HO_2) is favored, and the overall reaction rate decreases with increasing temperature. This is the classical negative temperature coefficient (NTC) behavior observed in the low-temperature oxidation of paraffinic fuels. At atmospheric pressure, the low-temperature regime is below 600 K, and the transition between the low- and intermediate-temperature regimes occurs between 650 and 700 K. At 10 atmospheres, the onset of the NTC chemistry does not occur until 750 to 800 K [Dryer (1991)]. As the temperature is increased further, the reaction rate again increases as more olefinic HCs and hydrogen peroxide (H_2O_2) are produced. Eventually, the temperature is increased high enough so that hydrogen-oxygen branching reactions control the reaction rate in the high-temperature regime.

The oxidation mechanisms of HCs in the high-temperature regime are conceptually the most easy to understand. The high-temperature oxidation of HCs can be described as a sequential three-step process:

1. After initiation, the parent fuel is converted to lower-molecular-weight HCs and H_2O with little energy release.

2. The intermediate HC species are further converted to produce CO and H_2O.

3. Carbon monoxide is oxidized into carbon dioxide (CO_2), and the large fraction of the energy is released from the combustion process.

The reaction sequence can be conceptualized as

$$RH \rightarrow R' + R'' + H_2O \rightarrow CO + H_2O \rightarrow CO_2 + \text{heat} \tag{3.1}$$

where RH is the parent fuel species, and R' and R'' represent lower-molecular-weight HC species. It is important to distinguish between the disappearance of the parent fuel and total HC consumption. The conversion of the initial fuel into lower-molecular-weight HCs is termed HC conversion, whereas the complete oxidation of all HCs into CO, CO_2, and H_2O is termed HC consumption. In terms of producing lower engine-out HC emissions during a cold-start, automotive engineers are primarily interested in HC consumption and not HC conversion.

At temperatures near the transition into the high-temperature regime (950 K at atmospheric pressure), the initiation reaction for paraffins is typically H atom abstraction by a radical. After the initial abstraction, the alkyl radicals can decompose via β-scission into smaller alkyl radicals and olefinic species. At high temperatures, this is an important pathway because it leads to rapid buildup of a radical pool. For olefinic fuels, the initiation reaction is typically radical addition at the double bond, which eventually is followed by fragmentation at the double-bond site. At temperatures well into the high-temperature regime, the initiation reaction can be unimolecular thermal decomposition of the parent fuel, or decomposition by collision with a third body that fragments the parent fuel molecule. As the reaction proceeds, radicals continue to abstract H atoms from the intermediate HC species; there is a cascading of the species into lower-molecular-weight alkyl radicals, formaldehyde (CH_2O), formyl radicals (CHO), and CO. The final reaction in all HC combustion systems is the oxidation of CO to CO_2. Because the radicals preferentially react with any HCs present in the mixture rather than with CO, CO oxidation is delayed until after most of the HCs are consumed.

Detailed reaction mechanisms for HC fuels have been under development for many years. These reaction mechanisms are extremely complex. Even for a simple fuel such as methane, the oxidation mechanism involves more than 63 species and 200 elementary reactions. At their current stage of development, detailed reaction mechanisms for large fuels such as n-heptane and iso-octane incorporate hundreds of species and thousands of reactions [Curran *et al.* (1998 and 2002)]. Although these models are too large to use in engine simulation programs, they have been used in simplified configurations that approximate the thermodynamic conditions and time scales existing within an engine in order to identify the chemical time scales of interest.

3.5.2 *Hydrocarbon Consumption Rates*

One of the objectives of developing detailed reaction mechanisms is to estimate the time scales required for HC consumption over the ranges of temperature and pressure that exist in an engine. Figure 3.13 is a plot of the normalized total HC mass as a function of time for iso-octane at conditions representative of the exhaust gas during a cold start. The initial concentration of iso-octane was 1000 ppm, and the background gas was composed of 9% CO_2, 14% H_2O, 4.5% CO, 2% H_2, 1% O_2, and the balance nitrogen. The composition of the background gas is the estimated composition for an engine operated at an equivalence ratio of $\Phi = 1.5$.

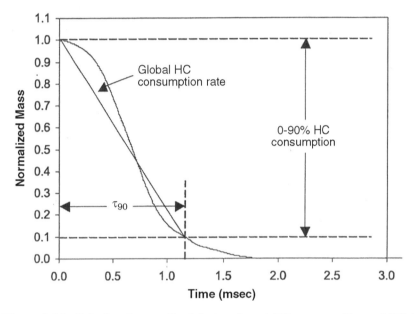

Figure 3.13 Calculated normalized fuel and total HC mass profiles at 1450 K, atmospheric pressure in an iso-octane/air system; 1000 ppm iso-octane, 9% CO_2, 14% H_2O, 4.5% CO, 2% H_2, 1% O_2, and the balance gas is nitrogen.

The chemical kinetic mechanism used for iso-octane was the one developed by Curran *et al.* (2002). This mechanism contains the low-temperature chemical kinetics that are known to be important for paraffinic fuels. For the data shown in Figure 3.13, the calculations were performed at 1450 K and atmospheric pressure. The mass fraction of the total HC mass in the system is normalized by the initial mass. At these conditions, the iso-octane disappears extremely rapidly and is completely consumed on a time scale of 0.01 msec. By comparison, the total HCs in the system are not consumed until 1.15 msec. The global consumption rate is defined as the time to consume 90% of the HC mass in the system.

Figure 3.14 is a plot of the calculated global HC consumption time as a function of temperature at 1, 5, and 10 atmospheres pressure. The curves through each set of data are exponential curve fits. As expected from chemical kinetics, the global consumption time decreases exponentially with increasing temperature. Also shown in the figure are the times for one engine cycle at engine speeds of 1000, 2000, and 3000 rpm. For the HCs to be consumed during the time scale of one engine cycle, the temperature must be in the range of 1200 to 1500 K. Calculations performed at the same conditions using n-heptane as the fuel show similar results. This is somewhat counterintuitive, because it might be expected that a fuel that exhibits significant low-temperature energy release such as n-heptane would demonstrate much more post-flame HC consumption. The reason for this is that because of the short time scales available for post-flame HC consumption within the engine, the temperature must be high enough that

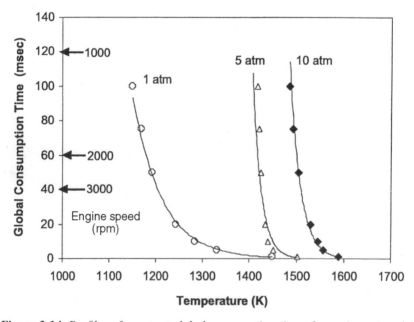

Figure 3.14 Profiles of constant global consumption times for an iso-octane/air system as a function of pressure and temperature; constant pressure and temperature calculations.

the reaction rates are dominated by high-temperature chemistry. Differences among the consumption rates of fuels are exhibited at temperatures lower than the 1300 to 1500 K range. Although the amount of HC consumption is expected to be similar among different fuels, the amount of fuel conversion into low-molecular-weight HC species will be different. For example, experiments performed by injecting different paraffinic tracer gases into the exhaust port of an engine operated on methane have shown that different amounts of the initial tracer gas disappear within the exhaust port [Yang *et al.* (2000)].

Post-flame HC consumption processes in engines occur in an environment where both the pressure and temperature are changing rapidly throughout the engine cycle. These calculations are intended to be strictly qualitative, but they illustrate the important point that to obtain HC consumption during the characteristic time scales available in engines, the temperature must be high enough that high-temperature kinetics dominate the consumption rate. As will be discussed in the next section, both experimental results and detailed computational modeling indicate that post-flame HC oxidation in engines takes place at temperatures higher than 1300 to 1500 K. The simplified calculations at constant pressure show the correct trends that are observed in the engine.

3.5.3 Post-Flame Hydrocarbon Consumption in Engines

The post-flame HC oxidation processes occurring in engines are extremely complex. Within the cylinder, the HC sources are non-uniformly distributed, and there are large temperature and concentration gradients near the cylinder walls. Within the exhaust port and manifold, the flow is highly turbulent, and the temperature changes drastically during exhaust blowdown. The physical nature of the process, and the rapid time and spatial variations in the cylinder and exhaust system, make it challenging to estimate HC consumption rates in engines.

Global post-flame HC consumption rates in engines have been estimated experimentally by Eng *et al.* (1997). A generalized HC consumption rate correlation was developed from engine experiments over a range of operating conditions using a wide range of paraffinic, olefinic, and aromatic fuels. The results indicated that post-flame HC consumption takes place at temperatures higher than 1500 K. For normal operating conditions with typical spark advance near the minimum spark advance for best torque (MBT), the post-flame HC oxidation is essentially completed within the cylinder before the exhaust valve opens. Similar results were obtained from exhaust port injection studies performed by Bian *et al.* (1998) and Yang *et al.* (2000). In these experiments, the engine was operated on methane, and different paraffinic fuels were injected into the exhaust port. The extent of oxidation of the injected tracer fuel was determined by speciating the exhaust gas HC. A range of C_3 to C_5 paraffinic and olefinic fuels was used as tracer fuels in the experiments. By varying the exhaust gas temperature by changing the intake temperature, they determined that the critical cutoff temperature for HC consumption was 1300 to 1500 K for all the fuels tested.

Experimental results obtained by Tamura *et al.* (2001) have provided further evidence of the essential features of post-flame HC consumption in engines and that it takes place at high temperatures. In these experiments, they performed simultaneous imaging of unburned fuel and hydroxyl (OH) radicals in an optical engine using a laser-induced fluorescence (LIF) technique. A quartz liner was used so that the entire cylinder could be observed during the expansion stroke. The engine was operated with methane as the fuel, and acetone was used as the tracer species to indicate the location of the unburned fuel within the cylinder. The results showed that unburned fuel was observed only after the disappearance of OH in the bulk gas late in the expansion. In some of the images, a thin layer of fuel was observed near the cylinder wall, and a thin intense region of OH was observed immediately next to it. Experiments with increased backpressure clearly showed an increase in OH after the crevice gases enter the cylinder, which is further evidence of post-flame HC oxidation taking place. These results indicate that the unburned HCs are able to generate and replenish the radical pool from their own oxidation. It is not necessary that there is an established radical pool within the burned gas to consume the HCs. All that is required is that the temperatures are sufficiently high to initiate the high-temperature H_2-O_2 chain branching chemistry. They concluded that with atmospheric exhaust backpressure, the reactions were quenched when the exhaust valve opened.

Several computational studies have been performed to investigate the effect of operating conditions and fuel type on post-flame HC consumption in engines. The models are for a simplified one-dimensional representation of the diffusion of a wall layer of HC into the high-temperature burned gas. These models are a continuation of the computational studies initiated by Adamczyk and Lavoie (1978) and Westbook *et al.* (1981) for the quenching of a flame on a surface. The initial results were obtained with a simple one-step kinetic model for methane [Min (1994)]. The results obtained using the one-step chemistry indicated that a large amount of fuel escaped oxidation during the expansion stroke. The shortfall of the one-step chemistry for these calculations is that the chemistry was developed for flame speed calculations and does not take into account the rapid thermal decomposition of the fuel at high temperatures in the absence of a radical pool. Subsequent calculations have been performed using detailed chemistry, and a submodel was included for the effect of turbulence in the bulk gas on the diffusion rates [Hochgreb and Oliveira (1999)]. These calculations were performed using detailed chemical kinetic mechanisms for both propane and iso-octane. The results of these calculations showed that the chemical reactions are effectively quenched at temperatures below 1400 to 1600 K for both fuels. The computational results obtained with a more realistic chemistry agree well with the experimental observations.

These computational studies have been extended to cold engine operation and specifically cold-start operation. The oxidation of a thin layer of liquid fuel along the cylinder wall was computationally investigated by Oliveira and Hochgreb (2000). In these studies, methanol was used for the surrogate fuel, and a detailed chemical kinetic mechanism was employed. A liquid fuel layer with initial thicknesses varying from 0.1 to 0.5 mm was deposited along the cylinder wall. The calculations were performed at an engine speed of 1500 rpm and an engine load of 375 kPa net mean effective pressure (NMEP) and MBT spark timing. The results indicated that regardless of the initial thickness of the liquid fuel layer, 97% of the fuel that was vaporized within a given engine cycle was consumed by the end of the engine cycle. Successive engine cycles were modeled to estimate the time it takes to completely vaporize the liquid fuel film. It was found that 500 cycles were required to consume a 0.5-mm-thick liquid fuel layer. This corresponds to a time period of 40 seconds, which is in general agreement with the time scales observed for differences in A/F ratio measured from the intake and exhaust gases [Shayler et al. (1999)]. The fuel vaporization rate within a cycle was dominated primarily by the flame arrival time, and neither the volatility of the fuel nor the wall temperature had a significant effect on the vaporization rate. These calculations are in good agreement with the experimental estimates, in that they both indicate that a significant amount of HC consumption is taking place within the cylinder and that most of the HC consumption occurs at temperatures higher than 1300 to 1500 K.

Secondary air injection into the exhaust has been used for many years to obtain reduced cold-start emissions. The principle behind using secondary air is to operate the engine overall fuel-rich and then inject secondary air into the exhaust manifold to provide an oxidizer to react the CO and HC over the catalyst. The use of secondary air is discussed in detail in Chapter 6. Although the main objective of secondary air is to react the CO and HC over the catalyst, it also is possible to obtain HC consumption within the exhaust port when the system is designed correctly. As the above results have shown, post-flame

HC consumption takes place at temperatures higher than 1300 to 1500 K. The chemical time scales increase considerably when the temperature is decreased below this range. The temperature of the exhaust gas changes significantly during an engine cycle, and immediately following EVO and exhaust blowdown, the exhaust gas temperature is within this critical 1300 to 1500 K window for a short period of time [Hernandez et al. (2002)]. Hydrocarbon oxidation can take place if the air can be mixed with the exhaust gas without significantly reducing the temperatures within the exhaust port during blowdown. Because the chemical time scale increases exponentially as the temperature decreases, it is expected that the mixing time between the secondary air and the exhaust gas will be critical to achieving HC consumption within the exhaust port.

The effects of secondary air injection location, exhaust manifold design, and engine operation on engine-out (or equivalently, converter-in) HC emissions were investigated by Borland and Zhao (2002). Experiments performed with the secondary air directed toward the exhaust valve did not show any evidence of exhaust port oxidation. This was attributed to poor mixing between the secondary air and the exhaust gas, and the reduced temperatures caused by the addition of secondary air. The pressure pulse from the exhaust gas during blowdown shuts off the secondary air momentarily when the airstream is directed toward the exhaust valve. However, results obtained with a sparger-type design, where the air was injected perpendicularly to the exhaust flow, showed evidence of HC consumption. With the sparger design, the secondary air is not shut off by the strong blowdown flow, and the air is entrained into the exhaust gas. A series of exhaust manifolds with different volumes was tested to obtain longer residence times within the manifold. By changing the manifolds, the exhaust gas temperature at 10 seconds after the cold start was increased from 380 to 850°C. This increase in temperature was caused by the energy release from HC oxidation within the exhaust port. Obtaining HC consumption within the exhaust port requires careful optimization of the mixing rate between the secondary air and the exhaust gas, and the exhaust port design.

3.6 Summary

After many years of extensive research at both the industrial and university levels, we currently have a good understanding of the important HC emissions mechanisms at steady-state conditions. Based on this prior research, and considering the important physics of each storage mechanism, the three most important cold-start HC emissions mechanisms are as follows:

1. Storage in crevices
2. Liquid fuel
3. Rich A/F operation

Quench layers have been shown not to be a significant source of HC emissions at warmed-up conditions. Estimates of the quench layer thickness and characteristic post-flame consumption rates at cold conditions indicate that quench layers are not an important HC emissions source during a cold start. The contribution to HC emissions from oil layer absorption is only ~5% at warmed-up operation. Based on the steady-state data and estimates of fuel solubility in oil at lower temperatures, it is estimated that oil layer absorption is not a significant contributor to cold-start HC emissions.

Although the important storage mechanisms have been identified, the relative contribution of each mechanism must be determined. This is no easy task. One of the most challenging problems with HC emissions research, at both steady-state and transient conditions, is that it is difficult to devise an experiment to determine the effect of one emissions mechanism without inadvertently changing another. An important requirement of every experiment is that the operating characteristics of the engine must not be changed in any significant way by the experiment. For example, if the piston ring-pack geometry is modified to investigate the contribution of piston crevices to cold-start HC emissions, the cylinder blowby flow rate also must be monitored to account for changes in blowby on the crevice mechanism. Another daunting problem with cold-start emissions research is that it is extremely difficult to obtain repeatable results due to the transient nature of the startup process.

Potential experiments that could be performed to investigate the effect of liquid fuel on cold-start HC emissions are to operate the engine on prevaporized gasoline or with a low-boiling-point single-component fuel such as iso-pentane. Some of these experiments have already been performed, but the influence of the liquid fuel relative to changes in the crevice mechanism and changes in the A/F ratio has not been determined. To investigate the effects of fuel oil absorption on HC emissions, the exhaust gas could be speciated during each cycle of the cold start to determine if there is evidence of preferential absorption of fuel components into the oil. Although initial work along these lines has already been performed, additional research must be performed to conclusively determine the influence of oil layer absorption on cold-start HC emissions.

3.7 References

1. Adamczyk, A.A. and Lavoie, G.A. (1978), "Laminar Head-On Quenching—A Theoretical Study," SAE Paper No. 780969, Society of Automotive Engineers, Warrendale, PA.

2. Alkidas, A.C. (1994), "The Effects of Fuel Preparation on Hydrocarbon Emissions of a SI Engine Operating Under Steady-State Conditions," SAE Paper No. 941959, Society of Automotive Engineers, Warrendale, PA.

3. Alkidas, A.C. (1997), "The Influence of Mixture Preparation on the HC Concentration Histories from an S.I. Engine Running Under Steady-State Conditions," SAE Paper No. 972981, Society of Automotive Engineers, Warrendale, PA.

4. Alkidas, A.C. (1999), "Combustion Chamber Crevices: The Major Source of Engine-Out Hydrocarbon Emissions Under Fully Warmed Conditions," *Prog. in Energy and Comb. Sci.,* Vol. 25.

5. Alkidas, A.C., Drews, R.J., and Miller, W.F. (1995), "Effects of Piston Crevice Geometry on the Steady-State Engine-Out Hydrocarbon Emissions of a S.I. Engine," SAE Paper No. 952537, Society of Automotive Engineers, Warrendale, PA.

6. Bian, X., Prabhu, S.K., Yang, W., Miller, D.L., and Cernansky, N. (1998), "Tracer Fuel Injection Studies on Exhaust Port Hydrocarbon Oxidation," SAE Paper No. 982559, Society of Automotive Engineers, Warrendale, PA.

7. Borland, M. and Zhao, Fuquan (2002), "Application of Secondary Air Injection for Simultaneously Reducing Converter-In Emissions and Improving Catalyst Light-Off Performance," SAE Paper No. 2002-01-2803, Society of Automotive Engineers, Warrendale, PA.

8. Bracco, F.V., Abraham, J., and Williams, F.A. (1985), "A Discussion of Turbulent Flame Structure in Premixed Charge Engines," SAE Paper No. 850345, Society of Automotive Engineers, Warrendale, PA.

9. Campbell, S., Clasen, E., Chang, C., and Rhee, K.T. (1996), "Flames and Liquid Fuel in an SI Engine Cylinder During Cold Start," SAE Paper No. 961153, Society of Automotive Engineers, Warrendale, PA.

10. Cheng, W.K., Hamrin, D., Heywood, J.B., Hochgreb, S., Min, K., and Norris, M. (1993), "An Overview of Hydrocarbon Emissions Mechanisms in Spark-Ignition Engines," SAE Paper No. 932708, Society of Automotive Engineers, Warrendale, PA.

11. Cheng, W.K. and Santoso, H. (2002), "Mixture Preparation and Hydrocarbon Emissions Behaviors in the First Cycle of SI Engine Cranking," SAE Paper No. 2002-01-2805, Society of Automotive Engineers, Warrendale, PA.

12. Curran, H.J., Gaffuri, P., and Pitz, W.J. (1998), "A Comprehensive Modeling Study of n-Heptane Oxidation," *Comb and Flame*, Vol. 114.

13. Curran, H.J., Gaffuri, P., and Pitz, W.J. (2002), "A Comprehensive Modeling Study of Iso-Octane Oxidation," *Comb and Flame*, Vol. 129.

14. Curtis, E., Russ, S., Aquino, C., Lavoie, G., and Trigui, N. (1998), "The Effects of Injector Targeting and Fuel Volatility on Fuel Dynamics in a PFI Engine During Warm-up: Part II—Modeling Results," SAE Paper No. 982519, Society of Automotive Engineers, Warrendale, PA.

15. Daniels, W.A. (1957), "Flame Quenching at the Walls of an Internal Combustion Engine," Sixth International Symposium on Combustion, The Combustion Institute, pp. 886–894.

16. Daniels, W.A. and Wentworth, J.T. (1962), "Exhaust Gas Hydrocarbons—Genesis and Exodus," SAE Paper No. 486B, Society of Automotive Engineers, Warrendale, PA.

17. Dent, J.C. and Lakshminarayanan, P.A. (1983), "A Model for Absorption and Desorption of Fuel Vapour by Cylinder Lubricating Oil Films and Its Contribution to Hydrocarbon Emissions," SAE Paper No. 830652, Society of Automotive Engineers, Warrendale, PA.

18. Dryer, F.L. (1991), "The Phenomenology of Modeling Combustion Chemistry," in *Fossil Fuel Combustion: A Source Book*, W. Bartok and A.F. Sarofim, eds., Wiley Interscience, New York, NY, pp. 121–213.

19. Eng, J.A., Leppard, W.R., Najt, P., and Dryer, F. (1997), "Experimental Hydrocarbon Consumption Rate Correlations from a Spark Ignition Engine," SAE Paper No. 972888, Society of Automotive Engineers, Warrendale, PA.

20. Fry, M., Nightingale, C., and Richardson, S. (1995), "High-Speed Photography and Image Analysis Techniques Applied to Study Droplet Motion Within the Porting and Cylinder of a 4-Valve SI Engine," SAE Paper No. 952525, Society of Automotive Engineers, Warrendale, PA.

21. Glassman, I. (1997), *Combustion*, 3rd Edition, Academic Press., San Diego, CA.

22. Gulati, S.T. (1999), "Thin Wall Ceramic Catalyst Supports," SAE Paper No. 1999-01-0269, Society of Automotive Engineers, Warrendale, PA.

23. Hernandez, J.L., Herding, G., Carstensen, A., and Spicher, U. (2002), "A Study of the Thermochemical Conditions in the Exhaust Manifold Using Secondary Air in a 2.0-L Engine," SAE Paper No. 2002-01-1676, Society of Automotive Engineers, Warrendale, PA.

24. Hochgreb, S. and Oliveira, I.B. (1999), "Effect of Operating Conditions and Fuel Type on Crevice HC Emissions: Model Results and Comparison with Experiments," SAE Paper No. 1999-01-3578, Society of Automotive Engineers, Warrendale, PA.

25. Ishizawa, S. and Takagi, Y. (1987), "A Study of HC Emissions from a Spark Ignition Engine—The Influence of Fuel Absorbed into Cylinder Lubricating Oil Film," *JSME*, Vol. 30, No. 260.

26. Kaiser, E.W., Siegl, W.O., and Russ, S.G. (1995), "Fuel Composition Effects on Hydrocarbon Emissions from a Spark-Ignition Engine—Is Fuel Absorption in Oil Significant?," SAE Paper No. 952542, Society of Automotive Engineers, Warrendale, PA.

27. Kaiser, E.W., Siegl, W.O., Lawson, G.P., Connolly, F.T., Cramer, C.F., Dobbins, K.L., Roth, P.W., and Smokovitz, M. (1996), "Effect of Fuel Preparation on Cold-Start Hydrocarbon Emissions from a Spark-Ignited Engine," SAE Paper No. 961957, Society of Automotive Engineers, Warrendale, PA.

28. Kidokoro, T., Hoshi, K., Hiraku, K., Satoya, K., Watanabe, T., Fujiwara, T., and Suzuki, H. (2003), "Development of PZEV Exhaust Emission Control System," SAE Paper No. 2003-01-0817, Society of Automotive Engineers, Warrendale, PA.

29. Kiehne, T.M., Matthews, R.D., and Wilson, D.E. (1986), "The Significance of Intermediate Hydrocarbons During Wall Quench of Propane Flames," 21st International Symposium on Combustion, The Combustion Institute, pp. 481–489.

30. Linna, J.R., Malberg, H., Bennett, P.J., Palmer, P.J., Tian, T., and Cheng, W.K. (1997), "Contribution of Oil Layer Mechanism to Hydrocarbon Emissions from Spark Ignition Engines," SAE Paper No. 972892, Society of Automotive Engineers, Warrendale, PA.

31. LoRusso, J.A., Kaiser, E.W., and Lavoie, G.A. (1983), "In-Cylinder Measurements of Wall Layer Hydrocarbons in a Spark Ignition Engine," *Comb. Sci. Tech.*, Vol. 33, p. 75.

32. Min, K. (1995), "Oxidation of the Piston Crevice Hydrocarbon During the Expansion Process in a Spark Ignition Engine," *Comb. Sci. Tech.*, Vol. 106.

33. Namazian, M. and Heywood, J.B. (1982), "Flow in the Piston-Cylinder-Ring Crevices of a Spark-Ignition Engine: Effect on Hydrocarbon Emissions, Efficiency, and Power," SAE Paper No. 820088, Society of Automotive Engineers, Warrendale, PA.

34. Oliveira, I.B. and Hochgreb, S. (2000), "Detailed Calculation of Heating, Evaporation, and Reaction Processes of a Thin Liquid Layer of Hydrocarbon Fuel," SAE Paper No. 2000-01-0959, Society of Automotive Engineers, Warrendale, PA.

35. Roberts, C.E. and Stanglmaier, R.H. (1999), "Investigation of Intake Timing Effects on the Cold-Start Behavior of a Spark-Ignition Engine," SAE Paper No. 1999-01-3622, Society of Automotive Engineers, Warrendale, PA.

36. Russ, S., Kaiser, E.W., and Siegl, W.O. (1995), "Effect of Cylinder Head and Engine Block Temperature on HC Emissions from a Single Cylinder Spark Ignition Engine," SAE Paper No. 952536, Society of Automotive Engineers, Warrendale, PA.

37. Russ, S., Thiel, M., and Lavoie, G.A. (1999), "SI Engine Operation with Retarded Ignition: Part 2—HC Emissions and Oxidation," SAE Paper No. 1999-01-3507, Society of Automotive Engineers, Warrendale, PA.

38. Sampson, M.J. and Heywood, J.B. (1995), "Analysis of Fuel Behavior in the Spark-Ignition Engine Start-Up Process," SAE Paper No. 950678, Society of Automotive Engineers, Warrendale, PA.

39. Shayler, P.J., Davies, M.T., and Scarisbrick, A. (1997), "Audit of Fuel Utilization During the Warm-Up of SI Engines," SAE Paper No. 971656, Society of Automotive Engineers, Warrendale, PA.

40. Shayler, P.J., Belton, C., and Scarisbrick, A. (1999), "Emissions and Fuel Utilization After Cold Starting Spark Ignition Engines," SAE Paper No. 1999-01-0220, Society of Automotive Engineers, Warrendale, PA.

41. Shin, Y., Cheng, W.K., and Heywood, J.B. (1994), "Liquid Gasoline Behavior in the Engine Cylinder of an SI Engine," SAE Paper No. 941872, Society of Automotive Engineers, Warrendale, PA.

42. Stanglmaier, R.H., Li, J., and Matthews, R.D. (1999), "The Effect of In-Cylinder Wall Wetting Location on the HC Emissions from SI Engines," SAE Paper No. 1999-01-0502, Society of Automotive Engineers, Warrendale, PA.

43. Sterlepper, J. and Spicher, U. (1993), "Flame Propagation into Top-Land Crevice and Hydrocarbon Emissions from an SI Engine During Warm-Up," SAE Paper No. 937101, Society of Automotive Engineers, Warrendale, PA.

44. Takeda, K., Yaegashi, T., Sekiguchi, K., Saito, K., and Imatake, N. (1995), "Mixture Preparation and HC Emissions of a 4-Valve Engine During Cold Starting and Warm-Up," SAE Paper No. 950074, Society of Automotive Engineers, Warrendale, PA.

45. Tamura, M., Sakurai, T., and Tai, H. (2001), "A Study of Crevice Flow in a Gas Engine Using Laser-Induced Fluorescence," SAE Paper No. 2001-01-0913, Society of Automotive Engineers, Warrendale, PA.

46. Wentworth, J.T. (1968), "Piston and Ring Variables Affect Exhaust Hydrocarbon Emissions," SAE Paper No. 680109, Society of Automotive Engineers, Warrendale, PA.

47. Westbrook, C.K., Adamczyk, A.A., and Lavoie, G.A. (1981), "A Numerical Study of Laminar Flame Wall Quenching," *Combustion and Flame*, Vol. 40, pp. 81–99.

48. Witze, P.O. (1999), "Diagnostics for the Study of Cold-Start Mixture Preparation in a Port Fuel-Injected Engine," SAE Paper No. 1999-01-1108, Society of Automotive Engineers, Warrendale, PA.

49. Yang, J., Kaiser, E.W., Siegl, W.O., and Anderson, R.W. (1993), "Effects of Port-Injection Timing and Fuel Droplet Size on Total and Speciated Exhaust Hydrocarbon Emissions," SAE Paper No. 930711, Society of Automotive Engineers, Warrendale, PA.

50. Yang, W., Miller, D.L., Zheng, J., and Cernansky, N.P. (2000), "Tracer Fuel Injection Studies on Exhaust Port Hydrocarbon Oxidation: Part II," SAE Paper No. 2000-01-1945, Society of Automotive Engineers, Warrendale, PA.

51. Yilmaz, E., Thirouard, B.P., Tian, T., Wong, V.W., and Heywood, J.B. (2001), "Analysis of Oil Consumption Behavior During Ramp Transients in a Production Spark Ignition Engine," SAE Paper No. 2001-01-3544, Society of Automotive Engineers, Warrendale, PA.

52. Yoshida, H., Kobayashi, H., Sato, A., and Tani, M. (1996), "Effect of Piston Second Land Shape on Oil Consumption," ASME Conference on Advances in Engine Design, ASME International, Fairfield, NJ.

CHAPTER 4

Characterization of Cold Engine Processes

Choongsik Bae
Korea Advanced Institute of Science and Technology

4.1 Introduction

Hazardous exhaust emissions, especially hydrocarbon (HC) emissions, under cold-start conditions in port fuel injected (PFI) spark ignition (SI) engines are determined by a chain of engine processes such as the spray characteristics of fuel injections, the interaction of the spray with the port wall and valve, the in-cylinder droplet field and associated mixture distribution, the flow/mixture state around the spark plug at ignition timing, and the flame/wall interaction.

A refined understanding of the combustion and pollutant formation process in SI engines is needed to evolve engines with lower emissions, especially HCs at cold start. As the requirements to reduce emissions become more stringent, there is an increasing need for clear measurement of realistic engine processes. This chapter is concerned with diagnostic techniques for engine process measurements, especially the laser-based *in-situ* diagnostics on the fuel sprays, mixture preparation, flame propagation, and pollutant formation, and the findings from various measurements.

The diagnostics on fuel sprays, mixture formation, and combustion can be divided into the classical ones of physical probing and remote sensing with non-intrusive optical diagnostics. Some optical diagnostics could still be considered classical (e.g., direct photography, absorption and emissions spectroscopy, Schlieren/shadowgraphy), although lasers have been introduced as more powerful sources of illumination in these techniques. Modern laser diagnostics have increased the insight into engine flow and combustion, but the old classical diagnostic tools are not completely obsolete. Laser-based measurements, undergoing continuous improvement, have received much research attention and have demonstrated their performance in mixture and combustion identification in the last couple of decades. The spectrum of diagnostic techniques recently was approached systematically and overviewed well by Zhao and Ladommatos (2001). These techniques usually are incorporated in optical engines for observation and measurement of the in-cylinder activity, in line with the need for improved understanding of the in-cylinder flow, mixing, combustion, and pollutant formation [Gold *et al.* (2000)]. Table 4.1 summarizes these diagnostic techniques.

This chapter introduces the measurement of fuel sprays, mixture distribution, combustion processes, and pollutant formation, with examples of probing and optical setups.

4.2 Fuel Injection Characteristics and Fuel Delivery into the Engine Cylinder

The fuel transport process plays an important role in engine cold-start operations. The fuel delivery into the engine cylinder initially is governed by fuel injection in the intake manifold, leading to mixture formation. The conventional PFI mixture formation processes have been well documented by Lenz (1992). The investigation of liquid fuel sprays interacted with highly turbulent flow has been possible by spray measurement, which gives a tip on the mixture formation process as shown in Figure 4.1 [Zhao *et al.* (1995)].

4.2.1 Fuel Sprays

Accurate determination of fuel spray characteristics is an important aspect in the development of mixture preparation systems. This includes spray patternation and fuel droplet sizing. Port fuel injection systems have evolved into electronic pulse-width-modulated systems utilizing sequentially timed individual injections into each port, offering significant advantages in engine transient response and HC emissions. Spray shapes are represented by spray angle and penetration development, which should be matched with intake manifold geometry and valve dynamics coupled with in-manifold flow field.

TABLE 4.1
DIAGNOSTICS TO CHARACTERIZE FUEL SPRAY, FLOW, MIXTURE CONCENTRATION, COMBUSTION PROCESSES, AND POLLUTANT FORMATION IN SI ENGINES

	Technique		Objectives and Application
Physical Probing/Analysis	Universal exhaust gas oxygen (UEGO) sensor; heated exhaust gas oxygen (HEGO) sensor		Air/fuel (A/F) measurement and wide-band HC detection in the exhaust stream
	Fast-response flame ionization detector (FFID/FRFID)		HC detection/HC concentration measurement in-cylinder or in the exhaust • Fuel film or residual gas estimation • Millisecond delay in detection
	Fast NOx/CO analyzer		NOx/CO measurement in cylinder or exhaust
Optical Measurement	Direct imaging		• Mie-scattered visualization/back-illuminated telescopic imaging (down to μm)/flame-illuminated imaging • Shapes, angle, and penetration of fuel spray/wall film imaging • Microscopic droplet imaging and sizing • Flame propagation and its structure
	Shadowgraphy, Schlieren photography		Vapor fields/density variation • Fuel evaporation/flame propagation
	Laser Diagnostics	Chemiluminescence	Radical measurement—A/F estimation
		Laser Doppler velocity (LDV)	Velocity measurement of a Mie scattering particle • Spray flow field/manifold/in-cylinder flow field • Point measurement ($10^{-27} - 10^{-8}$)*
		Particle image velocity (PIV)	2-D velocity measurement of Mie scattering particles • Spray/in-cylinder flow field in a plane • Point measurement ($10^{-27} - 10^{-8}$)*
		Phase Doppler anemometry (PDA)	Particle sizing of Mie scattering drops • Fuel droplet sizing in dispersed spray field • Point measurement ($10^{-27} - 10^{-8}$)*
		Diffraction method	Particle sizing of Mie scattering droplet cloud • Fuel droplet sizing through a volume of laser beam bundle in manifold/cylinder • Volume-averaged point measurement ($10^{-27} - 10^{-8}$)*
		Interferometric laser imaging	Interferometric laser imaging for drop sizing (ILIDS) • Fuel droplet sizing in a plane
		Laser extinction and absorption (LEA)	Vapor/liquid detection, droplet sizing
		Rayleigh scattering	Species measurement • Fuel vapor concentration in cylinder • Gaseous state ($10^{-28} - 10^{-25}$)*
		Raman scattering	Species concentration measurement in-cylinder • ($10^{-30} - 10^{-28}$)*
		Laser-induced fluorescence (LIF)	Liquid/gaseous radicals • Species concentration; combustible mixtures, intermediate radicals during combustion • Liquid/vapor detection • Fuel film measurement • ($10^{-25} - 10^{-20}$)*

* Sensitivity: relative scattering strength.

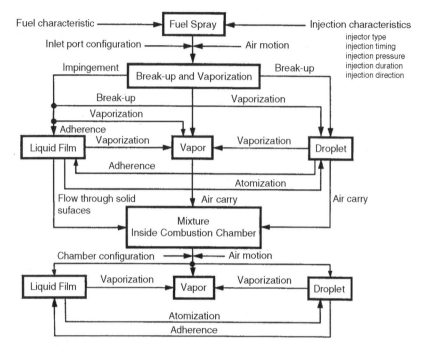

Figure 4.1 Mixture preparation process in PFI SI engines [Zhao et al. (1995)].

The degree of the preparation of a fuel/air mixture is characterized by fuel atomization and evaporation performance. Liquid phase spray is characterized by the droplet size distribution and droplet number density distribution. The portion of evaporated fuel also affects the mixture preparation process. The key parameters required to characterize a fuel spray from a port injector and their measurements have been reviewed by Zhao *et al.* (1995).

The direct imaging of sprays with front/back illumination can characterize fuel sprays, which enables the measurement of spray geometry and penetration depth. High-speed cinematography and holography have been used for spray patternation as well [Chigier (1991)]. The introduction of lasers improved the applicability of their techniques, in that lasers are considered more powerful sources of illumination.

It is important to know the droplet size distributions in the sprays, in addition to the general spray structure, because the droplet size distribution has a direct effect on the fuel atomization and vaporization processes. Classical methods of determining droplet size distribution, such as collection techniques, fractional separation techniques, and electrical methods, have evolved to optical methods, where either images of droplets or optical/physical properties are used for determining droplet sizes [Lenz (1992)]. The simplest optical droplet sizing technique is the direct imaging of fuel droplets by photographic films or solid-state cameras, which involves a spark light or a laser to freeze the fuel spray. The general structure of a fuel spray could be provided from the microscopic images. The droplet size distribution could be determined by counting and sizing the diameter of the droplets from the microscopic images in dilute regions of the droplet field. Detection of smaller droplets requires larger magnification using telescopic lenses, which can size down to a micrometer. Holographic recordings can extend the photographic techniques in three-dimensional measurements.

Advanced optical measurements using light scattering of fuel droplets can provide the temporal and spatial information on the microscopic spray fields, replacing tedious and time-consuming direct imaging. Light-scattering measurements could be divided into integral methods and particle counting. Fraunhofer diffraction is a typical integral method, measuring large numbers of droplets simultaneously. The scattering intensity comes from integration over space. The Fraunhofer diffraction pattern can be observed on a screen at a distance from the particle, when an opaque particle is illuminated by a beam of parallel monochromatic light. The diffraction pattern is unique for a collection of particles of different sizes, so that the size distribution can be deduced. The Fraunhofer diffraction method is implemented by a combination of a parallel beam of monochromatic light, typically from a laser, and an array of detectors. The light traverses a cloud of droplets or particles, and the transmitted and scattered light is collected by a lens. Its intensity distribution is observed by photodetectors in the focal plane. An apparatus developed by Swithenbank *et al.* (1977) has been adopted extensively and is available commercially. Droplets in a wide range of sizes can be measured using different light that is integrated over the measurement volume along the laser beam. It has limited spatial resolution, although an Abel inversion scheme can be used to procure spatially resolved measurements.

Laser phase Doppler anemometry (PDA) as a particle-counting method views one droplet at a time. The PDA technique is based on the detection of the phase of the light scattered from an individual particle. It requires an interference fringe pattern created within a small volume in the spray. A droplet moving this volume generates a scattered light modulated in time and space. The temporal modulation is related to the velocity of the droplet, while the spatial frequency is related to the droplet size. The light rays incident on a sphere of arbitrary size are partly reflected from the surface of the sphere (reflected ray) and are transmitted and refracted by the sphere (first-order refracted rays), while rays reflected from the internal surface are refracted in the backward direction (second-order refracted rays). The refracted ray undergoes a change of phase relative to the hypothetical wave. This phase difference is proportional to the diameter of the spherical particle.

The PDA configuration is similar to dual-beam laser Doppler velocimetry (LDV). The laser beam is split into two beams and then is focused to intersect using a transmitting lens. A particle passing through the beam intersection region produces a scattered interference fringe pattern that appears to move past the detectors. The frequency of Doppler burst signal from each detector is used to determine the particle velocity. The phase shift between two Doppler burst signals is used to calculate particle size. In practice, three detectors are used to determine the droplet size because these produce two pairs of redundant measurements that are compared to estimate the quality of measurements. The size distribution from PDA measurements is built up over time, leading to a temporal average, while the temporal distribution provides a measure of the mass flux of droplets of different sizes.

Ikeda et al. (1997) developed a high-data-rate PDA measurement to demonstrate the spray characteristics at each cycle and how each injection differed from the other, implying the mixture formation process. This technique proved to provide size-classified spray structure. The liquid fuel droplets also were studied in the intake port by phase Doppler interferometry, representing the intake port mixing process. The study results were correlated with the resulting combustion performance [Holthaus et al. (1997)].

Takagi and Skippon (1998) used interferometric laser imaging for droplet sizing (ILIDS) for in-cylinder spray characterization. This technique uses the angular oscillations in intensity observed in the wide-angle forward-scatter region, when droplets are illuminated by coherent radiation of a pulsed laser. Macrophotographic recording of a series of out-of-focus spots, corresponding droplets, onto fast monochrome film provides high spatial resolution in a wide field of view. Droplet size could be calculated using the principle that the spatial frequency of the fringes on each spot in the image plane is dependent on the diameter of the corresponding droplet. They found that the in-cylinder droplet size (SMD, or Sauter mean diameter) and the fuel volume in the droplet phase early in the intake are positively correlated with engine-out HC emissions and cyclic variability. These findings again indicate the advantages of enhanced fuel atomization.

Three-dimensional measurement of the spray structure of spray droplets was improved by laser holography [Anezaki et al. (2002)].

4.2.2 Wall Wetting

During a cold start, the temperature of the engine walls is too low to effectively vaporize the liquid fuel. Depending on the original size spectrum of the fuel droplets, the manifold geometry, and the temporal relation between fuel induction and the inlet process into the cylinder, a portion of the fuel droplets is deposited in the form of a fuel film along the manifold walls or on the inlet valve back surface. Another portion of the droplets either vaporizes in the airstream or enters the combustion chamber, still in droplet form. Therefore, liquid fuel can enter the cylinder during intake and can remain in the cylinder, both as droplets dispersed in the gas mixture and as liquid films on the walls. Some of the entering liquid fuel impinges on the top of the piston. Some of the vaporized fuel recondenses on the surfaces as well. The large droplets and the fuel film on the cylinder walls will then contribute to the liquid fuel, which could survive the combustion process into the exhaust process. The wall wetting of the combustion chamber was found to result in a significant increase in HC emissions [Stanglmaier et al. (1999)].

Knowing the quantities of liquid fuel film accumulated on the manifold walls under various load and operating conditions is not only important for judging the cold-start, warm-up, and response behavior of an engine, but moreover is the basis for analyzing the causes of mixture maldistribution. Most wall film measurements had been conducted indirectly by measuring the air/fuel (A/F) ratio, in-cylinder concentrations, or exhaust emissions of HCs, unless the geometry of the intake system was substantially interfered. An innovative experiment by Imatake et al. (1997), employing a flame ionization detector (FID) and electronically controlled valves, investigated the intake port and cylinder wall wetting. First, they froze an engine state during a transient time. Then hot air was admitted to the intake port (and combustion chamber), after which the HC-laden gases were removed through an FID to determine the intake port residual fuel, that is, wall wetting puddle masses. These results provided indications that fuel is delayed upon entering the combustion chamber and that puddles exist and build up in the intake port as the manifold pressure is increased. A similar setup of FID implementation was still used to confirm the contribution of fuel film flow that was intentionally simulated by a specially designed fuel probe [Landsberg et al. (2001)], and to find the fuel delivery efficiency [Santoso and Cheng (2002)]. Russ et al. (1998) used a universal exhaust gas oxygen (UEGO) sensor to detect A/F ratio deviations from stoichiometric caused by fuels of differing volatility and different injector targeting. They indirectly estimated the mass of the fuel film in the port.

An optical probing technique has been developed to interrogate the behavior of the port film. Evers and Jackson (1995) and Coste and Evers (1997) developed an optical sensor using the internally reflected light from the film surface. This probe can make point measurements in locations where the probe can gain optical access to the bottom of the fuel film. However, this is not suited yet to realistic fuel film measurement on the intake valve of a firing engine.

Laser-based optical techniques made a breakthrough to fuel film measurement on the back surface of the intake port and cylinder liner. Johnen and Haug (1995) used a fiber film optic-based laser-induced fluorescence (LIF) measurement technique to make fuel film thickness point measurements at several locations on the flat surface of a simplified model intake port.

Almkvist et al. (1995) made an LIF film thickness measurement along a line of laser light (formed from a laser sheet) focused into the inlet port of a running PFI engine. Other than slight modifications to allow for optical access, the port geometry was unmodified. They used 3-pentanone as a tracer for iso-octane because the vaporization properties of these two compounds are well matched. They examined the effect of cold versus warm engine operation, and the effect of intake valve open (IVO) versus intake valve closed (IVC) injection timing.

Felton et al. (1995) used an optical fiberscope to extract an LIF signal from a fuel film in the region of the valve guide and septum of a running engine. They interpreted the total normalized fluorescence intensity signal from the image to be an indicator of the total quantity of film stored in this region.

Senda et al. (1999) also used LIF to investigate the fuel film formed when a fuel spray impinges on a plate of glass at ambient conditions. For this simplified condition, they showed that the LIF technique is capable of making quantitative fuel film thickness measurements with good spatial and temporal resolution. Using this technique, they investigated the effect of fuel impingement direction and distance on the formation of adhered film.

Bruno et al. (2002) employed an LIF-based technique to examine the transient fuel film behavior of an intake valve during cold start of a PFI engine. Fluorescence from a tracer in the fuel was collected through a Borescope and imaged onto a charge-coupled device (CCD) camera, providing a two-dimensional image of the fuel film on the valve. Figure 4.2 shows the optical

Characterization of Cold Engine Processes

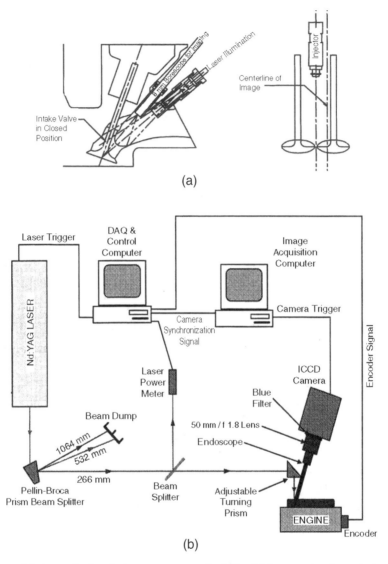

Figure 4.2 An optical technique to measure fuel film thickness using laser-induced fluorescence (LIF). (a) Optical access to the intake port and valve. (b) Schematic of the LIF intake valve and fuel film imaging experiment [Bruno et al. (2002)].

techniques using LIF. Ketone was doped in a mixture fuel of iso-octane and dimethylbutane. Ketone absorbs well at the incident wavelength of 266 nm and emits strongly with a peak fluorescence at approximately 430 nm. The average intensity of the fluorescence, over a region of interest, was taken to be proportional to the total amount of fuel present in the film during the cold-start transient. The information is component specific as well as temporally and spatially resolved, so that the observed differences in the behavior of the fuel film from different fuels could be related to the cold-start performance. The films of the low-volatility fuel (iso-octane) and the low-volatility component of a single two-component fuel were both found to grow initially for the first 5 to 15 cycles, and then slowly decline until the end of the experiment. These films were found to persist even after the intake valve opened. The high-volatility components (dimethylbutane) demonstrate a similar initial increase in fuel film, followed by a more rapid decline. Very little of the high-volatility fuel remained on the valve after the inlet valve opened.

As shown in Figure 4.3, two-dimensional visualization of liquid fuel films was provided by the application of LIF [Cho et al. (2001)]. The fluorescent intensity was related with the fuel film thickness on the quartz cylinder liner so that the cylinder wall wetting could be quantified with respect to injector angle. A single-cylinder research engine with an extended Bowditch piston also allowed the fuel film visualization by LIF [Witze (1999)].

4.2.3 Fuel Delivery into the Engine Cylinder

The fuel behavior at the intake port region was investigated by a high-speed direct imaging technique [Shin et al. (1995)]. The optical windows for illumination and observation were built on the intake runner so that the images of the charge intake process could provide the significant fuel delivery features. It was found that the reverse blowdown flow at the intake valve opening strip-atomizes the liquid film carrying the droplets away from the engine. Also, the forward flow strip-atomizes the film flow into the cylinder as droplets.

High-speed spectral infrared (IR) imaging was employed to identify some liquid fuel layers formed in the cylinder stemming from intake-port liquid fuel layers [Campbell et al. (1999)]. The locally reacting centers and luminous flows detected by IR imaging implied the existence of the liquid layers. The findings suggested that, even after the engine was well warmed, liquid fuel layers are formed over and in the vicinity of the intake valve. The sluggish consumption of those liquid layers, which were contained even until the exhaust valve opens, is expected to be one of the main emissions sources of unburned HCs.

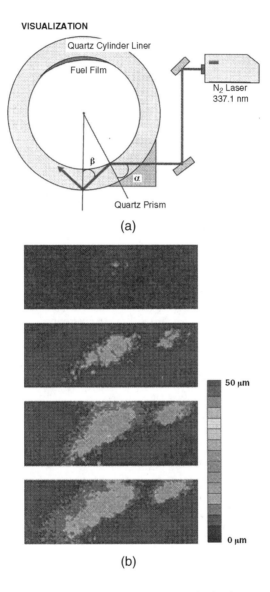

Figure 4.3 Fuel film visualization with LIF on the cylinder liner. (a) Principle of visualization. (b) Typical image of the wall fuel film thickness [Cho et al. (2001)].

The performance of different injection systems could be tested by investigating their spray and mixture formation by laser diagnostics. Standard PFI schemes have been shown to have an inherent problem associated with the use of closed-valve injection (CVI). They suffer from a so-called fuel "lag" due to the buildup of a liquid fuel film in the intake port. This makes precise A/F ratio control difficult and causes high levels of HCs during cold start. Another injection scheme is to use air-assisted port fuel injection (AAPFI) or air-forced port fuel injection (AFPFI) systems, which provide much better atomization and therefore can be used with an open-valve injection (OVI) strategy for better starting. Alkidas and Drews (1996) and Lee et al. (1999) reported clear benefits using AFPFI during cold start in terms of combustion stability and HC emissions. They investigated the in-cylinder mixture formation by planar laser-induced fluorescence (PLIF) and HC sampling with an FID, which provided the history of mixture formation and HC emissions as time passed from the start.

Richter et al. (2002) used a laser diffraction method for droplet sizing, a two-phase particle image velocimeter (PIV) for spray flow characterization, and visualization of the valve surface using LIF for fuel film monitoring. The standard production injector and a flash boiling injector were compared in terms of atomization and wall wetting. The spray generated by the flash boiling injector showed a significant reduction in droplet size (~10 μm compared to 100 μm for multipoint injection [MPI]) and a partial direct vaporization during the injection process by preheating the fuel inside the injector. However, this flash boiling injector led to an increase in fuel wall wetting under cold-start conditions, implying that there is a need for optimization between atomization and wall wetting in selecting the proper injection systems for better cold-start performance.

Measurements of the droplet velocity and size at various points in the optical engine cylinder were obtained as a function of injection timing corresponding to both CVI and OVI strategies [Chappius et al. (1997); Arcoumanis et al. (1998)]. The results confirmed that the mixture was fairly homogeneous for fuel injection with the valves closed and that some in-cylinder droplets were large—up to 110 μm (SMD)—generated by stripping of the liquid films from the valve and port surfaces and vaporized well before ignition. Evaporation of the fuel deposited on the hot surfaces of the inlet valves was enhanced by the backflow of hot residual gases that occurred in the early stages of intake. In contrast, injection toward the open valves leads to the deterioration of performance and high HC levels. This was attributed to the high concentration of small droplets of approximately 40 μm close to the liner on the exhaust valve side, resulting from impingement and secondary atomization on the liner.

Liquid fuel inflow into the cylinder was visualized by PLIF and PDA by Meyer and Heywood (1999a and 1999b), as shown in Figure 4.4. Zughyer *et al.* (2000) also found that most of the fuel under OVI conditions entered the cylinder as droplet mist, in contrast to CVI conditions. They observed that the combustion during the early cycles of cold start began with insufficiently vaporized fuel, followed by visible weak flame fronts developing to overall-rich combustion at a higher rate induced by late-vaporized fuel portions.

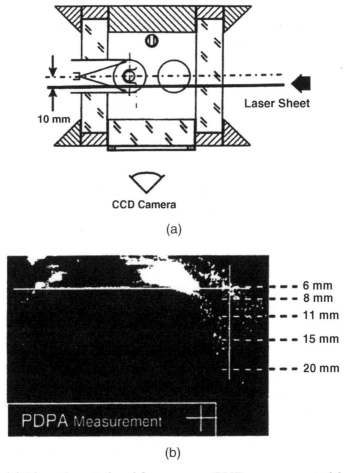

Figure 4.4 *Planar laser-induced fluorescence (PLIF) measurement of the liquid fuel inflow into a cylinder: (a) optical measurement plane, and (b) typical image of PLIF [Meyer and Heywood (1999b)].*

The absorption of unburned fuel into the engine cylinder wall oil film has been identified by Parks *et al.* (1998) by using an *in situ* measurement of the fuel/oil film interaction using LIF. Planar laser-induced fluorescence also was used to observe the in-cylinder transport of unburned fuel that, while trapped in the ring-land and ring-groove crevices, survives combustion in the propagating flame, eventually becoming one of the primary sources of unburned HC emissions [Green and Cloutman (1997); Swindal *et al.* (1997)].

4.3 Mixture Distribution and Its Interaction with Flow

Mixture distribution is associated with the interaction of the fuel spray with the port wall and valve, the in-cylinder droplet field, and the flow field, which consequently determines the flame propagation process and HC emissions, especially under cold-start conditions [Arcoumanis (1998)]. Liquid droplets of fuel spray may not be completely evaporated until the mixed charge from the intake manifold is inhaled into the cylinder, especially during cold start. The detection of the liquid portion of fuel and fuel vapor is necessary to identify the degree of fuel evaporation, which could be achieved by the laser-induced exciplex fluorescence (LIEF) method. The elastic scattering (also called Mie scattering) or fluorescence signal from the liquid phase is many orders of magnitude stronger than that from the fuel vapor, limiting the detection of liquid alone. The LIEF technique is versatile because the fluorescence from the liquid and the vapor phase can be separated in wavelength and detected separately. Laser extinction and absorption (LEA) is another technique for simultaneous detection of fuel vapor and liquid droplets, generating information about droplet size in the spray.

The combustion stability in SI engines basically depends on the A/F ratio. The realistic in-cylinder A/F ratio could be significantly different from the nominal value in the intake mixture during cold start due to the lag in the fuel transport and the loss of fuel to the crankcase [Shayler *et al.* (1997)]. The global A/F ratio in the engine process could be measured by an oxygen sensor (i.e., UEGO sensor) in the exhaust gas stream [Shayler *et al.* (2000); Winkler and Mueller (2001)], while the realistic in-cylinder A/F ratio could be monitored by high-speed sampling with a fast-response gas analyzer such as a fast-response flame ionization detector (FFID) and laser-based spectroscopic diagnostics, particularly the application of two-dimensional imaging of scattered laser light.

Universal exhaust gas oxygen sensors are widely used in engine exhaust streams to measure the recently burned A/F mixture ratio. In a wide-range

UEGO sensor, oxygen pumping and reference chambers are separated from the exhaust stream via a ceramic electrolyte (zirconia) and metallic electrodes. A voltage can be generated as a result of different oxygen partial pressures across the different chambers. However, UEGO sensor limitations can be significant due to the response times on the order of a few hundred milliseconds. Fast in-cylinder diagnostics on the combustible gas A/F ratio can be accomplished with an FFID, which can measure the HC concentration in-cylinder during the compression stroke. These data will be used to determine the A/F ratio [Ladommatos and Rose (1996); Cowart and Cheng (2000)]. An FFID usually measures the in-cylinder HC concentrations through the artificial channel into the engine cylinder, employing a sampling probe. The fast response time of some milliseconds in a total sampling/FFID measuring system could provide cycle-resolved data. Cowart (2002) compared the measuring performances of a UEGO sensor and an FFID. He found that the UEGO sensor effectively filters rich in-cylinder A/F ratio excursions and appropriately characterizes the transient fuel behavior, although it can significantly underestimate the extent of A/F excursions. However, the wall wetting parameter is not appropriately characterized by either the UEGO sensor (overestimated) or the FFID (underestimated).

It was found that high HC emissions that could be made during rapid throttle ramp rates (0.4 to 0.9 bar intake manifold air pressure [MAP] in three cycles) resulted from rich spikes during the fast transient operation, which was confirmed by FFID measurements [Cowart and Cheng (2000)]. This supposedly was due to the accumulated puddles during the transient operation. Fueling behavior is comparatively stable at stoichiometric condition with a slow throttle ramp rate (0.4 to 0.9 bar MAP in thirteen cycles). This implies that the throttle transient problem with high HC emissions might be solved by fast pressure measurement in the intake manifold, which could enable fast transient compensation.

Considering that even heated exhaust gas oxygen (HEGO) sensors cannot measure the exact A/F ratio of exhaust gas during cold-start conditions, Lee *et al.* (2002) developed a new estimator for the A/F ratio of the exhaust gas utilizing the exhaust gas temperature measured with a fast-response, fine-wire thermocouple (25-µm thickness). They generated A/F traces by the generalized regression neural network (GRNN) function approximation, which matched well with HC measurement by FFID. This method utilized the fact that the exhaust gas temperature is a function of the A/F ratio. Imatake *et al.* (1997) successfully utilized FID to investigate the intake port and cylinder wall through an innovative experiment, where the engine state was frozen by electronically controlled intake and exhaust valves, indicating the delay in the fuel entering the cylinder and the existence of puddles built up in the intake port. This implies that the exact measurement of exhaust gas temperature can provide information on the A/F ratio in compensation or replacement of FFID or UEGO measurements.

The A/F ratio of the combustion state could be estimated from the time-series chemiluminescence measurements of hydroxyl (OH), methylidyne (CH), and carbon dimer (C_2) radicals [Ikeda et al. (2001a)]. The comparison between chemiluminescence measurements in-cylinder and FFID sampling/analysis in the exhaust manifold proved the applicability of chemiluminescence in A/F measuring.

Mixture concentration near a spark plug could be measured by a laser IR absorption method [Nishiyama et al. (2003)]. An IR spark plug sensor with a double-pass measurement length was developed, as shown in Figure 4.5.

Another sampling and analysis technique was developed by Quader and Majkowski (1999) to measure the cycle-by-cycle equivalence ratio. They developed a diode laser-based spectroscopic technique incorporated with purpose-built sampling hardware to simultaneously measure the cycle-by-cycle fuel vapor-air equivalence ratio and residual gas carbon dioxide (CO_2) concentration inside the cylinder of an operating engine, as shown in Figure 4.6. A laser-based CO_2 absorption technique was used to analyze the samples, which were extracted once from each compression stroke. The stream of sample and carrier gas first passes through an optical cell (precatalyst cell) to measure the CO_2 level in the unburned sample gas. The sample was burned passing over a heated catalyst and then passed through a second optical cell (post-catalyst cell) to measure the CO_2 level again. The relative concentration of CO_2 in each sample was determined from the optical attenuation of a diode laser beam transmitted through the sample. From this sampling and analysis, it was found that combustion usually begins in the first cycle in which the fuel vapor-air equivalence ratio of the mixture inside the cylinder exceeds the lean flammability limit of the fuel at the desired temperature. The fuel vapor-air equivalence ratio for the start of combustion generally increased at lower temperatures. After the first cycle with combustion, dilution with residual gas can contribute to misfire in succeeding cycles. Hence, the fuel vapor-air equivalence ratio must be richer than the lean flammability limit for combustion to be sustained in succeeding cycles.

Characterization of Cold Engine Processes

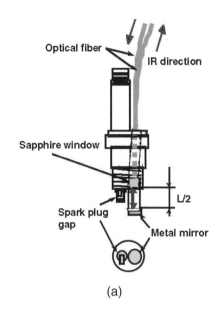

(a)

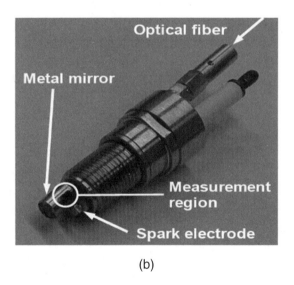

(b)

Figure 4.5 (a) Construction of an IR sensor in a plug, and (b) photograph of a spark plug sensor [Nishiyama et al. (2003)].

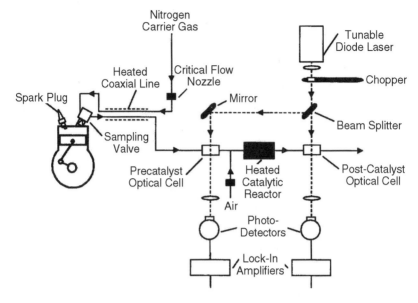

Figure 4.6 *Apparatus to analyze and measure CO_2 from samples [Quader and Majkowski (1999)].*

To better understand the mixture formation in port and engine cylinders, a number of non-intrusive laser-based techniques have been utilized. The three main laser-scattering methods are Rayleigh scattering, Raman scattering, and LIF.

Laser spectroscopic techniques involve the interaction and exchange of energy between a photon and a molecule. When a monochromatic radiation from a laser is passed through a transparent gas mixture, a small amount of the radiation energy is scattered. This scattered energy consists mostly of radiation of the incident wavelength (elastic scattering), so-called Rayleigh scattering, and a small amount of light at certain discrete wavelengths above and below that of the incident beam inelastic scattering referred to as Raman scattering. The scattered light signal, in general, is proportional to the gas density and a differential cross section that depends on the gas to be measured. For a binary mixture of fuel vapor and air, the mole fraction of fuel can be obtained from the Rayleigh scattering signal if calibration signals are measured in advance. Raman spectra also depend on the molecular structure of the molecules involved, so that the species concentration is measured from the Raman scattered signal.

The Rayleigh scattering technique has been used as an effective tool for fuel vapor concentration inside the engine combustion chamber [Arcoumanis et al. (1984); Zhao and Hiroyasu (1993)], while a spontaneous Raman scattering also was applied for in-cylinder mixture formation analysis [Knapp et al. (1997)]. Hinze and Miles (1999) used line imaging of Raman scattered light to simultaneously measure the mole fraction of carbon dioxide (CO_2), hydrogen oxide (water, H_2O), nitrogen (N_2), oxygen (O_2), and fuel (premixed propane, C_3H_8) in a four-valve PFI SI engine at idle (Figure 4.7). The measurements provided the level of fresh charge combustion residuals with the charge composition fluctuations in the vicinity of the spark plug and spatial mixing scales.

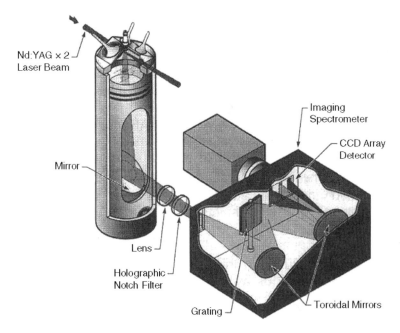

Figure 4.7 Raman scattered imaging in an optical research engine [Hinze and Miles (1999)].

Laser-induced fluorescence is the emission of light from an atom or molecule following excitation by a laser beam. The molecule is initially at a lower electronic state before it is excited to an upper electronic energy level by a laser source. The wavelength of the laser light must be chosen so that it coincides with the absorption wavelength. The LIF signal can be related to specific property concentrations of the absorbing species through modeling of the state-to-state transfer processes.

Laser-induced fluorescence has been used to characterize the fuel distribution inside the engine combustion chamber, including the differentiation between vapor and liquid. A pioneering work of Andresen *et al.* (1990) proved that LIF measured the density of fuel octane in the intake and in the cylinder. Two-dimensional LIF using planar laser sheet illumination (PLIF, or planar LIF) has been expanded to various species measurements to realize the detection of sub-parts-per-million levels, even under severe conditions in engines utilizing refined calibration methods. The use of fluorescent additive (tracer or dopant) into the fuel enhanced the sensitivity of fluorescent signal and the independence from quenching by oxygen. This method was named the LIEF (laser-induced exciplex fluorescence) technique. A variety of dopants have been tried in various fuels and engines: acetone in iso-octane [Wolff *et al.* (1994)], 3-pentanone in iso-octane [Neij *et al.* (1994); Johansson *et al.* (1995)], ethylmethylketone in gasoline [Lawrenz *et al.* (1992)], biacetyl in gasoline [Baritaud and Heinze (1992)], and nitrogen dioxide in gaseous fuel [Zhao *et al.* (1995)].

These LIF/PLIF/LIEF provided the physical insights into the in-cylinder mixture formation and the subsequent combustion processes. In most transient engine conditions, the mixing between the fuel and air is far from complete at induction bottom dead center (BDC), and the mixture tends to homogenize during compression, even with surviving heterogeneity. Injection timing can have an important effect on the mixture distribution, and its sensitivity is dependent on the flow pattern. The two-dimensional measurements on spatial fuel distributions by LIF provided the A/F maps that could give quantitative information on the mixing level to evaluate the correlation between the mixture quality and the combustion instability and subsequent pollutant formations. For example, fuel injection in the cold condition or in the intake stroke results in slow energy conversion and therefore low indicated mean effective pressure (IMEP) due to droplet penetration into the cylinder [Wolff *et al.* (1994)]. A planar laser-induced exciplex fluorescence (PLIEF) approach by Styron *et al.* (2000) enabled separate imaging of liquid and vapor fuel as a function of flow characteristics, as shown in Figure 4.8. The PLIEF measurement offered more evidence of non-uniform fuel distribution owing to liquid wall impingement with a high potential for producing increased HC emissions.

A locally rich region is created around the evaporating droplets, which is transported by the cylinder air motion so that it arrives in the vicinity of the spark

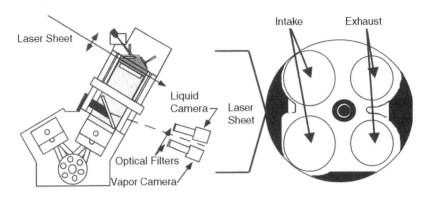

Figure 4.8 Simultaneous liquid and vapor imaging by a PLIEF with a horizontal laser sheet oriented perpendicular to the cylinder axis [Styron et al. (2000)].

plug at the time of ignition. It was found that the open valve and/or backflow injection induce the in-cylinder charge stratification, creating a region of relatively rich or less lean mixture close to the spark plug, in which the higher flame temperature enhances nitrogen oxide (NOx) formation [Takagi and Skippon (1998)]. While the bulk flow motion is governing the mixture formation on a large scale, the turbulent component of in-cylinder flow determines the speed of the combustion processes. Therefore, flow characteristics are interacted with flow in the port and cylinder affecting the fuel delivery into the cylinder, mixture formation/distribution, and the combustion process. This is why flow measurement is important.

Measurement of airflow has been obtained using LDV in both the port and the cylinder [Asanuma and Obotaka (1979); Arcoumanis and Whitelaw (1987)]. The PIV technique could give detailed information on the two-dimensional distribution inside the cylinder. Particle image velocimetry has been used to characterize the large-scale in-cylinder flows and even turbulence mostly in the engine cylinder [Stolz *et al.* (1992); Baby *et al.* (2002); Funk *et al.* (2002); Söderberg and Johansson (2002)]. Fuel fraction stratification is probably a good method for reducing CO_2 and NOx emissions. The in-cylinder flow must be optimized to obtain good fuel stratification.

Li *et al.* (2003) used PIV, as shown in Figure 4.9, to measure the strong tumble motion made by an optimized intake system. Holographic particle image velocimetry (HPIV) also could be used to measure the high-resolution velocity field [Konrath *et al.* (2001)].

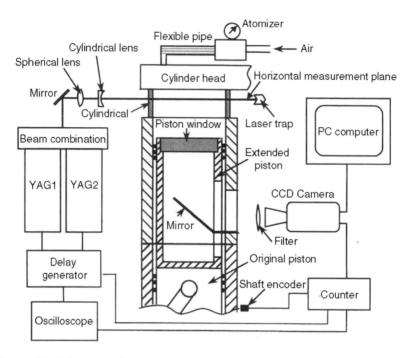

Figure 4.9 Schematic of an optical engine, PIV system, and the horizontal measurement plane [Li et al. (2003)].

The combination of flow measurement with LDV and flame imaging in-cylinder has been widely used [Arcoumanis *et al.* (1994)], which could be combined with fast spark-plug HC sampling, NOx/HC measurement with fast chemiluminescence, and FIDs/analyzers [Kampanis *et al.* (2001)].

4.4 Combustion Processes and Pollutant Formation

Laser diagnostics of flame characteristics have been improved significantly by high-speed cameras and powerful lasers in terms of spatial and temporal resolution for analyzing turbulent flow, ignition phenomena, and flame growth. Laser-induced fluorescence has been proved suitable to measure the species concentration of the transient free radicals in flames, which are intermediates in combustion chemistry. Applications of LIF to combustion have been concerned with the important and ubiquitous OH radicals in a flame. Two-dimensional imaging of LIF from the naturally occurring flame radical OH have been developed from a steady-state flame [Dyer and Crosley (1982)] and extended to optical engine applications [Felton et al. (1988); Andresen et al. (1990)]. Planar laser-induced fluorescence using laser sheet illumination, as shown in Figure 4.10, enabled the analysis of the flame propagation with fine spatial and temporal resolution [Foucher et al. (2001); Knaus et al. (1999)]. Laser-induced fluorescence also made the distribution of NO visible in the cylinder, although not during the combustion phase [Andresen et al. (1990); Berckmüller et al. (1997); Hildenbrand et al. (1998)].

Even the spark plug as an ion probe was proved useful as a combustion diagnostic tool [Yoshiyama and Tomita (2002)]. The ion current was well correlated with IMEP under idling condition, which shows the potential of combustion quality measurement. Two-dimensional Mie scattering visualization provided the refined turbulent flame contours in an optical engine [Ziegler et al. (1990)], where submicron-sized smoke particles, usually titanium dioxides, were added and scattered differently inside and outside the flame illuminated by a laser. Ikeda et al. (2001b) developed Cassegrin optics to measure local OH, CH, and C_2 radicals used with three CCD cameras incorporated in an optical engine. They found that CH and C_2 chemiluminescence signals can be a nice marker of flame front structure and its thickness, and the flame growth is pancake shaped with a hollow structure. Jansons et al. (2001) developed a high-speed spectral IR imaging method to investigate flame development, particularly for transient processes such as during cold start. This is a promising tool to improve the understanding of in-cylinder processes.

Liu and Wallace (1999) used an FFID to measure the in-cylinder HC concentrations on the post-flame period mainly to give a better understanding of the mechanism by which HC emissions form from crevices in SI engines. They found that the HC value is more dependent on the temperature and local mixing, and significant HC concentration could be found in the bulk gas at low-load conditions due to more HCs from the ring crevice, which reach the cylinder head and remain unburned during the post-flame period. An FFID also was used by Klein and Cheng (2002) to monitor behaviors of HC emissions in stopping and restarting. The HC measurements from two FFID probes at precatalyst and post-catalyst provided the information on the unburned HC behavior, leading to findings that HC in the shutoff process contained in the engine and exhaust system contributes to exhaust HCs in the restarting stage. Furthermore, an FFID also was used to confirm the beneficial performance of the recent advances in variable valve actuation (VVA) on the cold-start

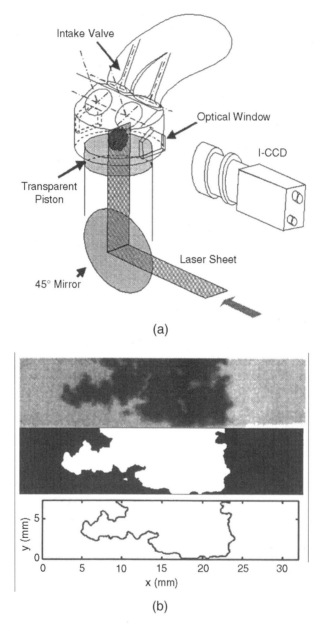

Figure 4.10 Flame visualization using PLIF. (a) Optical setup for PLIF imaging in a transparent engine lab. (b) Examples of instantaneous PLIF image, binary image, and flame contour [Foucher et al. (2001)].

period [Roberts and Stanglmaier (1999)]. Chemiluminescence measurements on some radicals could provide the information on the local flame front structure and its thickness. The FFID engine-out measurements directly after the exhaust valve have shown the instantaneous HC characterizations, proving the potential for simultaneous startup HC emissions reduction and driveability improvements by the optimized intake phasing, that is, retarded intake timing. The combination of FFID and flame visualization in an optical engine showed similar findings, that is, the reduction of HC emissions by spark timing retard [Choi et al. (2000)].

4.5 Summary

A refined understanding of the combustion and pollutant formation process in SI engines, especially HC emissions under cold-start conditions in PFI SI engines, could be achieved by diagnostic techniques, especially the laser-based *in-situ* diagnostics on the fuel sprays, mixture preparation, flame propagation, and pollutant formation. The diagnostic tools have been reviewed, which could be divided into the classical ones of physical probing and remote sensing with non-intrusive optical diagnostics, including laser diagnostic techniques.

Fuel transport process plays an important role in engine cold-start operations and could be characterized by fuel atomization and evaporation performances. The measurement of characteristics is enabled by the direct imaging of sprays with front/back illumination with the help of high-speed cinematography and holography. Classical methods determining droplet size distribution have evolved into optical methods, where either images of droplets or optical/physical properties are used for determining droplet sizes. Advanced optical measurements using light scattering of fuel droplets, such as Fraunhofer diffraction and laser PDA, can provide temporal and spatial information on the microscopic spray fields.

Wall wetting of the combustion chamber was found to result in a significant increase in HC emissions. Most wall film measurements during cold start had been conducted indirectly by measuring the A/F ratio, in-cylinder concentrations, or exhaust emissions of HCs. A UEGO sensor could be used to detect A/F ratios. Optical techniques such as optical probing and LIF made a breakthrough in fuel film measurement on the surface.

The detection of the liquid portion of fuel and fuel vapor is needed to identify the degree of fuel evaporation, which could be achieved by LIEF or LEA.

A realistic in-cylinder A/F ratio could be monitored by high-speed sampling with fast-response gas analyzers such as FFID and laser-based spectroscopic diagnostics. To overcome the limitation of UEGO sensor measurement during the cold-start condition, a new estimator for A/F of the exhaust gas also was developed utilizing the exhaust gas temperature measured with a fast-response fine-wire thermocouple, considering that the exhaust gas temperature is a function of the A/F ratio.

To better understand the mixture formation in port and engine cylinders, a number of non-intrusive laser-based techniques have been utilized. The three main laser-scattering methods are Rayleigh scattering, Raman scattering, and LIF. These measurements provide further evidence of non-uniform fuel distribution owing to liquid wall impingement with a high potential for producing increased HC emissions. Flow characteristics are interacted with fuel in the port and cylinder affecting the fuel delivery into the cylinder, mixture formation/distribution, and the combustion process. Measurement of airflow has been obtained using LDV and PIV techniques.

Laser diagnostics of flame characteristics have been improved significantly by high-speed cameras and powerful lasers. Laser-induced fluorescence has been proved suitable to measure the species concentration of the transient free radicals in flames. An FFID could be used to measure the in-cylinder HC concentrations of the post-flame period, mainly to give a better understanding of the mechanism by which HC emissions form from crevices in SI engines. A high-speed spectral IR imaging method also was developed to investigate flame development, particularly for transient processes such as during cold start. Development of diagnostic measurement techniques will enable the optimized design of clean engines with reduced emissions. This could be achieved by the refinement of optical and electronic components with the help of fast electronic control and data acquisition systems.

4.6 References

1. Alkidas, A.K. and Drews, R.J. (1996), "Effects of Mixture Preparation on HC Emissions of an SI Engine Operating Under Steady-State Cold Conditions," SAE Paper No. 961956, Society of Automotive Engineers, Warrendale, PA.

2. Almkvist, G., Denbratt, I., Josefsson, G., and Magnusson I. (1995), "Measurements of Fuel Film Thickness in the Inlet Port of an S.I. Engine by Laser

Induced Fluorescence," SAE Paper No. 952483, Society of Automotive Engineers, Warrendale, PA.

3. Andresen, P., Meijer, J., Schlüter, H., Voges, H., Koch, A., Hentschl, W., Oppermann, W., and Rothe, E. (1990), "Fluorescence Imaging Inside an Internal Combustion Engine Using Tunable Excimer Lasers," *Applied Optics*, Vol. 29, No. 16, pp. 2392–2404.

4. Anezaki, Y., Shirabe, N., Kanehara, K., and Sato, T. (2002), "3D Spray Measurement System for High Density Fields Using Laser Holography," SAE Paper No. 2002-01-0739, Society of Automotive Engineers, Warrendale, PA.

5. Arcoumanis, C. (1998), "Research Issues in Passenger Car Engines," 4th International Symposium on Diagnostics and Modeling of Combustion in Internal Combustion Engines (COMODIA 98), Kyoto, pp. 1–15.

6. Arcoumanis, C., Bae, C.S., and Hu, Z. (1994), "Flow and Combustion in a Four-Valve, Spark-Ignition Optical Engine," SAE Paper No. 940475, Society of Automotive Engineers, Warrendale, PA.

7. Arcoumanis, C., Gold, M.R., Whitelaw, J.H., Xu, H.M., Gaade, J.E., and Wallace, S. (1998), "Droplet Velocity/Size and Mixture Distribution in a Single-Cylinder Four-Valve Spark-Ignition Engine," SAE Paper No. 981186, Society of Automotive Engineers, Warrendale, PA.

8. Arcoumanis, C., Green, H.G., and Whitelaw, J.H. (1984), "Application of Laser Rayleigh Scattering to a Reciprocating Model Engine," SAE Paper No. 840376, Society of Automotive Engineers, Warrendale, PA.

9. Arcoumanis, C. and Whitelaw, J.H. (1987), "Fluid Mechanics of Internal Combustion Engines—A Review," *Proc. Instn. Mech. Engrs.*, Vol. 201, No. C1, pp. 57–74.

10. Asanuma, T. and Obokata, T. (1979), "Gas Velocity Measurements of a Motored and Firing Engine by Laser Anemometry," SAE Paper No. 790096, Society of Automotive Engineers, Warrendale, PA.

11. Baby, X., Dupont, A., Ahmde, A., Deslandes, W., Charnay, G., and Michard, M. (2002), "A New Methodology to Analyze Cycle-to-Cycle Aerodynamic Variation," SAE Paper No. 2002-01-2837, Society of Automotive Engineers, Warrendale, PA.

12. Baritaud, T.A. and Heinze, T.A. (1992), "Gasoline Distribution Measurements with PLIF in an SI Engine," SAE Paper No. 922355, Society of Automotive Engineers, Warrendale, PA.

13. Berckmüller, M., Tait, N.P., and Greenhalgh, D.A. (1997), "The Influence of Local Fuel Concentration on Cyclic Variability of a Lean Burn Stratified-Charge Engine," SAE Paper No. 970826, Society of Automotive Engineers, Warrendale, PA.

14. Bruno, B.A., Santavicca, D.A., and Zello, J.V. (2002), "LIF Characterization of Intake Valve Fuel Films During Cold Start in a PFI Engine," SAE Paper No. 2002-01-2751, Society of Automotive Engineers, Warrendale, PA.

15. Campbell, S., Lin, S., Jansons, M., and Rhee, K.T. (1999), "In-Cylinder Liquid Fuel Layers, Cause of Unburned Hydrocarbon and Deposit Formation in SI Engines," SAE Paper No. 1999-01-3579, Society of Automotive Engineers, Warrendale, PA.

16. Chappius, S., Cousyn, B., Posylkin, M., Vannobel, F., and Whitelaw, J.H. (1997), "Effects of Injection Timing on Performance and Droplet Characteristics of a Sixteen-Valve Four-Cylinder Engine," *Experiments in Fluids*, Vol. 22, pp. 336–344.

17. Chigier, N. (1991), "Optical Imaging of Sprays," *Prog. Energy Combust. Sci.*, Vol. 17, pp. 211–262.

18. Cho, H., Kim, M., and Min, K. (2001), "The Effect of Liquid Fuel on the Cylinder Liner on Engine-Out Hydrocarbon Emissions in SI Engines," SAE Paper No. 2001-01-3489, Society of Automotive Engineers, Warrendale, PA.

19. Choi, M.S., Sun, H.Y., Lee, C.H., Myung, C.L., Kim, W.T., and Choi, J.K. (2000), "The Study of HC Emission Characteristics and Combustion Stability with Spark Timing Retard at Cold Start in Gasoline Engine Vehicle," SAE Paper No. 2000-01-1082, Society of Automotive Engineers, Warrendale, PA.

20. Coste, T.L. and Evers, L.W. (1997), "An Optical Sensor for Measuring Fuel Film Dynamics of a Port-Injected Engine," SAE Paper No. 970869, Society of Automotive Engineers, Warrendale, PA.

21. Cowart, J. (2002), "A Comparison of Transient Air-Fuel Measurement Techniques," SAE Paper No. 2002-01-2753, Society of Automotive Engineers, Warrendale, PA.

22. Cowart, J.S. and Cheng, W.K. (2000), "Throttle Movement Rate Effects on Transient Fuel Compensation in a Port Fuel Injected SI Engine," SAE Paper No. 2000-01-1937, Society of Automotive Engineers, Warrendale, PA.

23. Dyer, M.J. and Crosley, D.R. (1982), "Two-Dimensional Imaging of OH Laser-Induced Fluorescence in a Flame," *Optics Letters*, Vol. 7, No. 8, pp. 382–384.

24. Evers, L.W. and Jackson, K.J. (1995), "Liquid Film Thickness Measurements by Means of Internally Reflected Light," SAE Paper No. 950002, Society of Automotive Engineers, Warrendale, PA.

25. Felton, P.G., Kyritsis, D.C., and Fulcher, S.K. (1995), "LIF Visualization of Liquid Fuel in the Intake Manifold During Cold Start," SAE Paper No. 952464, Society of Automotive Engineers, Warrendale, PA.

26. Felton, P.G., Mantzaras, J., Bomse, D.S., and Woodin, R.L. (1988), "Initial Two-Dimensional Laser Induced Fluorescence Measurements of OH Radicals in an Internal Combustion Engine," SAE Paper No. 881633, Society of Automotive Engineers, Warrendale, PA.

27. Foucher, F., Burnel, S., and Mounaïm-Rousselle, C. (2001), "Local Flame Front Structure in the Vicinity of the Piston in a Transparent SI Engine," SAE Paper No. 2001-01-1957, Society of Automotive Engineers, Warrendale, PA.

28. Funk, C., Sick, V., Reuss, D.L., and Dahm, W.J.A. (2002), "Turbulence Properties of High and Low Swirl In-Cylinder Flows," SAE Paper No. 2002-01-2841, Society of Automotive Engineers, Warrendale, PA.

29. Gold, M.R., Arcoumanis, C., Whitelaw, J.H., Gaade, J., and Wallace, S. (2000), "Mixture Preparation Strategies in an Optical Four-Valve Port-Injected Gasoline Engine," *Int. J. Engine Research*, Vol. 1, No. 1, pp. 41–56.

30. Green, R.M. and Cloutman, L.D. (1997), "Planar LIF Observations of Unburned Fuel Escaping the Upper Ring-Land Crevice in an SI Engine," SAE Paper No. 970823, Society of Automotive Engineers, Warrendale, PA.

31. Hildenbrand, F., Schulz, C., Sick, V., Josefsson, G., Magnusson, I., Andersson, Ö., and Aldén, M. (1998), "Laser Spectroscopic Investigation of Flow Fields and NO-Formation in a Realistic SI Engine," SAE Paper No. 980148, Society of Automotive Engineers, Warrendale, PA.

32. Hinze, P.C. and Miles, P.C. (1999), "Quantitative Measurements of Residual and Fresh Charge Mixing in a Modern SI Engine Using Spontaneous Raman Scattering," SAE Paper No. 1999-01-1106, Society of Automotive Engineers, Warrendale, PA.

33. Holthaus, B.E., Wagner, R.M., and Drallmeier, J.A. (1997), "Measurements of Intake Port Fuel/Air Mixture Preparation," SAE Paper No. 970867, Society of Automotive Engineers, Warrendale, PA.

34. Ikeda, Y., Hosokawa, S., Sekihara, F., and Nakajima, T. (1997), "Cycle-Resolved PDA Measurement of Size-Classified Spray Structure of Air-Assist Injector," SAE Paper No. 970631, Society of Automotive Engineers, Warrendale, PA.

35. Ikeda, Y., Kaneko, M., and Nakajima, T. (2001a), "Local A/F Measurement by Chemiluminescence OH*, CH* and C2* in SI Engine," SAE Paper No. 2001-01-0919, Society of Automotive Engineers, Warrendale, PA.

36. Ikeda, Y., Nishihara H., and Nakajima, T. (2001b), "Measurement of Flame Front Structure and Its Thickness by Planar and Local Chemiluminescence of OH*, CH* and C2*," SAE Paper No. 2001-01-0920, Society of Automotive Engineers, Warrendale, PA.

37. Imatake, N., Saito, K., Morishima, S., Kudo, S., and Ohhata, A. (1997), "Quantitative Analysis for Fuel Behavior in Port-Injection Gasoline Engines," SAE Paper No. 971639, Society of Automotive Engineers, Warrendale, PA.

38. Jansons, M., Lin, S., and Rhee, K.T. (2001), "High-Speed Images from Consecutive Cycles," SAE Paper No. 2001-01-3486, Society of Automotive Engineers, Warrendale, PA.

39. Johansson, B., Neij, H., Aldén, M., and Juhlin, G. (1995), "Investigations of the Influence of Mixture Preparation on Cyclic Variations in an SI Engine, Using Laser-Induced Fluorescence," SAE Paper No. 950108, Society of Automotive Engineers, Warrendale, PA.

40. Johnen, T. and Haug, M. (1995), "Spray Formation Observation and Fuel Film Development Measurements in the Intake of a Spark Ignition Engine," SAE Paper No. 950511, Society of Automotive Engineers, Warrendale, PA.

41. Kampanis, N., Arcoumanis, C., Kato, R., and Kometani, S. (2001), "Flow, Combustion and Emissions in a Five-Valve Research Gasoline Engine," SAE Paper No. 2001-01-3556, Society of Automotive Engineers, Warrendale, PA.

42. Klein, D. and Cheng, W.K. (2002), "Spark Ignition Engine Hydrocarbon Emissions Behaviors in Stopping and Restarting," SAE Paper No. 2002-01-2804, Society of Automotive Engineers, Warrendale, PA.

43. Knapp, M., Beushausen, V., Hentschel, W., Manz, P., Grünefeld, G., and Anderson, P. (1997), "In-Cylinder Mixture Formation Analysis with Spontaneous Raman Scattering Applied to a Mass-Production SI Engine," SAE Paper No. 970827, Society of Automotive Engineers, Warrendale, PA.

44. Knaus, D.A., Gouldin, F.C., Hinze, P.C., and Miles, P.C. (1999), "Measurement of Instantaneous Flamelet Surface Normals and the Burning Rate in an SI Engine," SAE Paper No. 1999-01-3543, Society of Automotive Engineers, Warrendale, PA.

45. Konrath, R., Schröder, W., and Limberg, W. (2001), "Three-Dimensional Flow Measurements Within the Cylinder of a Motored Four-Valve Engine Using Holographic Particle-Image Velocimetry," SAE Paper No. 2001-01-3493, Society of Automotive Engineers, Warrendale, PA.

46. Ladommatos, N. and Rose, D. (1996), "On the Cause of In-Cylinder Air-Fuel Ratio Excursions During Load and Fuelling Transients in Port-Injected Spark-Ignition Engines," SAE Paper No. 960466, Society of Automotive Engineers, Warrendale, PA.

47. Landsberg, G.B., Heywood, J.B., and Cheng, W.K. (2001), "Contribution of Liquid Fuel to Hydrocarbon Emissions Spark Ignition Engines," SAE Paper No. 2001-01-3587, Society of Automotive Engineers, Warrendale, PA.

48. Lawrenz, W., Köhler, J., Meier, F., Stolz, W., Wirth, R., Bloss, W.H., Maly, R.R., Wagner, E., and Zahn, M. (1992), "Quantitative 2D LIF Measurements of Air/Fuel Ratios During the Intake Stroke in a Transparent SI Engine," SAE Paper No. 922320, Society of Automotive Engineers, Warrendale, PA.

49. Lee, S., McGee, J.M., Quay, B.D., and Santavicca, D.A. (1999), "A Comparison of Fuel Distribution and Combustion During Engine Cold Start for Direct and Port Fuel Injection Systems," SAE Paper No. 1999-01-1490, Society of Automotive Engineers, Warrendale, PA.

50. Lee, T., Bae, C., Bohac, S.V., and Assanis, D. (2002), "Estimation of Air/Fuel Ratio of an SI Engine from Exhaust Gas Temperature at Cold Start Condition," SAE Paper No. 2002-01-1667, Society of Automotive Engineers, Warrendale, PA.

51. Lenz, H.P. (1992), *Mixture Formation in Spark-Ignition Engines*, Society of Automotive Engineers, Warrendale, PA.

52. Li, Y., Zhao, H., Leach, B., Ma, T., and Ladommatos, N. (2003), "Optimisation of In-Cylinder Flow for Fuel Stratification in a Three-Valve Twin-Spark-Plug SI Engine," SAE Paper No. 2003-01-0635, Society of Automotive Engineers, Warrendale, PA.

53. Liu, H. and Wallace, J.S. (1999), "Instantaneous In-Cylinder Hydrocarbon Concentration Measurement During the Post-Flame Period in an SI Engine," SAE Paper No. 1999-01-3577, Society of Automotive Engineers, Warrendale, PA.

54. Meyer, R. and Heywood, J.B. (1999a), "Effect of Engine and Fuel Variables on Liquid Fuel Transport into the Cylinder in Port-Injected SI Engine," SAE Paper No. 1999-01-0563, Society of Automotive Engineers, Warrendale, PA.

55. Meyer, R. and Heywood, J.B. (1999b), "Evaporation of In-Cylinder Liquid Fuel Droplets in an SI Engine: A Diagnostic-Based Modeling Study," SAE Paper No. 1999-01-0567, Society of Automotive Engineers, Warrendale, PA.

56. Neij, H., Johansson, B., and Aldén, M. (1994), "Development and Demonstration of 2D-LIF for Studies of Mixture Preparation in SI Engines," *Combustion and Flame*, Vol. 99, pp. 449–457.

57. Nishiyama, A., Kawahara, N., and Tomita, E. (2003), "In-Situ Concentration Measurement Near Spark Plug by 3.392 μm Infrared Absorption Method—Application to Spark Ignition Engine," SAE Paper No. 2003-01-1109, Society of Automotive Engineers, Warrendale, PA.

58. Parks, J., Armfield, J., Storey, J., Barber, T., and Wachter, E. (1998), "*In Situ* Measurement of Fuel Absorption into the Cylinder Wall Oil Film During Engine Cold Start," SAE Paper No. 981054, Society of Automotive Engineers, Warrendale, PA.

59. Quader, A.A. and Majkowski, R.F. (1999), "Cycle-by-Cycle Mixture Strength and Residual-Gas Measurements During Cold Starting," SAE Paper No. 1999-01-1107, Society of Automotive Engineers, Warrendale, PA.

60. Richter, B., Dullenkopf, K., Wittig, S., Tribulowski, J., and Spicher, U. (2002), "Influence of Atomization Quality on Mixture Formation, Combustion and Emissions in an MPI-Engine Under Cold-Start Conditions, Part I," SAE Paper No. 2002-01-2807, Society of Automotive Engineers, Warrendale, PA.

61. Roberts, C.E. and Stanglmaier, R.H. (1999), "Investigation of Intake Timing Effects on the Cold-Start Behavior of a Spark-Ignition Engine," SAE Paper No. 1999-01-3622, Society of Automotive Engineers, Warrendale, PA.

62. Russ, S., Stevens, J., Aquino, C., Curtis, E., and Fry, J. (1998), "The Effects of Injector Targeting and Fuel Volatility on Fuel Dynamics in a PFI Engine During Engine Warm-Up: Part 1—Experimental Results," SAE Paper No. 982518, Society of Automotive Engineers, Warrendale, PA.

63. Santoso, H. and Cheng, W.K. (2002), "Mixture Preparation and Hydrocarbon Emissions Behaviors in the First Cycle of SI Engine Cranking," SAE Paper No. 2002-01-2805, Society of Automotive Engineers, Warrendale, PA.

64. Senda, J., Ohnishi, M., Takahashi, T., Fujimoto, H., Utsunomiya, A., and Wakatabe, M. (1999), "Measurement and Modeling on Wall Wetted Fuel Film Profile and Mixture Preparation in Intake Port of SI Engine," SAE Paper No. 1999-01-0798, Society of Automotive Engineers, Warrendale, PA.

65. Shayler, P.J., Davies, M.T., and Scarisbrick, A. (1997), "Audit of Fuel Utilisation During the Warm-Up of SI Engines," SAE Paper No. 971656, Society of Automotive Engineers, Warrendale, PA.

66. Shayler, P.J., Winborn, L.D., Hill, M.J., and Eade, D. (2000), "The Influence of Gas/Fuel Ratio on Combustion Stability and Misfire Limits of Spark Ignition Engines," SAE Paper No. 2000-01-1208, Society of Automotive Engineers, Warrendale, PA.

67. Shin, Y., Min, K., and Cheng, W.K. (1995), "Visualization of Mixture Preparation in a Port Fuel Injection Engine During Engine Warm-Up," SAE Paper No. 952481, Society of Automotive Engineers, Warrendale, PA.

68. Söderberg, F. and Johansson, B. (2002), "Particle Image Velocimetry Flow Measurements and Heat-Release Analysis in a Cross-Flow Cylinder Head," SAE Paper 2002-01-2840, Society of Automotive Engineers, Warrendale, PA.

69. Stanglmaier, R.H., Li, J., and Matthews, R.D. (1999), "The Effect of In-Cylinder Wall Wetting Location on the HC Emissions from SI Engines," SAE Paper No. 1999-01-0502, Society of Automotive Engineers, Warrendale, PA.

70. Stolz, W., Köhler, J., Lawrenz, W., Meier, F., Bloss, W.H., Maly, R.R., Herweg, R., and Zahn, M. (1992), "Cycle Resolved Flow Field Measurements Using a PIV Movie Technique in an SI Engine," SAE Paper No. 922354, Society of Automotive Engineers, Warrendale, PA.

71. Styron, J.P., Kelly-Zion, P.L., Lee, C.F., Peters, J.E., White, R.A., and Lucht, R.P. (2000), "Multicomponent Liquid and Vapor Fuel Distribution Measurements in the Cylinder of a Port-Injected, Spark-Ignition Engine," SAE Paper No. 2000-01-0243, Society of Automotive Engineers, Warrendale, PA.

72. Swindal, J.C., Furman, P.A., Loiodice, M.E., Stevens, R.W., Liu, P.C., and Acker, W.P. (1997), "Fuel Distillation Effects on the Outgassing from a Simulated Crevice in an SI Engine Measured by Planar Laser-Induced Fluorescence," SAE Paper No. 970825, Society of Automotive Engineers, Warrendale, PA.

73. Swithenbank, J., Beer, J.M., Taylor, D.S., Abbott, D., and McGreath, G.C. (1977), "A Laser Diagnostic Technique for the Measurement of Droplet and Particle Size Distribution," *Prog. Astronaut. Aeronaut.*, Vol. 53, pp. 421–427.

74. Takagi, Y. and Skippon, S.M. (1998), "Effects of In-Cylinder Fuel Spray Formation on Emissions and Cyclic Variability in a Lean-Burn Engine, Part 1: Background and Methodology," SAE Paper No. 982618, Society of Automotive Engineers, Warrendale, PA.

75. Winkler, M.A. and Mueller, E. (2001), "Determination of the Air/Fuel Ratio of an SI Engine During Transients with a Standard UEGO Sensor," SAE Paper No. 2001-01-1955, Society of Automotive Engineers, Warrendale, PA.

76. Witze, P.O. (1999), "Diagnostics for the Study of Cold-Start Mixture Preparation in a Port Fuel-Injected Engine," SAE Paper No. 1999-01-1108, Society of Automotive Engineers, Warrendale, PA.

77. Wolff, D., Beushausen, V., Schlüter, H., Andresen, P., Hentschel, W., Manz, P., and Arndt, S. (1994), "Quantitative 2D-Mixture Fraction Imaging Inside an Internal Combustion Engine Using Acetone-Fluorescence," 3rd International Symposium on Diagnostics and Modeling of Combustion in Internal Combustion Engines (COMODIA 94), Yokohama, pp. 445–451.

78. Yoshiyama, S. and Tomita, E. (2002), "Combustion Diagnostics of a Spark Ignition Engine Using a Spark Plug as an Ion Probe," SAE Paper No. 2002-01-2838, Society of Automotive Engineers, Warrendale, PA.

79. Zhao, F., Lai, M., and Harrington, D.L. (1995), "The Spray Characteristics of Automotive Port Fuel Injection—A Critical Review," SAE Paper No. 950506, Society of Automotive Engineers, Warrendale, PA.

80. Zhao, F-Q. and Hiroyasu, H. (1993), "The Applications of Laser Rayleigh Scattering to Combustion Diagnostics," *Progress in Energy and Combustion Science*, Vol. 19, pp. 447–485.

81. Zhao, F.Q., Taketomi, M., Nishida, K., and Hiroyasu, H. (1994), "PLIF Measurements of the Cyclic Variation of Mixture Concentration in an SI Engine," SAE Paper No. 940988, Society of Automotive Engineers, Warrendale, PA.

82. Zhao, H. and Ladommatos, N. (2001), *Engine Combustion Instrumentation and Diagnostics*, Society of Automotive Engineers, Warrendale, PA.

83. Ziegler, G.F.W., Meinhardt, P., Herweg, R., and Maly, R. (1990), "Cycle-Resolved Flame Structure Analysis of Turbulent Premixed Engine Flames," SAE Paper No. 905001, Society of Automotive Engineers, Warrendale, PA.

84. Zughyer, J.R., Zhao, F., Lai, M., and Lee, K. (2000), "A Visualization Study of Liquid Fuel Distribution and Combustion Inside a Port-Injected Gasoline Engine Under Different Start Conditions," SAE Paper No. 2000-01-0242, Society of Automotive Engineers, Warrendale, PA.

CHAPTER 5

Spark Retardation for Improving Catalyst Light-Off Performance

Stephen Russ
Ford Motor Company

5.1 Introduction

Meeting stringent emissions standards requires the catalyst system to achieve operating temperature and high conversion efficiency (light-off) as soon as possible following a cold start. To meet this goal, sophisticated catalyst heating devices have been proposed, including electrically heated substrates [Socha and Thompson (1992)], exhaust system combustion devices [Ma *et al.* (1992)] and secondary air injection into the exhaust [Kollmann *et al.* (1994)]. With the exception of secondary air injection, these exhaust heating devices have not found widespread application due to the added cost and complexity accompanying these systems. More rapid catalyst light-off typically is achieved by packaging the catalyst as close as possible to the engine to minimize heat losses from the exhaust gases and by optimizing the cold-start calibration. This chapter will focus on calibration actions that result in faster catalyst light-off and what can be done with the base engine design to facilitate stable engine operation under these conditions.

5.2 Calibration Actions for Improving Catalyst Light-Off

Cold-start calibrations for rapid catalyst light-off typically include three actions:

1. **Increased idle engine speed**—Increased cold-idle revolutions per minute (rpm) increases both the exhaust gas temperature and the mass flow rate to provide more heat to the catalyst at cold start. The impact of the cold-idle revolutions per minute on the mass flow rate and exhaust gas temperature is shown in Figures 5.1 and 5.2 from Ueno (2000). The work of Kaiser *et al.* (1995) also shows that hydrocarbon (HC) oxidation increases at higher engine speeds, presumably due to increased exhaust gas temperatures and improved mixing.

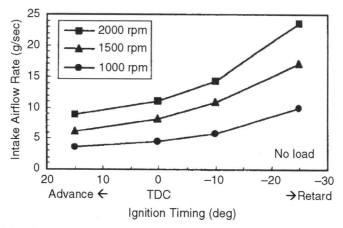

Figure 5.1 *Effect of engine speed and ignition timing on airflow [Ueno (2000)].*

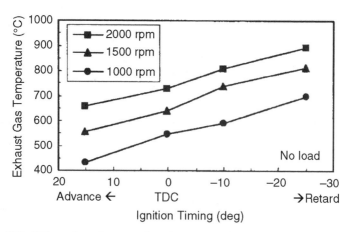

Figure 5.2 *Effect of engine speed and ignition timing on exhaust temperature [Ueno (2000)].*

2. **Lean air/fuel ratio**—A slightly lean air/fuel (A/F) ratio results in a small amount of excess oxygen to improve HC oxidation [Kaiser *et al.* (1995)]. This lowers the feedgas HC emissions during the portion of the cycle when the catalyst is inactive. Additionally, a slightly lean A/F ratio reduces the temperature at which the catalyst becomes active (i.e., the light-off temperature) by up to 80°C, as shown by Kubo *et al.* (1997). This is due to the excess oxygen, as well as an increased percentage of reactive low-molecular-weight olefins in the exhaust.

3. **Retarded ignition timing**—Retarding ignition timing results in increased exhaust gas temperature, as well as increased exhaust mass flow due to the decrease in engine thermal efficiency. The impact of spark retard on the exhaust gas flow rate and temperature are shown in Figures 5.1 and 5.2 from Ueno (2000). This dramatically increases the warm-up rate of the catalyst, as well as increasing the oxidation of the feedgas HC emissions after leaving the cylinder.

Examples of these actions used to decrease cold-start emissions can be found in many recent studies. The work of Nakayama *et al.* (1994) utilized leaner A/F ratios and retarded ignition to decrease cold-start HC emissions and to increase catalyst temperatures. Kaiser *et al.* (1994) describe a cold-start calibration utilizing all three of these actions. The engine speed was increased by 200 rpm, the A/F ratio was changed from stoichiometric to 5% lean, and the spark was retarded up to 25° from minimum spark advance for best torque (MBT). This calibration change produced approximately 50% lower total HC emissions with an increased percentage of olefins, which are more reactive in the catalyst. The work of Chan and Zhu (1996) illustrates the large impact of spark retard alone on catalyst light-off. The spark was retarded by 26°, improving the catalyst light-off time to less than 50 seconds from an initial value of more than 200 seconds. Research done by Mandokoro *et al.* (1997) on a single-cylinder engine reports that enleanment (10% lean) and the use of spark retard can reduce engine-out emissions by 50%, with further benefits from improved in-cylinder swirl motion. This work also illustrates an increased percentage of low-molecular-weight olefins in the exhaust for these operating conditions.

There are many examples of the use of these calibration changes to meet stringent emissions standards. Takahashi *et al.* (1998) utilized optimization of the ignition timing and A/F ratio to improve the light-off time of the catalyst, with catalyst formulation actions to reduce the HC emissions in the first phase of the Federal Test Procedure (FTP). A 60 to 70% reduction was achieved, enabling a V6 vehicle to meet low emissions vehicle (LEV) and almost ultra-low emissions vehicle (ULEV) emissions standards. The quick warm-up control system described by Ueno (2000) uses a large amount of spark retard to quickly light-off the catalyst and halves the HC emissions during the FTP drive cycle. This control system, with dramatically retarded ignition timing, was used by Kitagawa (2000), with a lean A/F ratio to achieve super ultra-low emissions vehicle (SULEV) emissions standards. Kidokoro *et al.* (2003) used 15° of additional spark retard and changed the A/F ratio from 14 to 15.5 to reduce HC emissions from cold start (including fast idling) by approximately 40 to 50% to meet SULEV tailpipe standards.

5.3 Engine Operation with Retarded Ignition

The use of spark retard increases exhaust gas temperatures because the burned gas is not ideally expanded and does not perform as much work on the piston as the MBT case. In fact, studies by Russ *et al.* (1999) and Chen *et al.* (2001) illustrate that with sufficient spark retard, combustion may not even be complete by the time of exhaust valve opening. Subsequent reactions in the exhaust port and manifold release heat to increase the exhaust gas temperature. The use of aggressive spark retard has two main drawbacks. The first is the obvious loss in efficiency, which results in a small fuel economy penalty. This penalty is not large because aggressive spark retard typically is used for less than 1 minute during engine warm-up. The second drawback is that spark retard increases cycle-to-cycle indicated mean effective pressure (IMEP) fluctuations, causing engine roughness (usually expressed in terms of standard deviation of IMEP). Figure 5.3 illustrates the HC emissions and exhaust gas temperature benefits, with the increase in standard deviation of indicated mean effective pressure (SDIMEP).

Increased engine roughness is a critical factor in determining the amount of ignition retard that can be implemented in the cold-start calibration. The following is a more detailed explanation of the source of the increased cycle-to-cycle IMEP variations.

Figure 5.4 presents a plot of the individual cycle IMEP versus the individual cycle burn rate/combustion phasing (location of 50% mass fraction burned [MFB]) for typical cold retarded spark operation, from Russ *et al.* (1999). This plot shows the strong correlation of individual cycle IMEP with the burn rate that is not present at normal (MBT) ignition timing.

Spark Retardation for Improving Catalyst Light-Off Performance

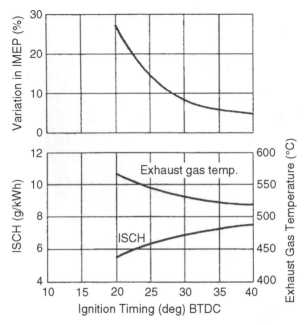

Figure 5.3 *Influence of ignition timing on variation in IMEP, indicated specific hydrocarbon (ISHC) emissions, and exhaust temperature under fast idle condition; 1500 rpm, A/F = 13.0, water temperature = 30°C [Nakayama et al. (1994)].*

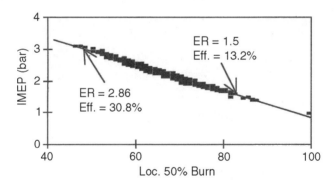

Figure 5.4 *Effect of combustion phasing on cyclic IMEP; 1200 rpm, 1.0-bar brake mean effective pressure (BMEP), 14.6 A/F, 5° above top dead center (ATDC) spark, 20°C coolant and oil [Russ et al. (1999)].*

An explanation of this strong correlation is found in Figure 5.5, a plot of the normalized heat release profile for the same operating conditions, along with the ideal Otto cycle thermal efficiency (assuming $\gamma = 1.35$) calculated from the expansion ratio from various crank angles to the exhaust valve opening (EVO). Figures 5.4 and 5.5 illustrate that at these retarded spark conditions, the expansion ratio and therefore the thermal efficiency rapidly decrease during the combustion event. The heat release (and pressure rise) from the first gas to burn is accompanied by a volume expansion and therefore produces useful work. The energy release from the end of the combustion event, occurring close to EVO, has little accompanying increase in volume and therefore does little expansion work and does not contribute to the IMEP of the cycle. Most of the heat is released near the 50% MFB location; therefore, the 50% MFB location does a good job of representing the average location of the heat release and thus the expansion ratio for each cycle. Note the strong correlation between the IMEP and theoretical efficiency for the fast-burn cycle compared to the IMEP and theoretical efficiency for the slow-burn cycle highlighted by the arrows in Figure 5.4.

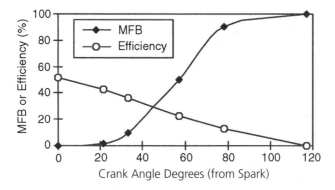

Figure 5.5 Burn rate and ideal efficiency profile; 1200 rpm, 1.0-bar BMEP, 5° ATDC spark, 14.6 A/F, 20°C fluids) [Russ et al. (1999)].

These plots clearly illustrate that the primary cause of cyclic variations in IMEP for retarded spark operation is variations in the combustion phasing (location of 50% MFB). The expansion ratio decreases rapidly during combustion for retarded spark timing; therefore, the combustion phasing of an individual cycle determines the cycle thermal efficiency and the IMEP produced. Solutions to enable additional spark retard at cold start therefore must focus on making the combustion event repeatable (i.e., similar burn rate cycle to cycle) to decrease engine roughness.

In addition to operation with retarded ignition, a slightly lean A/F ratio typically is calibrated as outlined in the previous section. The engine must be capable of stable operation at this slightly lean calibrated A/F, in addition to having acceptable driveability with less volatile (i.e., heavier, higher driveability index [DI]) fuels. These fuels can produce leaner A/F excursions due to additional fuel held up in liquid films on the intake port and valve surfaces.

5.4 Approaches for More Robust Operation with Ignition Retard

Operation with retarded ignition and lean A/F can cause increased engine roughness, as discussed in Section 5.3. Typically, base engine design changes are made for low emissions applications to improve the engine combustion stability to enable more aggressive use of spark retard and lean A/F while ensuring good driveability with less volatile fuels. This section focuses on two of these actions used to make the combustion system more robust.

5.4.1 Enhanced Charge Motion

A variable valve timing and lift electronic control (VTEC) system has been used to help stabilize lean, retarded spark cold-engine operation [Nakayama *et al.* (1994)]. The more stable lean engine operation at the low valve lift was due to a reduction in valve overlap, as well as increased intake air velocities that improve fuel atomization and create a faster burn rate. The leaner A/F ratio, in addition to retarded ignition timing, produced a 45% HC reduction. The VTEC system also improved the driveability with less volatile fuel. Enhanced charge motion through the application of swirl control valves also has been shown to improve combustion stability with retarded ignition timing, as illustrated in Figure 5.6 for a LEV emissions level V6 [Takahashi *et al.* (1998)]. In this study, several shapes of plate were investigated and indicate a potential improvement of more than 40°C in exhaust gas temperature and 10% reduction in HC mass flow. A subsequent study utilizing this type of swirl control valve for a SULEV emissions level L4 illustrates an improvement of more than 10° ignition retard for stoichiometric operation, as shown in Figure 5.7 [Webster *et al.* (2000)].

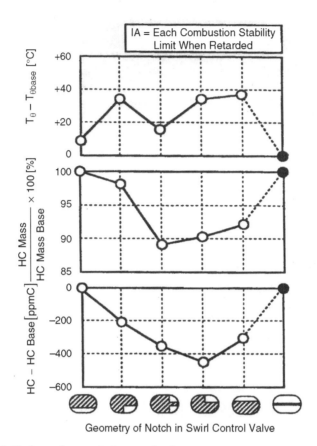

Figure 5.6 Hydrocarbon emissions and exhaust temperature versus swirl control valve (1600 rpm, 390 kPa BMEP, stoichiometric A/F, 40°C coolant, ignition at stability limit) [Takahashi et al. (1998)].

An intake air control valve producing tumble flow in the cylinder is shown in Figure 5.8 [Kidokoro et al. (2003)]. This valve improved spark retard and enleanment capability at a fixed combustion stability limit of 1 Nm torque fluctuation, as shown by the lines in Figure 5.9. In addition to the improvement in combustion stability, the higher air velocities and turbulence levels in the intake port improved the fuel wetting and mixture formation. Figure 5.10 shows a 24% reduction in the required fuel injection by the presence of the intake air control valve. This reduces HC emissions and improves driveability. The use

Spark Retardation for Improving Catalyst Light-Off Performance

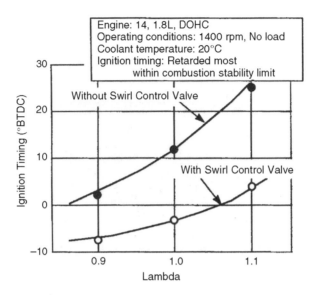

Figure 5.7 Expansion of ignition retard limit by swirl control valve [Webster et al. (2000)].

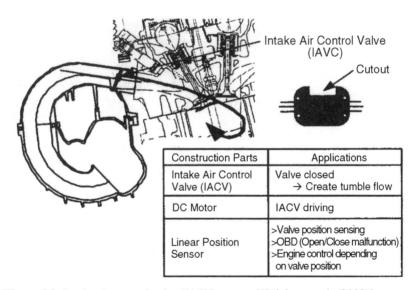

Figure 5.8 Intake air control valve (IACV) system [Kidokoro et al. (2003)].

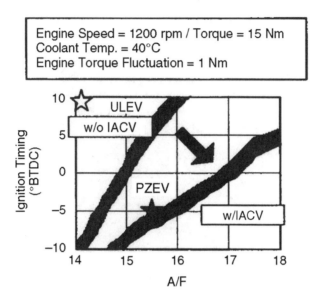

Figure 5.9 Combustion improvement with the IACV system [Kidokoro et al. (2003)].

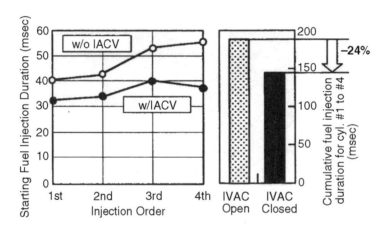

Figure 5.10 Required fuel amount for the starting cycle [Kidokoro et al. (2003)].

of the intake air control valve (IACV) enabled the cold-start calibration to be 1.5 A/F leaner with 15° of additional spark retard, as shown in Figure 5.9.

5.4.2 Dual Spark Ignition

Multiple spark plugs produce a faster burn rate and more stable combustion due to the additional flame area produced by igniting in several locations. Both Honda [Ogawa *et al.* (2003)] and Chrysler [Lee *et al.* (2002)] have recently put dual-spark-plug engines into production. On the Chrysler engine, a significant improvement in the combustion stability was reported as the coefficient of variation (COV) in the IMEP at idle was reduced from 12 to 8% with the use of dual ignition.

The impact of dual spark plugs on cold-start combustion stability was studied by Russ *et al.* (1999a) and compared to enhanced charge motion. In this study, a single-cylinder research engine with a relatively slow burn rate with single-plug ignition (1500 rpm, 3.8-bar IMEP, MBT spark ~ 32°) was used. The burn rate improved with either the use of a charge motion control valve (CMCV) or igniting with twin spark plugs (MBT spark ~ 26° for both cases). The combination of dual spark ignition with the CMCV closed produced a very fast burn (19° MBT spark) with good mixing. Figure 5.11 shows the combustion stability results for cold operation for the four different engine configurations. For single-plug operation with the CMCV open, the SDIMEP was above 0.25 bar at 150° spark and degraded with further ignition retard. The enhanced in-cylinder charge motion with CMCV closed improved both the burn rate and the SDIMEP substantially. Although the exhaust gas temperatures and HC emissions were not improved at fixed spark timing, the additional combustion stability allowed for additional spark retard, which did show improvements relative to the base engine. The dual-plug configuration with the CMCV open produced almost identical exhaust gas temperatures as the single plug with the CMCV closed at fixed spark timing due to similar burn rates. However, note that the dual plug produces lower SDIMEP levels. The improved stability was presumably due to the fact that any inhomogeneities in the chamber were "averaged out" by the two spark locations. The case with the dual spark plugs firing and the CMCV closed produced the fastest burn rate and the lowest SDIMEP levels.

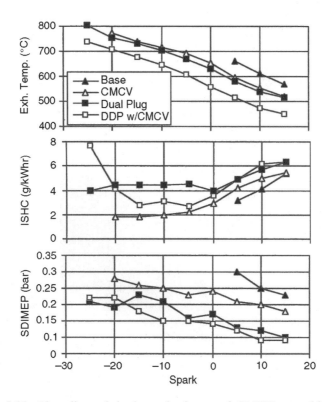

Figure 5.11 The effect of dual spark plugs and CMCV on cold operation; 1200 rpm, 2.5-bar IMEP, 15.5:1 A/F, 20°C fluids) [Russ et al. (1999a)].

5.5 Summary

Calibration changes to improve catalyst light-off typically involve increased idle speed, a leaner A/F, and significantly retarded spark timing. These calibration changes, with design actions to improve catalyst light-off such as close-coupled packaging, low thermal mass exhaust systems, and thin-wall substrates, have enabled low emissions standards to be met without sophisticated catalyst heating systems. The use of retarded spark timing causes an increase in engine roughness (SDIMEP) due to an increased sensitivity to cyclic burn rate differences. In addition, engine operation with lean A/F ratios increases engine roughness and can cause driveability concerns with less volatile fuels. Base engine design changes such as the addition of swirl control valves or VTEC systems can improve the engine combustion system robustness to lean engine operation with retarded ignition due to increased turbulence levels and improved in-cylinder mixing. These charge motion control devices also can improve the fuel preparation due to higher intake port air velocity and turbulence levels reducing the amount of injected fuel required for cold start. Dual spark ignition also can improve the engine combustion system robustness with retarded ignition due to the fast burn rate and the fact that the flame initiation is more repeatable because it is averaged over two locations.

5.6 References

1. Chan, S.H. and Zhu, J. (1996), "The Significance of High Value of Ignition Retard Control on the Catalyst Lightoff," SAE Paper No. 962077, Society of Automotive Engineers, Warrendale, PA.

2. Chen, Y., Wang, J.-X., Zhuang, R.-J., and Yang, T. (2001), "Analysis of Combustion Behavior During Cold-Start and Warm-Up Process of SI Gasoline Engine," SAE Paper No. 2001-01-3557, Society of Automotive Engineers, Warrendale, PA.

3. Kaiser, E.W., Siegl, W.O., Baidas, L.M., Lawson, G.P., Cramer, C.F., Dobbins, K.L., Roth, P.W., and Smokovitz, M. (1994), "Time-Resolved Measurement of Speciated Hydrocarbon Emissions During Cold Start of a Spark-Ignited Engine," SAE Paper No. 940963, Society of Automotive Engineers, Warrendale, PA.

4. Kaiser, E.W., Siegl, W.O., Trinker, F.H., Cotton, D.F., Cheng, W.K., and Drobot, K. (1995), "Effect of Engine Operating Parameters on Hydrocarbon Oxidation in the Exhaust Port and Runner of a Spark-Ignited Engine," SAE Paper No. 950159, Society of Automotive Engineers, Warrendale, PA.

5. Kidokoro, T., Hoshi, K., Hiraku, K., Satoya, K., Watanabe, T., Fujiwara, T., and Suzuki, H. (2003), "Development of PZEV Exhaust Emission Control System," SAE Paper No. 2003-01-0817, Society of Automotive Engineers, Warrendale, PA.

6. Kitagawa, H. (2000), "L4-Engine Development for a Super-Ultra-Low Emissions Vehicle," SAE Paper No. 2000-01-0887, Society of Automotive Engineers, Warrendale, PA.

7. Kollmann, K., Abthoff, J., Zahn, W., Bischof, H., and Giarhre, J. (1994), "Secondary Air Injection with a New Developed Electrical Blower for

Reduced Exhaust Emissions," SAE Paper No. 940472, Society of Automotive Engineers, Warrendale, PA.

8. Kubo, S., Mandokoro, Y., Taki, M., Takeda, K., and Murai, T. (1997), "Reduction in Cold Hydrocarbon Mass Emissions by Combustion Control of Gasoline Engine: Part 2—The Effect of Equivalence Ratio on Catalytic Reactivity," Paper No. 9737428 (in Japanese), 14th JSME/SAEJ Internal Combustion Engine Symposium.

9. Lee, R.E., Winship, M., Hartman, P., MacFarlane, G., Maru, D.B., Pannone, G., Martinez, T., and Cruz, J. (2002), "The New DaimlerChrysler Corporation 5.7L Hemi V8 Engine," SAE Paper No. 2002-01-2815, Society of Automotive Engineers, Warrendale, PA.

10. Ma, T., Collings, N., and Hands, T. (1992), "Exhaust Gas Ignition (EGI)—A New Concept for Rapid Light-Off of Automotive Exhaust Catalysts," SAE Paper No. 920400, Society of Automotive Engineers, Warrendale, PA.

11. Mandokoro, Y., Kubo, S., Taki, M., Ban, H., Takeda, K., and Murai, T. (1997), "Reduction in Cold Hydrocarbon Mass Emissions by Combustion Control of Gasoline Engine: Part 1—The Effect of In-Cylinder Gas Flow on Reactivity Promotion of Hydrocarbons," Paper No. 9737419 (in Japanese), 14th JSME/SAEJ Internal Combustion Engine Symposium.

12. Nakayama, Y., Maruya, T., Oikawa, T., Fujiwara, M., and Kawamata, M. (1994), "Reduction of HC Emission from VTEC Engine During Cold-Start Condition," SAE Paper No. 940481, Society of Automotive Engineers, Warrendale, PA.

13. Ogawa, H., Matsuki, M., and Eguchi, T. (2003), "Development of a Power Train for the Hybrid Automobile—The Civic Hybrid," SAE Paper No. 2003-01-0083, Society of Automotive Engineers, Warrendale, PA.

14. Russ, S., Lavoia, G.A., and Dai, W. (1999a), "SI Engine Operation with Retarded Ignition: Part 1—Cyclic Variations," SAE Paper No. 1999-01-3506, Society of Automotive Engineers, Warrendale, PA.

15. Russ, S., Thiel, M., and Lavoie, G.A. (1999b), "SI Engine Operation with Retarded Ignition: Part 2—HC Emissions and Oxidation," SAE Paper No. 1999-01-3507, Society of Automotive Engineers, Warrendale, PA.

16. Socha, L.S. Jr. and Thompson, D.F. (1992), "Electrically Heated Extruded Metal Converters for Low Emission Vehicles," SAE Paper No. 920093, Society of Automotive Engineers, Warrendale, PA.

17. Takahashi, H., Momoshima, S., Ishizuka, Y., Tomita, M., and Nishizawa, K. (1998), "Engine-Out and Tail-Pipe Emission Reduction Technologies of V-6 LEVs," SAE Paper No. 980674, Society of Automotive Engineers, Warrendale, PA.

18. Ueno, M. (2000), "A Quick Warm-Up System During Engine Start-Up Period Using Adaptive Control of Intake Air and Ignition Timing," SAE Paper No. 2000-01-0551, Society of Automotive Engineers, Warrendale, PA.

19. Webster, L., Nishizawa, K., Momoshima, S., and Koga, M. (2000), "Nissan's Gasoline SULEV Technology," SAE Paper No. 2000-01-1583, Society of Automotive Engineers, Warrendale, PA.

CHAPTER 6

Secondary Air Injection for Improving Catalyst Light-Off Performance

Fuquan (Frank) Zhao
Brilliance Jinbei Automobile Corporation

Mark Borland
DaimlerChrysler Corporation

6.1 Introduction

Improving catalyst light-off performance during cold start and reducing engine-out (more accurately referred to as converter-in) emissions prior to catalyst light-off have been regarded as the keys to meeting future stringent emissions regulations. As a result, most development efforts have focused on the cold-start period. Accelerating the heating process of the catalytic converter is commonly known as the most effective approach to drastically reduce cold-start hydrocarbon (HC) emissions. Because it generally takes some time to light off the catalyst (the time varies, depending on the emissions goal), reducing converter-in HC emissions prior to catalyst light-off is as important as lighting off the catalyst faster, both of which must be obtained simultaneously. Many technologies and control strategies have been proposed, and some have already been incorporated into production. Some of the technologies are applied independently, and some must be combined with other approaches to maximize the benefits. Among these, secondary air injection into the exhaust port in combination with rich engine operation received a lot of attention due to its robust and consistent performance to meet the development goal and its relative ease of implementation [Borland and Zhao (2002); Kollmann et al. (1994)]. Coupling the ultra-fuel-rich engine combustion with the injection of precisely metered air into the exhaust port results in an exothermic reaction in the port and exhaust manifold before the exhaust gases reach the converter. This can effectively reduce HC emissions inside the exhaust manifold and simultaneously accelerate the heating process of the converter following a cold start of gasoline engines. When compared with other approaches, secondary air injection can be implemented relatively easily with today's engine system, without requiring a major design change.

The thermal and chemical processes associated with secondary air injection inside the exhaust system are complex. All key design and operating parameters such as secondary air injection location, exhaust manifold design, spark retardation, engine-enrichment level, and secondary air flow rate that will affect the success for implementing the secondary air injection strategy must be understood and optimized to maximize the simultaneous benefit of improving catalyst light-off performance and reducing converter-in emissions. This chapter provides a general description of the physical and chemical processes associated with secondary air injection. A comparison of thermal oxidation and catalytic oxidation is presented to highlight the different functions of HC oxidation in the exhaust manifold and inside the catalytic converter. The role of mixing and temperature in enhancing the thermal and catalytic oxidation and the effects of engine enrichment and secondary air injection quantity are outlined to sort out the key issues that must be addressed in practical implementation of this technology. Other considerations associated with practical application also are mentioned briefly at the end of this chapter.

6.2 Principle and System Layout of Secondary Air Injection

Engine-out exhaust gas composition varies substantially with the engine-fueled air/fuel (A/F) ratio. Figure 6.1 plots exhaust gas compositions and their concentrations as a function of fuel/air equivalence ratio [D'Alleva and Lovell (1936); Stivender (1971); Harrington and Shishu (1973); Spindt (1965); Heywood (1988)]. When the engine is fueled with a richer-than-stoichiometric mixture, hydrogen (H_2) and carbon monoxide (CO) will be produced and exhausted into the exhaust stream. The concentrations of CO and H_2 rise steadily as the mixture becomes richer. A similar trend is true with engine-out

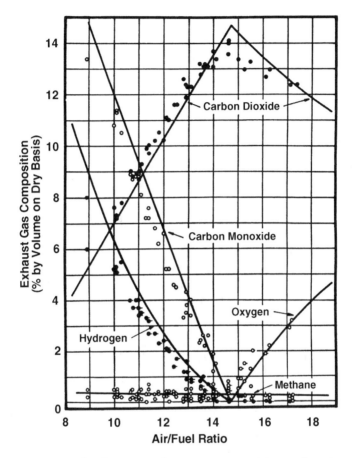

Figure 6.1 Relationship between exhaust gas composition and engine air/fuel (A/F) ratio for spark ignition (SI) engines [D'Alleva et al. (1936)].

HC emissions [Heywood (1988)]. Figure 6.2 presents the ratio of CO and H_2 for a typical HC fuel. Injecting secondary air into the exhaust port, which allows the secondary air to mix and react with these reactants (H_2, CO, and HC) produced during engine-rich operation, leads to an exothermic reaction in the exhaust system. This reaction can effectively consume HCs inside the exhaust manifold and thus lower converter-in HCs, which otherwise will break through the converter and be emitted as HC emissions before catalyst light-off. Moreover, this reaction can simultaneously accelerate the heating process of the converter following a cold start of gasoline engines. The engine-out nitrogen

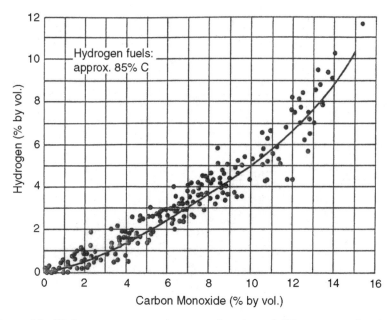

Figure 6.2 Hydrogen concentration as a function of CO concentration in the engine-out exhaust for SI engines [Leonard (1961)].

oxides (NOx) also can be reduced due to the lack of oxygen for NOx formation when operating the engine richer, in addition to the evaporative charge cooling effect [Heywood (1988)].

However, the physical and chemical processes associated with secondary air injection are complicated, and its benefit strongly depends on the system optimization, which usually varies with the specific application. All design and operating parameters that determine engine-out emissions, exhaust gas temperature, mixing process of reactants with the secondary air, and the mixture residence time inside the exhaust port and manifold will affect the level of success of the secondary air injection strategy. All these key elements will be discussed in detail in the following sections.

Figure 6.3 illustrates the generic layout of the secondary air injection system [Kollmann *et al.* (1994)]. The most important element in the secondary air system is the electrical secondary air pump with both inlet filter and silencer. The blower sucks the air from the environment and pumps it, via the secondary

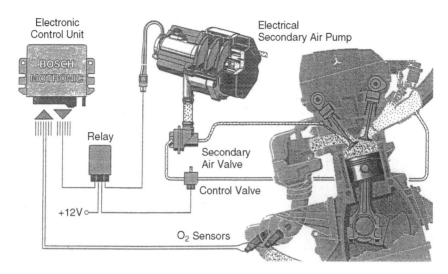

Figure 6.3 Configuration of the secondary air injection system [Kollmann et al. (1994)].

air valve, very close to the outlet valves of the engine. The injection tube is integrated into the cylinder head. The system operates during warm-up of the catalyst and, perhaps, during brief diagnostic cycles when the engine is warm. The duration of operation is dependent on the engine temperature. After the A/F ratio control is active, the secondary air system is deactivated by the engine controller. The secondary air valve is opened only during air pump operation and prevents backflow of the exhaust gas at all other times. With the relay, the engine controller simultaneously activates the blower and the control valve. The system also must meet the onboard diagnostics (OBD) requirements.

Figure 6.4 shows the typical events that occur during cold start and the time histories of temperatures and HC emissions with secondary air injection. Clearly, the engine is started just after the clock time of 5 seconds. Immediately, there is a spike in the HC concentration associated with the fuel enrichment required for good startability. Then, at approximately 8 seconds, the thermal oxidation begins to dominate after the secondary air injection is turned on, the HC concentration drops dramatically, and both the exhaust temperature and the catalyst temperature increase as the secondary injection air continues. This trend continues throughout the idle until approximately 38 seconds in the clock, when the secondary air and excess fuel enrichment are cut off.

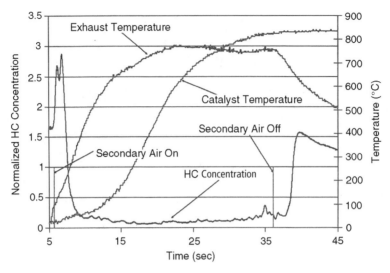

Figure 6.4 *Typical events and time histories of temperatures and HC concentrations with secondary air injection [Borland and Zhao (2002)].*

6.3 Thermal Oxidation Versus Catalytic Oxidation

In general, the chemical reactions occurring in the exhaust system with secondary air injection can be categorized into two types of oxidation of reactants. These two types of oxidation are both exothermic, but the temperatures required for these to occur rapidly are dramatically different. One is the oxidation process that happens inside the exhaust port and manifold prior to the catalytic converter, which in this chapter is referred to as thermal oxidation. The other is the oxidation of HC and CO that occurs inside the catalytic converter, which is significantly different from that of the thermal oxidation due to the presence of a catalyst and is referred to as catalytic oxidation. The thermal oxidation usually requires a temperature of 750°C or higher to become significant, depending on the exhaust species and residence time [Heywood (1988)]. It is believed that the production of H_2 in engine combustion can significantly enhance the thermal oxidation process inside the exhaust port and manifold, and CO and HC can well sustain the process once it is established [Crane *et al.* (1997)]. In contrast, the conversion of HC and CO can start at a similar rate with a much lower temperature (200°C) on the surface of certain types of catalysts [Heywood (1988)].

Secondary Air Injection for Improving Catalyst Light-Off Performance

Figures 6.5 and 6.6 show a comparison of thermal and catalytic reactivity for H_2, CO, and HCs. Clearly, HCs are more reactive in the thermal oxidation process, which is the opposite inside the catalytic converter where CO has a

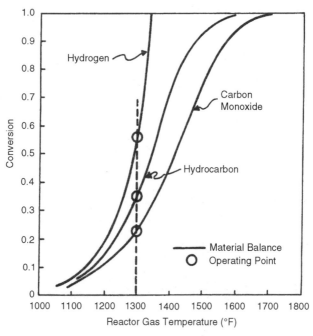

Figure 6.5 *Thermal reactivity of H_2, HCs, and CO [Lord et al. (1973)].*

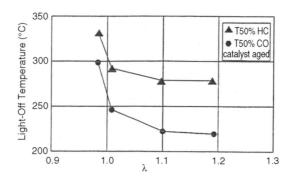

Figure 6.6 *Light-off temperature for HCs and CO with an aged catalyst [Kollmann et al. (1994)].*

lower light-off temperature. Both thermal oxidation and catalytic oxidation contribute to heating the catalyst for shortening the light-off time. By comparison, catalytic oxidation is more effective in improving the catalyst light-off by heating the catalyst directly when it reaches a certain temperature threshold, which must be provided by cylinder-out exhaust heat plus upstream thermal oxidation. Part of the heat produced via thermal oxidation will be dissipated to manifold walls; as a result, its impact on catalyst light-off will be slightly degraded. However, thermal oxidation can be effective in removing engine-out HC emissions, if adequate temperatures are sustained. This is extremely important prior to catalyst light-off. Therefore, efforts must be directed toward reducing the heat loss along the exhaust system, such as using the thin dual-wall exhaust manifold.

Rich engine operation generally produces higher engine-out HC emissions compared to that of lean operation, which demands a further reduction in converter-in emissions. With the addition of secondary air, the converter-in emissions can be reduced significantly due to the enhanced oxidation in the manifold. As will be discussed in the next section, a certain level of temperature is necessary to activate the chemistry and sustain the reactions. It has been demonstrated that adding air too far downstream from the exhaust valve cannot benefit from this emissions reduction due to the lack of thermal oxidation, although it may produce an improvement in catalyst light-off performance [Borland and Zhao (2002)].

The effect of exhaust chemistry on the catalyst tends to be ignored in most of the analysis or data interpretation when dealing with secondary air injection. Note that the catalyst is exposed not only to the hot exhaust gases, but also to high chemical loads from the leftover HC and CO emissions. The introduction of these chemical constituents at certain concentrations is important to realize self-sustaining chain reactions. Therefore, it is strongly believed that the combination of this chain reaction and thermal loading allows the catalyst to ultimately sustain the high conversion efficiencies following cold start. This chemical impact should not be ignored when locating the catalyst, designing the exhaust system to maximize the benefit of secondary air injection, or calibrating the secondary air injection system. A good compromise between the exhaust manifold thermal oxidation and the catalytic oxidation of CO and HC emissions is necessary. The former provides the heat necessary to initiate the exothermic reaction of CO and HC on the catalyst surface and to reduce the HC and CO loading on the catalyst to minimize any emissions breakthrough, although the latter has a direct heating effect on the catalyst surface and requires a certain level of HC and CO emissions to achieve this. If the thermal oxidation consumes too much engine-out reactants, which is important in minimizing HC emissions prior to catalyst light-off during cold start, the catalyst light-off process may be slowed due to a reduced energy release directly inside the catalytic converter. In contrast, if the converter is overdosed with the reactants due to poor thermal oxidation, there may be more emissions breakthrough, even though the catalyst may be lit off slightly faster. Therefore, the system must be optimized to balance these two oxidation processes to maximize the benefits.

6.4 Role of Temperature and Mixing in Enhancing the Thermal Oxidation Process

It is easy to understand that engine-out reactants can be oxidized in the exhaust manifold only when oxygen is present at a sufficiently high temperature. To maximize the thermal oxidation rate in the manifold, the pathway from the exhaust valve to the converter inlet must provide an environment with a high temperature and a high oxygen concentration over a sufficient period of time. The reaction rate increases exponentially with the temperature. A small increase in the initial temperature on the order of 10 to 20°K can result in the oxidation time of the mixture being cut in half [Hernandez (2002)]. Any design and operating parameters such as spark retardation and early exhaust valve opening that can increase engine-out gas temperature will dramatically enhance the thermal oxidation process. Therefore, the early initiation of thermal oxidation around the periphery of the exhaust valve is crucial to the success of secondary air injection.

Air injection must be as close as possible to the exhaust valve where the exhaust gases are hottest. As shown in Figure 6.7, air injection location has a significant impact on HC emissions. A minimal amount of oxygen is needed to react with the exhaust gases, and a minimal amount of reactants (e.g., HC, CO, H_2) and radicals (e.g., OH, O, H) must be available to start the reaction. When the secondary air is in an extreme excess ratio or the mixing of exhaust gases with secondary air occurs too rapidly, the thermal reaction will be quenched [Kadlec et al. (1972); Son et al. (1999)]. Once the reaction is quenched, it will be difficult to initiate again. On the other hand, if this mixing occurs too slowly, the heat loss to the wall may eventually lower the temperature below that necessary for retaining the thermal oxidation. This leads to the belief that promoting the mixing inside the port during the blowdown process is critical.

Figures 6.8 through 6.11 show the results derived from an experiment of secondary air injection with different airflow rates where the A/F ratio is the

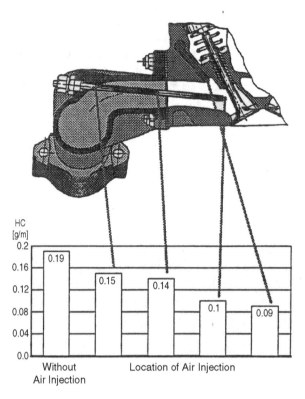

Figure 6.7 *Effect of secondary air injection location on HC emissions reduction [Kollmann et al. (1994)].*

exhaust A/F ratio measured with secondary air injection, exhaust temperature at the manifold is measured near the exhaust port that is slightly upstream of the secondary air injection, and the inlet temperature and the center temperature are measured at the inlet of the converter and at the center of the converter, respectively. Clearly, with an increase in the amount of secondary air injected, the A/F ratio gradually shifts toward the lean side. Because the exhaust temperature at the manifold was measured at the exhaust port and near the location of the secondary air introduction, the exhaust gas was cooled significantly with an increase in the flow rate of secondary air. However, there is almost no difference in the converter inlet temperature among different airflow rates, which indicates that the exothermic reactions inside the exhaust manifold are sufficient

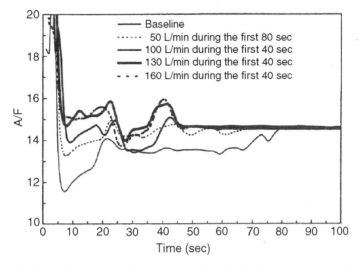

Figure 6.8 *Time histories of exhaust A/F ratio with different flow rates of secondary air injection [Son et al. (1999)].*

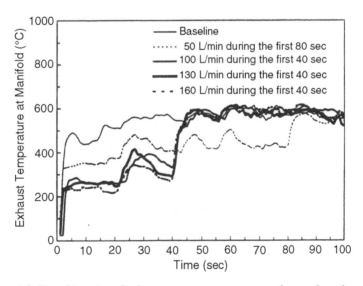

Figure 6.9 *Time histories of exhaust temperature measured near the exhaust port with different flow rates of secondary air injection [Son et al. (1999)].*

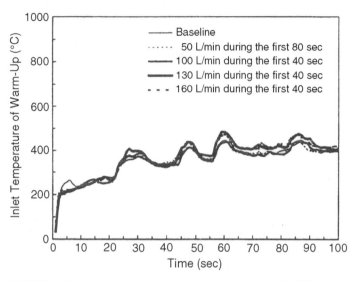

Figure 6.10 Time histories of converter inlet temperatures with different flow rates of secondary air injection [Son et al. (1999)].

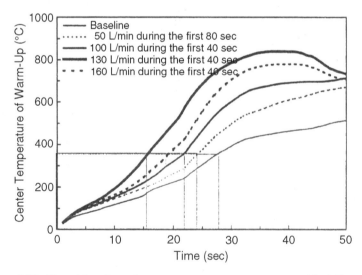

Figure 6.11 Time histories of converter center temperature with different flow rates of secondary air injection [Son et al. (1999)].

to heat the cold air. When examining the converter center temperature, it is clear that there is an optimal amount of secondary air that produces the best catalyst light-off benefit and emits the fewest HC and CO emissions.

The temperature and mass flow rate at the exhaust port vary significantly through the cycle. Ideally, the amount of air delivered should be in phase with the exhaust gas flow rate to maximize the benefit of secondary air injection. Figure 6.12 illustrates the time history of exhaust gas flow rate through the cycle, and the ideal airflow rate is plotted schematically. To realize this ideal secondary air delivery rate, a pulse-controlled secondary air delivery system is required. In reality, all the applications use an open-ended tube to deliver secondary air. This makes the secondary air delivery out of phase with the exhaust gas flow process, namely, a lot of air will be accumulated inside the exhaust port when the exhaust valve is closed, and the secondary air flow may be shut off due to the high backpressure when the exhaust valve opens. This will complicate the process of optimally utilizing the secondary air.

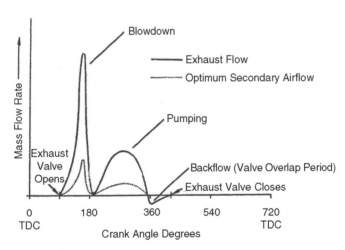

Figure 6.12 Time history of exhaust gas flow rate through the cycle and the ideal secondary air delivery rate in phase with the engine exhausting process [Herrin (1975)].

Immediately after the exhaust valve opens, the temperature initially drops rapidly due to the fact that a large amount of air delivered before the exhaust valve opens is accumulated there to mix with the exhaust gases. Then the temperature increases quickly due to the exhaust gas blowdown. During blowdown, exhaust gases exit the engine and flow through the exhaust system at a higher pressure, which prevents the exhaust gases from mixing with the secondary air. At the exhaust port, exhaust blowdown can start thermal oxidation as long as a small amount of secondary air is available. The capability of being able to deliver some secondary air and get it entrained into the exhaust gases is important in utilizing the heat available during the blowdown phase. The mixing between secondary air and exhaust gases also is occurring during this blowdown phase, which prepares the thermal condition for the oxidation once the secondary air and exhaust gases become mixed. Because the pressure from the exhaust gas during the blowdown phase of the exhaust process was high enough to shut off the flow of air out the tube and leave the exhaust gases largely unmixed with secondary air, it was found that a sparger-type design can produce better results compared to the open-ended tube-type design [Herrin (1975); Borland and Zhao (2002)] due to the fact that the sparger-type design allows the secondary air and exhaust gases to better mix during the blowdown phase. Figure 6.13 illustrates the schematic of the open-ended and sparger-type injection tubes for delivering secondary air.

After the blowdown, the exhaust gases have lower temperatures but a better mixing rate and longer residence time, which can certainly initiate thermal oxidation if a minimum temperature is maintained. Visualization of the exhaust port revealed that after the blowdown phase, a blue luminous zone is observed. This blue zone moves downstream with the exhaust stream and can reach the exhaust manifold collector area [Koehlen *et al.* (2002)]. The last part of the mixture from the cylinder flowing through the exhaust port has an even lower temperature, which may prevent any thermal oxidation. After the exhaust valve closes, only secondary air is supplied into the residual gas in the port at a relatively lower temperature until the next engine cycle.

In the pathway of the exhaust gases from the exhaust valve to the catalytic converter, the exhaust gases will be heated by the reaction occurring, lose heat to the walls, and either be heated or cooled by the surrounding gas. At the manifold collector, the temperature of the mixture decreases due to the heat losses to the wall and the mixing of the exhaust gases with secondary air coming from other cylinders. The worst case for mixing occurs when the secondary air is forced out the exhaust port during the exhaust blowdown. However, the exhaust gas blowdown also provides relatively hotter exhaust gases. Any

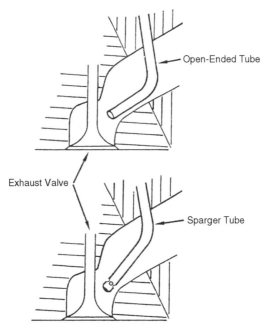

Figure 6.13 Schematic of open-ended and sparger-type tubes for secondary air injection [Herrin (1975)].

design and operating parameters that can prevent heat losses to the walls and improve mixing with a longer residence time in the manifold are effective in enhancing the completion of the thermal oxidation process. The mixing of exhaust gases between cylinders is as critical as that within each cylinder event [Yamamoto et al. (2001)]. The detailed design features of the exhaust manifold such as runner length, runner orientation, and manifold volume are important parameters for optimization.

Note that the effort to improve mixing around this region is crucial, both from the point of thermal oxidation and of catalytic oxidation because the catalytic oxidation requires a homogeneous mixture inside the catalytic converter where the mixing ceases. Figure 6.14 shows the H_2 conversion efficiency as a function of reactor temperature with different levels of mixing (indicated by mixing index I_m). Clearly, when the reactants are not well mixed with the oxygen (lower I_m), the conversion rate levels off after a certain threshold of reactor temperature. Below this temperature threshold, the H_2 conversion is kinetics limited, and the conversion rate is not strongly dependent on the mixing level.

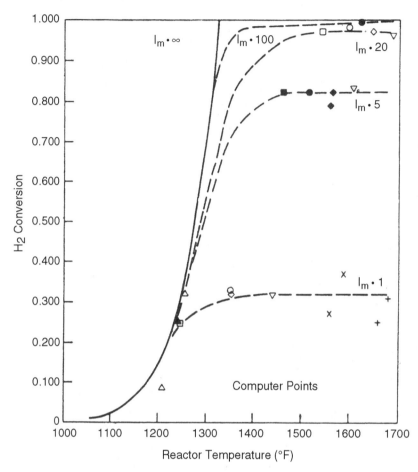

Figure 6.14 *Hydrogen conversion efficiency as a function of reactor temperature with various levels of mixing [Lord et al. (1973)].*

Once the reactor temperature is above this threshold, the H_2 conversion rate is no longer limited by the chemical kinetics and shows a strong dependence on the mixing level. This clearly highlights the importance of enhancing the mixing process of reactants and secondary air inside the manifold prior to the catalytic converter.

When considering the residence time for enhancing thermal oxidation in the manifold, the position for a close-coupled catalyst may no longer be as critical as that for the non-secondary air system. Instead, placing the catalyst slightly distant from the manifold collector will provide more time for oxidation to occur before reaching the catalyst. This also will alleviate the concern of fast thermal deterioration with the closed-coupled location, extend the life of the catalyst, and relax the packaging constraints. To study the effect of exhaust manifold design on secondary air injection, several configurations of exhaust manifold were fabricated to emphasize the impact of mixing (M), volume (V) for extended residence time, and a combination of both mixing and volume (H). Figures 6.15 and 6.16 show the time histories of HC emissions and catalyst temperatures with various exhaust manifolds. It is evident that the exhaust manifold design can significantly affect the HC emissions and catalyst light-off performance due to its enhanced mixing and extended residence time of reactants inside the exhaust manifold.

Spark retardation is a widely used technique for raising exhaust gas temperatures. Higher exhaust gas temperature enhances the initiation of the thermal oxidation reaction. Figure 6.17 plots the exhaust gas and catalyst temperatures measured using various levels of spark retardation. These values are the average of temperatures once they reach their steady-state values. Clearly, the advantage with the spark retardation is that hotter exhaust gases, resulting from thermal oxidation, heat the catalyst more quickly to the catalyst light-off temperature. The points in Figure 6.17 with 0° of retardation show the exhaust temperatures

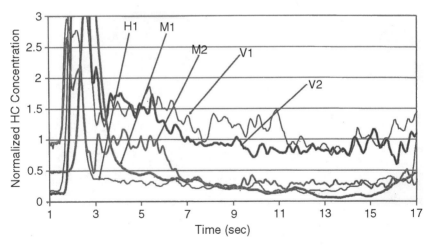

Figure 6.15 Time histories of HC emissions with various exhaust manifolds [Borland and Zhao (2002)].

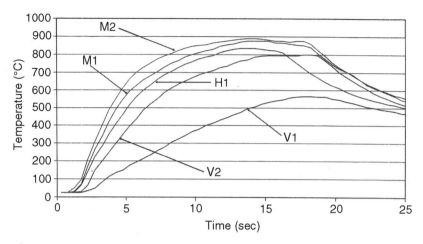

Figure 6.16 Time histories of catalyst temperatures with various exhaust manifolds [Borland and Zhao (2002)].

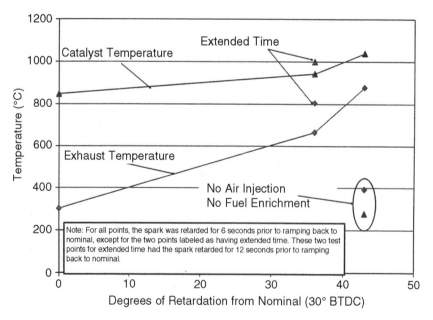

Figure 6.17 Effect of spark retardation on exhaust gas and catalyst temperatures [Borland and Zhao (2002)].

with air injection and fuel enrichment, but without spark retardation. In this case, the thermal oxidation reaction does not seem to be initiated in the exhaust gases. The data show that the catalytic oxidation reaction was initiated, and the air injection still has a dramatic effect on the catalyst temperature. To study the benefit of the period of spark retardation on catalyst light-off, two cases of spark retardation for 36° were examined. Figure 6.17 shows both cases, with each case holding the spark retardation for 6 and 12 seconds, respectively, after which the spark is ramped back toward the normal idle value around 30° before top dead center (BTDC). Using the exhaust gas temperature as an indicator of the thermal oxidation reaction, it can be seen that the additional thermal energy added to the exhaust by holding the retarded spark longer does enhance the rate of the oxidation reaction. It also was found that the extended use of spark retardation has a rather smaller benefit on the catalyst temperature increase. At the spark retardation of 43°, the retardation was held for 6 seconds before being ramped back. The extra spark retardation raises the peak exhaust gas temperature and, to a much lesser degree, the catalyst temperature. It can be concluded that some spark retardation must be used with secondary air injection to initiate the thermal oxidation reaction in the manifold. The benefit in improving the reaction rate must be examined carefully against the concerns with combustion stability resulting from extreme spark retardation.

6.5 Requirements for Engine Enrichment and Secondary Air Injection Quantity

Engine enrichment level and secondary air injection quantity are two more determining parameters that must be examined carefully to maximize the secondary air benefits. The level of engine fueling enrichment determines the composition of engine-out emissions (reactants) and their concentrations. It was reported that mass emissions in Bag I of the Federal Test Procedure (FTP) test were consistently higher with less engine enrichment [Crane *et al.* (1997)]. For a leaner mixture, a relatively warmer exhaust gas can be obtained, but it contains fewer reactants. As a result, the energy conversion is low, even though the oxidation is thermally favorable. Therefore, a relatively leaner in-cylinder mixture is not the favored strategy for secondary air injection [Hernandez (2002)]. Note that over-enriching the engine may jeopardize combustion stability, which may lead to an engine misfire and subsequently a sharp increase in HC emissions.

Regarding the quantity of secondary air, there are two opposite effects. On one hand, a certain quantity of secondary air is necessary to maximize the reactions in front of the catalyst, which will result in increased temperature. On the other hand, an excess amount of secondary air will cool the exhaust, which may slow the reactions or, in the worst case, quench the thermal oxidation inside the exhaust system.

Figure 6.18 shows the time-averaged steady-state values of HC concentration and temperature as a function of exhaust lambda. Here, the air mass injected was held constant, and the exhaust lambda was varied by varying the engine lambda proportionately. Clearly, the exhaust and catalyst temperatures increase linearly as the exhaust lambda decreases. As the decrease of exhaust lambda results from a decrease in engine lambda, more reactants (engine-out emittants) are available for enhancing temperature increases. The effect of exhaust lambda on HC emissions is definitely nonlinear, showing a minimum around the lambda value of 1.25. For a mixture leaner than the optimum, there are not enough reactants produced to match the injected air. For a richer mixture than the optimum, the engine-out HC emissions dominate the trend.

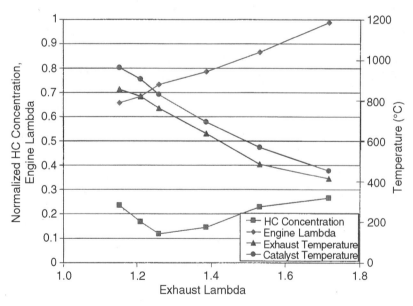

Figure 6.18 Effect of exhaust lambda on exhaust temperature, catalyst temperature, and HC concentration with a fixed amount of secondary air injected [Borland and Zhao (2002)].

To study the effect of the airflow rate, a test was conducted with a constant exhaust lambda of 1.25, and the engine lambda was varied with the change in airflow rates. Figure 6.19 shows the results derived from this study. The HC concentration steadily increases as the airflow rate decreases. It is speculated that the trend toward a lower HC concentration with increasing airflow will continue to a point where the engine lambda required to maintain the exhaust lambda is low enough to cause rich misfire. The temperatures do not show the same linear response to the airflow increase. As the airflow increases, the temperatures increase as expected, up to a point, and then begin to decrease. This decrease is likely due to the fact that the extra air mass injected to maintain the constant exhaust lambda cools the mixture and slows the reaction.

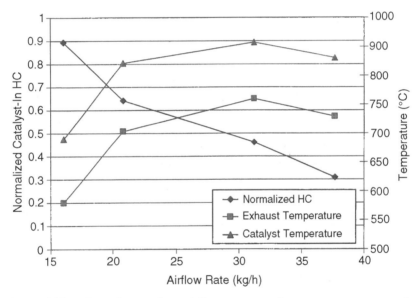

Figure 6.19 *Effect of secondary airflow rate on exhaust temperature, catalyst temperature, and HC concentration at a fixed exhaust lambda of 1.25 [Borland and Zhao (2002)].*

6.6 Secondary Air Injection Control and Onboard Diagnostics

As already discussed, there exists an optimal A/F ratio entering the catalyst that minimizes the converter-in HCs and maximizes the converter-in exhaust gas temperature. Because secondary air injection is used on a cold start when the catalyst is not lit off, heating the catalyst as rapidly as possible while minimizing the emissions that pass through is the primary goal. This optimal A/F ratio is affected by the mass of air flowing through the engine, the mass of air being pumped by the secondary air pump, and the amount of fuel being injected into the engine. Variations in any of these values will cause deviations from the targeted converter-in A/F ratio. Additionally, if the A/F ratio is too rich, the catalyst temperature can easily exceed 1000°C during idle, which will significantly degrade catalyst durability. There are a number of strategies for controlling the A/F ratio. These strategies range from simple open-loop control, which is widely used today, to more complex strategies, including various types of closed-loop control.

6.6.1 Open-Loop Control

Open-loop control generally relies on the air pump injecting a certain amount of air. A premapped amount of fuel is injected to try to obtain the goal A/F ratio. This type of control system typically could not account for changes in the system that may take place over time, such as clogged air filters, changes in air temperature, changes in altitude, or changes in the battery voltage. Some of these factors could be addressed if a suitable model of the air pump flow were developed. This model for various environmental conditions could achieve the goal A/F ratio more precisely. Although this type of control is inexpensive and works well under optimum conditions, it is still not sufficient to meet the most stringent emissions regulations when considering the variability of production and the variability due to aging.

6.6.2 Closed-Loop Control

As emissions standards become more stringent, the use of closed-loop control algorithms for the control of secondary air injection becomes more popular. Because the goal is to control the A/F ratio entering the catalyst, three variables can be used to modify this value: (1) the amount of airflow from the air pump, (2) the amount of airflow from the engine, and (3) the A/F ratio of the exhaust gases leaving the engine (prior to mixing with the secondary air). Theoretically, all these factors could be controlled; however, practically, an active control of the engine airflow will probably be unacceptable to the customer. Several sensors are available that can be used to provide the feedback information required.

- **Airflow control**—One way of controlling the A/F ratio is to control the amount of air flowing into the exhaust system. This air can come from either the engine or the air pump. The amount of air from the engine can easily be changed by varying either the engine speed or the spark advance. However, customer satisfaction problems could be presented by this approach. Varying the engine speed or load at idle could be perceived by customers as a driveability problem. Varying the amount of air pumped by the air pump has been proposed by Golden *et al.* (2001). Their approach would apply an extra control module to supply pulse width modulation (PWM) control of the air pump motor in response to changes in operating conditions to vary the amount of air being supplied. While matching the engine conditions with the air pump seems to have the minimum impact on the existing engine control system, this type of system requires the expense of an extra control module. In addition, given the inertia of the spinning impeller in the pump, it is difficult to change the airflow rate quickly by controlling the air pump. To accurately maintain the optimum A/F ratio during transients, it would seem that some sort of fuel compensation also would be required. The advantage of this type of system is the better matching of the air pump flow to the engine flow. Because there is a minimum A/F ratio at which an engine can operate without the danger of misfire from being overly rich, the air pump flow must be matched with the expected engine flow so that the optimum A/F ratio is achievable. For a partial zero emissions vehicle (PZEV), the air pump typically will be used only at idle, and the air pump is sized for that engine airflow rate. For other emissions test cycles where the vehicle is more lightly loaded early in the test, it may be necessary to operate the air pump for a longer period of time. In this case, the engine airflow can vary widely. If the air pump flow were fixed, that would limit the range of attainable A/F ratios.

- **Fueling control**—The other option for controlling the A/F ratio is to control the amount of fuel being injected. If one knows the amount of air from the two sources, the amount of fuel required to reach the target A/F ratio can easily be calculated. Because all modern engines have direct control over the amount of fuel injected into the engine, this is an actuator that is available at no extra cost and with few customer satisfaction ramifications. Modification of fueling also is generally quicker to respond than the modifications of the airflow rate previously discussed.

6.6.3 Sensors for Feedback Control

To implement the closed-loop control scheme, a sensor must be available to sense what is being controlled in order to modify the control inputs accordingly. Some of the candidate sensors are outlined as follows: .

- **Mass airflow sensor**—The most direct measurement of airflow is through the use of a mass airflow (MAF) sensor. This type of sensor is quite common and can be found in most modern cars. The MAF sensor can be used to measure only the pump airflow. Because most engine control schemes require the amount of air flowing through the engine to either be calculated or be measured for A/F ratio control, the amount of air flowing through the engine generally is known. The additional information of the amount of air coming from the air pump allows the calculation of the quantity of fuel required to reach the optimum A/F ratio. This method is easy to add to the existing control algorithms. Additionally, this sensor may be used to satisfy some of the requirements for diagnostic monitoring as required by the OBD II regulations. The major drawback of this approach is the extra expense of the additional components and the space required to install the components. Mass airflow sensors warm up quickly, so they can be used during the entire operating period. An MAF sensor alone probably would not be sufficient for OBD because where the air was pumped is still unknown.

- **Wide-range oxygen sensor**—A wide-range oxygen (WRO_2, or universal exhaust gas oxygen [UEGO]) sensor is another common sensor used in automotive control systems. It is used to sense the A/F ratio of the exhaust from the engine. If this sensor were exposed to both sources of air as well as the fuel, it could yield a direct reading of the A/F ratio in the exhaust. These sensors are common, but they are more expensive than the more widely used switching oxygen sensors. The other drawback of this method is the time required for the sensor to reach its operating temperature. Typically, the WRO_2 sensor is heated when the engine is started. It can take 10 to 20 seconds (or longer) from this time for the sensor to reach its operating temperature. Even with a light-off time of 10 seconds, half of the idle time on the FTP has passed before closed-loop control can begin with this sensor. In practical operation, the situation could be worse if the typical driver does not idle for 20 seconds after a cold start. Therefore, this type of sensor would, most likely, need to be combined with some type of model for satisfactory control during the entire run time. A UEGO sensor may be sufficient for OBD purposes but may have difficulty meeting the rate-based OBD requirements due to the long warm-up time.

- **Temperature sensor**—It has been shown that a relationship exists between the exhaust gas temperature and the A/F ratio when secondary air is introduced under a condition with thermal oxidation in the exhaust gases. A temperature sensor could be used to sense the temperature, and, if a relationship can be well established, it could be used as a measure of the A/F ratio. Because a durable sensing element probably will demonstrate a considerable lag, given the rapid increase in exhaust gas temperatures at the onset of the thermal oxidation reaction, it probably would not be possible to use this reading directly. Instead, this signal would likely have to be manipulated by some algorithm that tried to predict the temperature while taking into account the time constant of the sensing element. This type of algorithm would require some logic so that it would not predict the temperature incorrectly if operating conditions change. This type of device should be sufficient for OBD.

6.7 Other Application Considerations for Secondary Air Injection

6.7.1 Application of Secondary Air Injection to Vee Engines

The application of secondary air injection to vee engines requires some unique considerations in both control and OBD. These considerations include the method of secondary air introduction, namely, one bank or both, and the number of secondary air pumps required.

Where to introduce the secondary air is largely influenced by the exhaust system architecture and catalytic converter locations. If there is one catalyst per engine bank, then the system would certainly benefit most from the secondary air injection into each bank. If all exhaust flows through a single catalyst at some point in the exhaust system, air injection into only one single bank of the engine may be able to accomplish the required emissions reduction.

With the air introduction into the single bank, the fuel enrichment and spark retardation required for optimum emissions reduction would be applied only to that bank. The other engine bank then would be operated under a condition for an optimal catalyst light-off within the given idle stability constraints.

The number of air pumps required will be determined by the emissions reduction target. As has been discussed, the mass flow rate of secondary air injection has an influence on the exhaust and catalyst temperatures, as well as the amount of converter-in HC reduction. If there is an air pump for each bank (two air pumps or a single air pump supplying only one bank), then the system can be controlled and diagnosed the same way as an independent in-line engine. When a single air pump is used to supply both banks of a vee engine, several issues may arise. The most critical issue is how to ensure that the ratio of air to each bank remains constant through the useful life of the vehicle. For a system with a feedback control of the airflow to adjust the fuel enrichment, this will require at least two measurements of the secondary airflow (either one for each bank, or one for the system airflow and another measurement for one of the banks). Onboard diagnosis for the case of one air pump and two banks also may require two separate measurements. It is easy to imagine a severe degradation in emissions performance of an exhaust system in which the air flowing to one bank is blocked, and there is no indication to the control system of this occurrence.

6.7.2 Application of Secondary Air Injection to Turbocharged Engines

The application of secondary air injection for tailpipe emissions reduction on turbocharged engines has been realized in production for several years. Many of the design challenges for the turbocharged engine are similar to those for the naturally aspirated engine. The unique elements of the turbocharged engine are the sensitivity of transient performance to the exhaust manifold design and the effect of the turbochargers on catalyst light-off.

Inherent in any turbocharged engine is a lag in boost pressure during a throttle transient. One way engine designers attempt to minimize this is through the reduction of volume between the exhaust ports and the turbine blades of the turbocharger. It has been reported that increasing the exhaust manifold volume has a beneficial effect on the converter-in HC reduction due to the increased residence time of the exhaust gases. Therefore, clearly on the turbocharged engine, there must be a tradeoff between the converter-in HCs and the transient performance of the turbocharger.

The delay in catalyst light-off due to the turbocharger is another challenge to low emissions vehicle development. The turbocharger housing and turbine blades represent added thermal mass in the exhaust system. Naturally, turbine blades are designed to effectively extract energy from the exhaust, which can further lower the exhaust temperatures and slow the catalyst light-off process. Even if the thermal oxidation reaction is initiated in the exhaust gases, the reaction will largely be quenched as it passes through the turbocharger. Testing has shown a drop of up to 230°C from the inlet of the turbocharger to its outlet during an idle after cold start on a four-cylinder engine.

Some manufacturers have tried to avoid the catalyst light-off penalty by placing a portion of the exhaust system catalyst volume in front of the turbocharger. This leads to improved catalyst light-off and lower tailpipe HC emissions, but this design also contributes to a large volume upstream of the turbocharger and the associated turbo lag. Other manufacturers have sought to bypass the exhaust gases around the turbocharger for a period of time after cold start. Using a suitable actuator, the waste gate or other bypass is held open. In fact, one manufacturer also tried to integrate a second bypass that is much larger than the waste gate into the turbocharger housing, for the sole purpose of bypassing exhaust gases during cold start. This bypass even has an integral mini-catalyst built in to aid in catalyst light-off. Some suppliers are beginning to address the catalyst light-off penalty that is associated with turbochargers by designing a thermally insulated turbine housing, integrating exhaust bypass passages, or attaching catalysts directly to the turbine housing outlet.

6.7.3 Other Application Issues

Other concerns such as the air pump noise level, its volume for packaging, weight, system cost, and system durability must be carefully examined before implementing this technology. Because the engine is fueled with an ultra-rich mixture and a large amount of liquid fuel may enter the cylinder directly, particularly during cold start, some of the rich-combustion-related issues must be thoroughly evaluated. Those include soot loading on the combustion chamber and catalyst, likely oil dilution via repeated and extended fuel-rich operation, and spark plug fouling concerns. In addition, thermal loading in the exhaust system and catalyst must be examined to determine the optimal design of the exhaust system and the catalyst location. Cold-start fuel consumption should be minimized.

6.8 Summary

Both thermal and catalytic oxidation reactions of CO, HC, and H_2 with secondary air injection affect the catalytic converter light-off performance. Catalytic oxidation alone can be used to improve catalyst light-off performance, but this strategy gives up many of the benefits that come with thermal oxidation. Promoting thermal oxidation in the manifold is effective in raising exhaust gas temperature and in lowering the converter-in HC concentration. An improvement in the thermal oxidation reaction may degrade the catalytic oxidation reaction due to the increased consumption of reactants upstream of the catalyst. There is a tradeoff between the performance of thermal oxidation and catalytic oxidation reactions. Mixing is the key to success of the secondary air injection strategy. Proper mixing immediately downstream of the air injection location is critical. The optimal secondary air quantity is highly dependent on the mixing quality. The exhaust manifold design has a major influence on the performance of the thermal oxidation reaction and the subsequent catalytic oxidation. Enhanced mixing and increased residence time are the design goals. The exhaust lambda must be optimized to minimize the emissions entering the catalyst. The amount of air injected influences the catalyst-in emissions and temperatures by increasing engine enrichment (and thus increased production of reactants) while maintaining the optimum exhaust lambda.

The thermal and chemical processes associated with secondary air injection inside the exhaust system are complex and must be well understood to maximize the simultaneous benefit of improving catalyst light-off performance and reducing converter-in emissions. Some of the key design and operating parameters are summarized as follows:

- Engine-out emissions (reactants) concentrations and temperature
    - Engine A/F ratio (enrichment level)
    - Spark timing (level of spark retardation)
    - Idle speed
    - Exhaust cam timing
- Mixing of secondary air with the reactants and mixture residence time
    - Air injection location
    - Air injection orientation

- Airflow rate
- Capability for delivering secondary air during the exhaust blowdown phase
- Exhaust port design
- Exhaust valve opening timing (blowdown flow versus displacement flow)
- Exhaust manifold design for enhanced mixing (manifold runner length, runner orientation of each cylinder, plenum volume, collector location, cross talk between cylinders)

• Thermal conditions of the exhaust system
- Exhaust port geometry (affecting heat rejection to the port wall)
- Exhaust manifold design and material (affecting heat loss to the manifold wall, such as a dual thin-wall exhaust manifold)

• Reaction inside the converter
- Mixing quality of the mixture prior to entering the converter
- Composition of the mixture

6.9 References

1. Borland, M. and Zhao, F. (2002), "Application of Secondary Air Injection for Simultaneously Reducing Converter-In Emissions and Improving Catalyst Light-Off Performance," SAE Paper No. 2002-01-2803, Society of Automotive Engineers, Warrendale, PA.

2. Crane, M., Dodge, L.G., Thring, R.H., and Podnar, D.J. (1997), "Reduced Cold-Start Emissions Using Rapid Exhaust Port Oxidation (REPO) in a Spark-Ignition Engine," SAE Paper No. 970264, Society of Automotive Engineers, Warrendale, PA.

3. D'Alleva, B.A. and Lovell, W.G. (1936), "Relation of Exhaust Gas Composition to Air-Fuel Ratio," *SAE Transactions*, Vol. 38, No. 3, pp. 90–96.

4. Golden, J.E., Kochs, M., Kloda, M., and Van De Venne, G. (2001), "Innovative Secondary Air Injection Systems," SAE Paper No. 2001-01-0658, Society of Automotive Engineers, Warrendale, PA.

5. Harrington, J.A. and Shishu, R.C. (1973), "A Single-Cylinder Engine Study of the Effects of Fuel Type, Fuel Stoichiometry and Hydrogen-to-Carbon Ratio on CO, NO, and HC Exhaust Emissions," SAE Paper No. 730476, Society of Automotive Engineers, Warrendale, PA.

6. Hernandez, J.L., Herding, G., Carstensen, A., and Spicher, U. (2002), "A Study of the Thermochemical Conditions in the Exhaust Manifold Using Secondary Air in a 2.0-L Engine," SAE Paper No. 2002-01-1676, Society of Automotive Engineers, Warrendale, PA.

7. Herrin, R.J. (1975), "The Importance of Secondary Air Mixing in Exhaust Thermal Reactor Systems," SAE Paper No. 750174, Society of Automotive Engineers, Warrendale, PA.

8. Heywood, J. (1988), *Internal Combustion Engine Fundamentals*, McGraw-Hill, New York.

9. Kadlec, R., Sondreal, E.A., Patterson, D.J., and Graves, M.W. Jr. (1972), "Limiting Factors on Steady-State Thermal Reactor Performance," SAE Paper No. 730202, Society of Automotive Engineers, Warrendale, PA.

10. Koehlen, C., Holder, E., and Vent, G. (2002), "Investigation of Post Oxidation and Its Dependence on Engine Combustion and Exhaust Manifold Geometry," SAE Paper No. 2002-01-0744, Society of Automotive Engineers, Warrendale, PA.

11. Kollmann, K., Abthoff, J., Zahn, W., Bischof, H., and Giarhre, J. (1994), "Secondary Air Injection with a New Developed Electrical Blower for Reduced Exhaust Emissions," SAE Paper No. 940472, Society of Automotive Engineers, Warrendale, PA.

12. Leonard, L.S. (1961), "Fuel Distribution by Exhaust Gas Analysis," SAE Paper No. 379A, Society of Automotive Engineers, Warrendale, PA.

13. Lord, H., Sondreal, E.A., Kadlec, R.H., and Patterson, D.J. (1973), "Reactor Studies for Exhaust Oxidation Rates," SAE Paper No. 730203, Society of Automotive Engineers, Warrendale, PA.

14. Son, G.-S., Kim, D.J., Lee, K.Y., and Choi, E.-R. (1999), "A Study on the Practicability of a Secondary Air Injection for Emission Reduction," SAE Paper No. 1999-01-1540. Society of Automotive Engineers, Warrendale, PA.

15. Spindt, R.S. (1965), "Air-Fuel Ratios from Exhaust Gas Analysis," SAE Paper No. 650507, Society of Automotive Engineers, Warrendale, PA.

16. Stivender, D.L. (1971), "Development of a Fuel-Based Mass Emission Measurement Procedure," SAE Paper No. 710604, Society of Automotive Engineers, Warrendale, PA.

17. Yamamoto, S., Tanaka, D., Kuwahara, K., and Ando, H. (2001), "Useful Combustion in Cylinder During Exhaust Stroke and in Exhaust Port with Gasoline Direct Injection," Proceedings of the 5th International Symposium on Diagnostics and Modeling of Combustion in Internal Combustion Engines (COMODIA), pp. 187–192.

CHAPTER 7

Effects of Fuel Properties and Fuel Reforming on Cold-Start Hydrocarbon Emissions and Catalyst Light-Off

James A. Eng
General Motors Research

7.1 Introduction

The main reason for the increase in engine-out hydrocarbon (HC) emissions during a cold start is that only a fraction of the fuel vaporizes at low temperature. Equilibrium calculations estimate that only 10 to 20% of the fuel vaporizes during the first few cycles of a cold start [Boyle *et al.* (1993); Cheng and Santoso (2002)]. As a result, the engine must be significantly over-fueled to create a combustible mixture within the cylinder. This over-fueling results in large amounts of liquid fuel entering the cylinder, which can be a major source of HC emissions. Significant reductions in HC emissions could be achieved if the engine could be operated closer to stoichiometry during cold start.

Fuel vaporization and mixture preparation in a port fuel injected (PFI) spark ignition (SI) engine during a cold start depend critically on both the ambient temperature and the engine coolant temperature. In particular, the temperature of the intake valve is important for obtaining vaporization during the first engine cycle. After the first successful fired cycle, backflow of hot residual gases into the intake port aids in vaporizing the fuel on subsequent cycles [Kaiser *et al.* (1996); Kidokoro *et al.* (2003)]. During the first engine cycle of a cold start, the intake port and valves are at low temperature, and there is no residual gas backflow into the intake port to assist fuel vaporization. The fuel injector has a large effect on mixture preparation during cold start. Although improved fuel atomization does not eliminate the problem of fuel vaporization, better atomizing injectors dramatically reduce the amount of fuel that is deposited on the walls of the intake port. By maintaining more of the fuel in the intake air, rather than on the intake port wall, the engine can be operated with less over-fueling. Droplet diameters smaller than an estimated 10 μm are required to have the fuel remain airborne and follow the airflow [Zimmermann *et al.* (1999)]. All manufacturers have incorporated improved fuel atomization injectors into their certified super ultra-low emissions vehicles (SULEVs) [Kidokoro *et al.* (2003); Oguma *et al.* (2003)].

Engine warm-up tests have demonstrated that the effect of fuel on HC emissions is similar in magnitude to the effect of the fuel injector [Takeda *et al.* (1995); Shayler *et al.* (1996)]. Shayler *et al.* performed warm-up tests at a constant engine speed and a fixed air/fuel (A/F) ratio of 13.5 using fuels with a range of volatilities and fuel injectors with different atomization characteristics. Variations in fuel volatility resulted in changes in HC emissions that were comparable to the changes achieved with different fuel injectors. The effects of fuel variability on cold-start driveability and HC emissions must be considered during vehicle development programs so that the engine and emissions control system are robust to variations in fuel quality. One approach that has been used to compensate for changes in fuel volatility on combustion performance during cold engine operation is to monitor crankshaft speed fluctuations and adjust the engine fueling to maintain acceptable performance with low-volatility fuels [Kishi *et al.* (1998)].

One potential strategy to reduce cold-start HC emissions is to modify the fuel onboard the vehicle in some manner that will reduce tailpipe emissions during cold start. Technologies that have been developed along these lines are onboard fuel reformers and distillation systems. Fuel reformers convert the liquid fuel into a mixture of carbon monoxide (CO) and hydrogen (H_2), with only trace amounts of HCs, that is used to fuel the engine during cold start. Fuel distillation systems fractionate the fuel into high-volatility and low-volatility fuel streams. The engine is cold started with the high-volatility fuel to obtain better mixture preparation. Fuel reformer systems also have been proposed to generate hydrogen onboard to obtain faster catalyst light-off during cold start.

7.2 Gasoline Properties

Because the increase in HC emissions during a cold start is due in a large part to the poor volatility characteristics of gasoline, it is appropriate to review those properties of gasoline that are relevant to cold engine operation. For more information on gasoline composition and general aspects of gasoline, the interested reader can refer to the *Automotive Fuels Reference Book* by Owen and Coley (1995) or to the review papers by Gibbs (1993 and 1996) on gasoline history and development.

7.2.1 Composition

A fully blended gasoline is composed of hundreds of HC compounds with different molecular structures and boiling points. The composition of gasoline varies widely around the world and both regionally and seasonally within the United States. In the United States, a typical non-oxygenated gasoline is composed of roughly 60% paraffins (single C–C bond), 30% aromatics (benzene ring), and 10% olefins (double C=C bond) [Leppard et al. (1992)]. The paraffins are composed of low-molecular-weight compounds such as n-butane, 2-methyl pentane, and 2,2,4-trimethyl pentane (iso-octane). The aromatic species are composed of 4 to 8% toluene (methylbenzene) and roughly 8% xylenes (1,3-dimethylbenzene and 1,4-dimethylbenzene). The olefinic content in gasoline has been reduced significantly in the last 15 years in response to environmental concerns and to minimize the reactivity of the exhaust gas to generate ozone in the atmosphere, as well as minimizing gum formation in the fuel system. European gasoline typically contains up to 20% olefins and has a higher aromatic content than a typical U.S. gasoline [Kopasz et al. (2001)].

Oxygenates have been blended into gasoline for many years to improve the octane rating of gasoline. The effect of the oxygenate on the octane quality of a fuel depends on the detailed composition of the fuel, because the octane behavior of a given component in a fuel blend is modified by the other components in the fuel. For example, the Research Octane Number (RON) for neat methanol is 109, and the Motor Octane Number (MON) is 89; however, in gasoline, the blending RON for methanol is 127 to 136, and the blending MON is 99 to 104 [Owen and Coley (1995)]. Federal regulations mandate that reformulated gasolines (RFGs) contain oxygen levels between a minimum of 1.5% to a maximum of 2.7% by weight.

The addition of oxygenates to gasoline also causes an emissions reduction in older vehicles without closed-loop emissions control and exhaust gas aftertreatment [Leppard et al. (1995)]. Although the largest emissions reduction obtained from the addition of oxygenates is in CO emissions, there is also an HC emissions benefit. Both alcohols (R–O–H where R is an HC group) and ethers (R–O–R) are blended into gasoline. Alcohols that are added to gasoline are methanol (MeOH), ethanol (EtOH), iso-propanol (IPA), and t-butanol (TBA). Because of the poor solubility of methanol in gasoline when water is present, it must be used with TBA added as a cosolvent. Ethers that are added are methyl-tertiary butyl ether (MTBE), tertiary amyl butyl ether (TAME), and ethyl-tertiary butyl ether (ETBE). Methyl-tertiary butyl ether is the most common oxygenate added to gasoline [Chevron (1996)]. However, MTBE is highly soluble in water, and unacceptably high levels have been detected in the groundwater in many areas of the United States, leading to it being banned in the state of California as a gasoline additive [Porse (2002)]. Although MTBE is still blended into gasoline in other states, it is likely that it will be completely banned in the future.

Simplified species models for gasoline can be created by grouping the many different HC species into representative classes of compounds [Chen et al. (1994)]. These simplified models for gasoline can be empirically developed to have vaporization characteristics that closely mimic those of gasoline. Table 7.1 lists the composition of a simplified model for California Phase II RFG [Cheng and Santoso (2002)]. Also listed in the table are the molecular weights and boiling points of the representative species. The boiling points for the pure compounds range from 0 to near 220°C.

7.2.2 Volatility

The volatility of a fuel is characterized by its vapor pressure and boiling range, or distillation curve. The volatility of a fuel is commonly characterized by two different standardized tests. The most common means to quantify fuel volatility are the ASTM distillation curve and the Reid vapor pressure (RVP) [Owen and Coley (1995)]. The RVP is determined at standard conditions and follows ASTM Procedure D323. In this test procedure, liquid fuel at 0°C is connected to a vapor chamber filled with air at 37.8°C (100°F). The air volume is four times larger than the fuel volume. The entire apparatus is immersed in a water

TABLE 7.1
MODEL FUEL COMPOSITION FOR CALIFORNIA PHASE II REFORMULATED GASOLINE [CHENG AND SANTOSO (2002)]

Compound	Volume Percent	Boiling Point (°C)
n-butane	2.0	−1
Iso-pentane	16.6	28
Iso-hexane	10.6	60
2,3 dimethyl pentane	11.9	79
2,2,4 trimethyl pentane	16.1	100
2,2 dimethyl heptane	2.6	132
n-decane	0.7	175
n-undecane	0.4	196
n-dodecane	0.3	216
Benzene	1.2	80
Toluene (methylbenzene)	7.4	110
m-xylene	10.1	139
Iso-propyl benzene	4.7	152
Iso-butyl benzene	2.4	172
MTBE	13.0	58

bath that is maintained at a temperature of 37.8°C. The fuel RVP is equal to the equilibrium vapor pressure and is measured with a standard pressure gauge. U.S. Environmental Protection Agency (EPA) regulations established in 1992 restrict the maximum summer RVP in northern states to 62 kPa (9 psi) and in the southern tier states to 54 kPa (7.8 psi) [Gibbs (1993)]. The EPA does not regulate RVP during winter months.

The distillation curve for a fuel is determined using ASTM Procedure D86. In this procedure, a 100-mL sample is distilled under specified conditions at atmospheric pressure using a controlled temperature. The fuel is placed in a flask and is heated using a specified temperature-time history to vaporize the fuel. The vaporized fuel passes through a cooling tube, and the volume of condensate is recorded as a function of temperature. At the conclusion of the test, some residual that did not vaporize is left within the flask. The difference between the initial sample volume and the sum of the amount of condensate recovered and the residual volume is equal to the losses from the low-boiling-point compounds. The low-boiling-point compounds typically are butanes (C_4) and pentanes (C_5). The distillation curve is shifted accordingly to subtract the losses. Figure 7.1 shows a typical distillation curve for a summer-grade gasoline [Jorgensen et al. (1999)]. Also shown are the boiling points for some of the compounds listed in Table 7.1. Important parameters determined from the distillation curve are the initial and final boiling points, and the temperatures for 10, 50, and 90% evaporation of the fuel. These temperatures are denoted in Figure 7.1 as T_{10}, T_{50}, and T_{90}, respectively. The initial boiling point is the location where liquid fuel first starts to condense. Typical ranges for T_{10} are 40 to 60°C, T_{50} from 90 to 110°C, and T_{90} from 160 to 180°C [Gibbs (1993)].

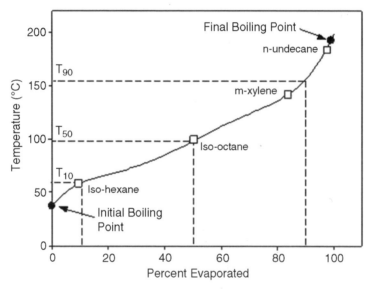

Figure 7.1 Distillation curve for a typical summer-grade gasoline [Jorgensen et al. (1999)].

Although the RVP and ASTM distillation curves are used almost universally to characterize gasoline volatility, neither is an accurate representation of the vaporization processes of gasoline in an engine. In an engine, the manifold pressure is below atmospheric pressure, and this low pressure effectively reduces the temperatures on the distillation curve. For example, the T_{50} temperature for an average gasoline is reduced from 102°C at atmospheric pressure to 77°C at 50 kPa pressure [Curtis et al. (1998)]. It also is known that flash boiling can occur with gasoline in a PFI engine [Aquino et al. (1998)]. This has the effect of significantly changing the spray pattern from the injector and altering the wall-wetting tendency of the fuel spray. Regarding the RVP test, note that at the temperature at which the test is performed (37.8°C), only the lowest-molecular-weight compounds have vaporized, principally HCs smaller than C_5 or C_6. Thus, the RVP test greatly emphasizes the front end of the distillation curve. For example, the initial boiling point for the fuel shown in Figure 7.1 was 40°C, which is higher than the temperature at which the RVP test is performed. In essence, the RVP is used to characterize the volatility of the fuel fraction that is largely lost from the ASTM distillation test.

The addition of oxygenates to gasoline dramatically changes the volatility. The blending RVP of an oxygenate depends on the oxygenate type, the amount present, and the RVP of the base fuel. The addition of methanol causes a significant increase in RVP. Adding 5% MeOH to gasoline increases the RVP from 56 to 90 kPa (8 to 13 psi) [Furey (1985)]. By comparison, the effect of ethers on RVP is much smaller. The blending RVP of MTBE is below that of gasoline, which means the refiner can add other low-boiling-point compounds to the fuel to increase the RVP. The net effect of oxygenates on the distillation curve is to shift the front end of the curve to lower temperatures.

Fuel volatility has a large effect on volatile organic compound (VOC) evaporative emissions from vehicles. Evaporative emissions from vehicles are effectively restricted to be zero for the partial zero emissions vehicle (PZEV) standard (zero being defined as 350 mg over a three-day diurnal test). Extensive testing has shown that the only fuel property that affects evaporative emissions is RVP [Owen and Coley (1995)]. It also has been established that oxygenates in gasoline do not have an effect on evaporative emissions other than the increase in RVP. The desire to decrease VOC emissions, particularly from older vehicles, was the main reason that the EPA established limits on RVP.

7.2.3 Driveability

The cold-weather driveability of a gasoline is defined in terms of whether the vehicle will start easily, idle smoothly, and have good acceleration with no hesitation under cold ambient conditions [Owen and Coley (1995)]. Acceptable volatility of gasoline under cold conditions is crucial for avoiding driveability problems. However, too much vaporization under hot ambient conditions can result in hot-weather driveability problems. In particular, it is important to design the fuel system to maintain the fuel temperature below the low-end boiling point so that the fuel does not vaporize upstream of the injector at the fuel rail pressure. The driveability of a fuel is the result of complex interactions among the vehicle, engine, fuel handling system, and volatility characteristics of the fuel. Thus, although a fuel may exhibit good cold-weather driveability in most vehicles, problems may be experienced in other vehicles, depending on the engine design. This vehicle/fuel interaction makes identifying the root causes of driveability problems extremely challenging.

A number of procedures have been developed to characterize the cold-weather driveability of a fuel. In the United States, the most commonly used procedure is the Coordinating Research Council (CRC) Procedure 598 [Jorgenson et al. (1999)]. The procedure incorporates a series of light, moderate, and wide open throttle (WOT) maneuvers, followed by a number of idles. The vehicle is rated during the test in terms of stalls, hesitations, and general problems in vehicle transient response. Each malfunction is rated in terms of being moderate, severe, or harsh, and a demerit rating is built up for the tests using weighting factors. The procedure is designed to be completed before the vehicle is totally warmed up, so that the test emphasizes the cold driveability of the fuel. The driveability of a fuel under cold operating conditions has a direct influence on cold-start HC emissions under real-world operating conditions.

The fuel parameters that influence driveability are not simple and can vary widely from vehicle to vehicle. Many studies have been performed to elucidate the effects of fuels on vehicle driveability, and several volatility indices have been developed to correlate driveability with fuel properties. In the United States, one of the most commonly used is the driveability index (DI) [Barker et al. (1988)]. The DI is calculated from the ASTM distillation curve as

$$DI = 1.5\, T_{10} + 3\, T_{50} + T_{90} \qquad (7.1)$$

The units on the DI are the same as those used for the ASTM D86 distillation temperatures. The DI for a typical U.S. gasoline ranges from 540 to 700°C (1000 to 1300°F) [Gibbs (1993)]. The driveability characteristic of a fuel varies nonlinearly with the DI, with a higher DI resulting in more cold-start-related driveability problems [Jorgenson et al. (1996a)]. Clearly, the DI depends heavily on the T_{50} temperature, and it is known that this temperature is critical for HC emissions during a cold start.

An increase in cold-weather driveability problems has been observed with oxygenated gasoline. Although these problems have been observed, there is no consensus about the root cause of the effect of oxygenates on driveability. Oxygenated gasoline will tend to lean the mixture A/F ratio slightly on older vehicles without closed-loop control, which is expected to cause an increase in driveability problems due to leaner mixtures. However, a CRC study of fuels at cold operating conditions in late-model vehicles employing PFI fuel injection systems also exhibited an effect of oxygenates on driveability [Jorgenson *et al.* (1999)]. The recommended DI for oxygenated fuels developed by the CRC is

$$DI = 1.5\, T_{10} + 3\, T_{50} + T_{90} + 7.2\, \delta_{MTBE} + 30\, \delta_{EtOH} \qquad (7.2)$$

where δ_{MTBE} and δ_{EtOH} are equal to 1 if the oxygenate is present, and 0 if it is not. Note that the units on Eq. 7.2 are in degrees Celsius. Using a typical range of DI for non-oxygenated fuels of 540 to 700°C, the percent increase in DI from the addition of oxygenates ranges from 5 to 7%.

Other driveability indices have been developed to characterize the fuel by the amount of fuel vaporized at a specified temperature, otherwise known as the evaporation driveability index (EDI). For refinery blending calculations, it is better to use the percent evaporated at a given temperature (E_{XX}), as opposed to the temperature at which a given amount evaporates (T_{XX}), because the levels of the percent evaporated at a given temperature blend linearly and it is easier for the refinery to control that number [Owen and Coley (1995)]. The CRC developed the following EDI to correlate driveability with the weighted demerits, including an oxygenate offset effect:

$$EDI = E_{70} + 1.44\, E_{100} + 1.6\, E_{140} - 15\, \delta_{MTBE} - 41\, \delta_{EtOH} \qquad (7.3)$$

where E_{70}, E_{100}, and E_{140} are the percentages of fuel evaporated at 70, 100, and 140°C, respectively, and δ_{MTBE} and δ_{EtOH} are defined as previously in Eq. 7.2. Because the E index is the inverse of the T index, higher values of EDI are a result of improved vaporization and consequently better driveability. Typical values for EDI range from 100 to 250. European and Japanese automobile manufacturers typically favor using driveability indices based on percent evaporated.

7.2.4 Reformulated Gasoline

Reformulated gasoline (RFG) was introduced into the market to reduce both tailpipe and evaporative emissions from vehicles. Reformulated gasoline was first introduced by ARCO in September 1989 in southern California. It was intended to reduce the emissions from older vehicles that were not equipped with catalytic converters. The emissions reductions were obtained by lowering the RVP to reduce evaporative emissions, and by reducing aromatics and olefins to reduce the ozone-forming potential and generate lower levels of toxins such as benzene in the exhaust gas [Chevron (1996)]. Oxygenates were added to the fuel to reduce CO and HC emissions.

Soon after ARCO introduced its RFG, other companies developed similar products for market distribution in high-pollution areas around the country. Because development of the new fuels was voluntary, there were wide variations in the composition of RFG from different manufacturers. The 1990 Clean Air Act Amendments included a regulation for the production of RFG to be marketed in non-ozone emission attainment regions in the country. Nine large U.S. cities fall into the ozone non-attainment region and consume roughly 30% of the fuel in the country [Porse (2002)]. The EPA established a "regulation-negotiation" (Reg-Neg) for RFG composition. Table 7.2 [Chevron (1996)] lists selected parameters for both California Phase II and federal RFG from the Reg-Neg agreement.

**TABLE 7.2
RFG REGULATION-NEGOTIATION AGREEMENT
[OWEN AND COLEY (1995)]**

Specification Item (Max.)	California	Federal
RVP	48 kPa (7.0 psi)	<48 kPa
T_{50} distillation temperature	95° (220°F)	Not specified
T_{90}	167°C (330°F)	Less than 1990 refinery average
Olefins (% volume)	10	Less than 1990 refinery average
Aromatics (% volume)	30	Not specified
Benzene (% volume)	1.20	<0.95 average, 1.3% max.
Oxygen (% mass)	2.7	1.5 min.–2.7 max.
Sulfur (ppm mass)	80	Less than 1990 refinery average

The oxygenates were mandated to be included in RFG primarily to obtain an emissions reduction from older, open-loop-control vehicles by effectively forcing a lean shift with the oxygenate. The Reg-Neg agreement does not specify the type of oxygenate that can be used. Methyl-tertiary butyl ether currently accounts for roughly 87% of all oxygenates in RFG [Porse (2002)]. Many areas throughout the country have reported groundwater contamination with MTBE, which led California to ban the use of MTBE in RFG. Other states have followed suit and have banned MTBE, and it is likely that a federal ban will be implemented in the future. The sulfur level in RFG gasoline is regulated because sulfur is known to poison the catalyst and reduce catalyst efficiency.

7.3 Fuel Effects on Hydrocarbon Emissions

Vehicle tailpipe emissions are a complex interaction among the engine design, emissions control system, ambient temperature, and fuel properties. The overall impact of the fuel on emissions is small, relative to the effects of vehicle design and ambient temperature [Heiken *et al.* (2001)]. To demonstrate the relative effects of vehicle design, temperature, and fuel on emissions, Bazzani *et al.* (2000) performed a matrix of tests using ten different vehicles, and six different fuels at ambient temperatures of 10°C and –5°C. The tests were performed on a modified Coordinating European Council (CEC) cold-weather driveability cycle. As such, the HC emissions changes are not directly applicable to the changes that would be expected over the U.S. Federal Test Procedure (FTP) cycle. Statistical analysis of the data indicated that the single largest effect on HC emissions was vehicle design. The difference between the highest-emitting vehicle and the lowest-emitting vehicle averaged over all fuels was 140%. By comparison, the effect of ambient temperature on HC emissions was 110% (averaged over all fuels and vehicles), and the effect of fuel type on HC emissions was roughly 20%. In other words, the effect of vehicle design is seven times larger than that of the fuel. This is good from the standpoint of vehicle and engine design because it suggests that significant reductions can be made in vehicle emissions without having to develop exotic fuels. From a clean air standpoint, the 20% reduction due to fuel changes is important because this change applies to all vehicles that are currently on the road.

To minimize vehicle emissions under real-world conditions, the engine control system must have a mechanism to account for changes in fuel quality during starting and warm-up. Early during cold start, the oxygen sensor for closed-loop control is not operational, and the engine is operated in open-loop mode.

Consequently, significant errors in fueling can exist when the fuel volatility is significantly different than that used to develop the open-loop engine calibration. For example, variations of up to three A/F ratios were measured in the exhaust gas early during a cold-start test when an ultra-low emissions vehicle (ULEV) certified engine was operated with a fuel with a low volatility (700°C DI) [Kishi et al. (1998)] when compared to a fuel with a higher volatility (616°C DI). Figure 7.2 shows the A/F ratio measured in the exhaust gas with a wide-range air/fuel (WRAF) sensor during the first 20 seconds of the FTP.

As shown in Figure 7.2, the A/F ratio was leaner with the high DI fuel, such that the resulting A/F ratio in the exhaust gas was lean of stoichiometric. At these lean A/F ratios, the combustion was unstable and the engine would misfire, leading to significant increases in HC emissions. A straightforward approach to dealing with variations in fuel volatility is to develop the engine calibration so that acceptable combustion occurs with the lowest expected volatility fuel.

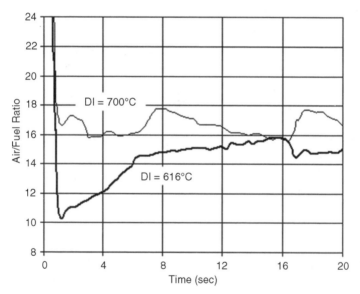

Figure 7.2 *Effect of fuel DI on measured exhaust A/F ratio during a cold start [Kishi et al. (1998)].*

The negative effect of this brute-force approach is that the engine is forced to run richer than required for fuels with higher volatility, which leads to an increase in cold-operation HC emissions. More innovative approaches have been used to accommodate a range of fuel volatility. For example, Honda developed a control system to account for fuel volatility by monitoring crankshaft speed fluctuations during cold start. When the crankshaft fluctuations become too large, the control system increases the fueling level to maintain acceptable combustion stability. Using this control system, the tailpipe HC emissions obtained using a 650°C DI fuel over the FTP cycle were decreased by nearly 50%, compared to emissions without the control system active [Kishi *et al.* (1998)]. Because the exhaust gas temperature is a strong function of in-cylinder A/F ratio, another approach that has been proposed to account for poor fuel volatility is to monitor exhaust gas temperature during cold start [Ferguson and Griffin (2000)]. When using this approach, the fueling level is modified to obtain a desired exhaust gas temperature.

Jorgenson *et al.* (1996b) developed a correlation between fuel DI and tailpipe HC emissions. Tailpipe HC emissions increase nonlinearly with the fuel DI, as defined in Eq. 7.1. While HC emissions are essentially independent of DI for levels below 600°C, for DI increases above this amount, the HC emissions increase more or less linearly. When the DI was increased from 650 to 732°C, there was a 35% increase in HC emissions. Recognizing the effect of volatility and DI on HC emissions, it has been proposed that the EPA should develop legislation to limit DI values to below 600°C [Heiken *et al.* (2001)]. The work of Jorgensen *et al.* was performed on 1990 to 1994 vintage vehicles that met Federal Tier I and Tier II emissions standards. As the HC emission standards are lowered to ULEV and SULEV levels, the effect of fuels on emissions is expected to increase. In FTP tests performed with a ULEV certified vehicle, the first cycle HC emissions were increased by 45% by changing the fuel from California Phase II gasoline with a DI index of 621°C to a 693°C DI fuel [Kirwan *et al.* (1999)]. The amount of HC emissions produced during the first cycle accounted for 95 to 98% of the total emissions. Similarly, Heiken *et al.* (2001) measured a 50% increase in FTP emissions when the fuel DI was changed over the same range of temperatures. Figure 7.3 shows the effect of fuel DI on FTP tailpipe HC emissions from a ULEV certified vehicle.

Fuel volatility has an effect not only on the HC emissions produced during cold start but also on HC emissions from transients during warmed-up operation. The engine control system must maintain tight A/F control of the engine to have the catalyst operate at high conversion efficiency. When the volatility of the fuel is changed, the dynamics of the port wall film layer are altered, and significant A/F ratio excursions during transient operation can result. Curtis *et al.* (1998) performed a set of fuel shutoff experiments during engine warm-up to investigate the effects of fuel volatility on fueling variations during engine warm-up

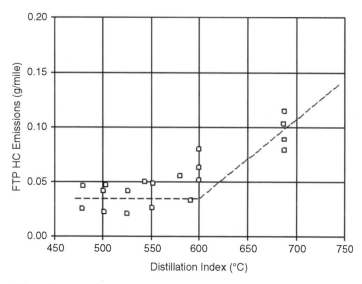

Figure 7.3 Effect of fuel DI on FTP tailpipe HC emissions (g/mile) [Heiken et al. (2001)].

and transients. Using a fuel with a 638°C DI, the resulting A/F variations had a first-order time constant of 10 seconds. By comparison, experiments performed with a 678°C DI fuel produced variations of up to four A/F ratios that had a time decay constant of 25 seconds. If the engine control and feedback system is not designed to be robust to these changes in fuel quality, the HC emissions will increase throughout the entire operating regime.

The effect of fuels on vehicle emissions was extensively investigated by the Auto/Oil Air Quality Improvement Research Program (AQIRP) in the early and mid-1990s. The AQIRP was a cooperative test program initiated by three domestic automotive manufacturers and fourteen petroleum companies. The goal of the program was to evaluate the effect of fuels on vehicle emissions from vehicles with different levels of emissions control technology. The program evaluated the effects of fuel distillation temperature, RVP, and olefin, aromatic, and sulfur concentrations in the fuel. The results that are relevant to the effects on cold-start HC emissions are reviewed briefly here. The interested reader can refer to the many SAE special publications generated from the AQIRP program for more details and for discussion of fuel effects on emissions.

The results of the AQIRP program indicated that the fuel parameters that have the largest effect on FTP tailpipe HC emissions are the T_{50} and T_{90} distillation temperatures and the fuel sulfur level [Leppard *et al.* (1995)]. Detailed analysis of the FTP modal data showed that the high emissions produced from fuels with high T_{50} and T_{90} distillation temperatures resulted from these fuels producing significantly higher engine-out emissions during cold start and the first cycle of the FTP. During the remainder of the FTP cycle, the engine-out emissions from these fuels showed only a small increase in emissions relative to fuels with lower T_{50} and T_{90} temperatures. Although the level of sulfur in the fuel had no effect on the engine-out HC emissions, increasing sulfur levels resulted in higher tailpipe HC emissions due to decreased catalytic converter efficiency. The reduced converter efficiency was especially pronounced during the first test cycle. The results indicated that the single largest fuel effect of California Phase II RFG on tailpipe HC emissions was the reduced sulfur level in the fuel.

Fuel additives can have a detrimental impact on both engine-out and tailpipe HC emissions. Methylcyclopentadienyl manganese tricarbonyl (MMT) is used as an octane improver in Canada and in certain regions of the United States. Studies have shown that its use can lead to numerous vehicle problems such as catalyst fouling, spark plug and engine deposits, and changes in O_2 sensor performance and onboard diagnostics (OBD) catalyst monitoring [Benson and Dana (2002)]. The Alliance of Automobile Manufacturers (AAM) conducted an extensive test program on certified low emissions vehicles (LEV) over an extended test of 75,000 to 100,000 miles. One fleet of vehicles was fueled with a clear base fuel, while the other vehicle fleet was fueled with the same base fuel with MMT addition. The two test fleets were identical and were composed of a range of vehicle makes and models. At the conclusion of the test, the HC, CO, nitrogen oxide (NOx), and carbon dioxide (CO_2) emissions were all statistically higher for the MMT-fueled vehicles. The FTP tests performed by the manufacturer on one pair of test vehicles showed a 38% increase in engine-out HC emissions from the MMT-fueled vehicle, whereas the tailpipe HC emissions had increased by 120% [McCabe *et al.* (2004)]. Systematic swapping of the parts between the two vehicles indicated that roughly half of the increase in tailpipe HC emissions was caused by deposits on the cylinder head and spark plugs, and the other half was due to degradation of catalyst performance.

7.4 Onboard Fuel Reformers

The poor volatility of gasoline at low temperatures represents a significant barrier to reducing cold-start HC emissions. All of the control strategies and improved fuel injection systems that have been developed are means of dealing with the problem of obtaining a combustible mixture in spite of the poor vaporization characteristics of gasoline at low temperature. As an alternative, if the fuel could be prevaporized before entering the engine, it is expected that significant reductions in engine-out HC emissions could be achieved. For example, Boyle *et al.* (1993) performed cold-start tests with gasoline using an experimental prevaporizer system. Depending on the engine and control system hardware, engine-out cold-start HC emissions were decreased by 45 to 65% with the prevaporizer system. Heated tip injectors have been developed to obtain better fuel preparation [Zimmermann *et al.* (1999)]. Engine tests performed with heated tip injectors have demonstrated a 20 to 35% reduction in engine-out HC emissions 20 seconds after cold start.

Fuel reformers are an alternative approach to prevaporizing systems that are being developed to produce lower cold-start emissions. Fuel reformers partially oxidize gasoline into reformate containing high concentrations of CO and H_2, with only small concentrations of HC. The reformate is used to cold start the engine rather than gasoline. After the engine is started and the catalyst reaches its light-off temperature, the engine is transitioned into operation on gasoline. An additional potential use of fuel reformers is to extend the dilution limit of the engine during part-load operation by adding hydrogen-rich reformate to the intake [Quader *et al.* (2003)]. The higher dilution limits for the engine result in a fuel economy improvement due to reduced pumping losses. Recently, there has been increased interest in using fuel reformers as a means to produce hydrogen for fuel cell vehicles. In the absence of a hydrogen infrastructure, fuel reformers represent a potential near-term solution to generate hydrogen from gasoline. The side benefit of this to SI engine development is that many of the resulting technologies from the fuel cell development work can be applied directly to gasoline-fueled SI engines.

Figure 7.4 shows a representative layout of the different steps that take place within a fuel reformer unit. Fuel reformers reform a liquid HC fuel using one of the following three major processes:

1. Steam reforming
2. Partial oxidation (POx) reforming
3. Autothermal reforming

Effects of Fuel Properties and Fuel Reforming

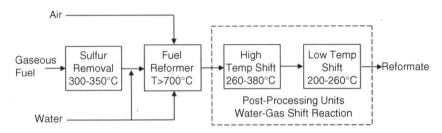

Figure 7.4 *Typical fuel reformer process diagram and temperature ranges* [Fuel Cell Handbook *(2000)*].

Depending on the type of fuel reforming process used in the reformer unit, there will be an external supply of air, water, or a combination of both to the reformer reactor.

To speed the reforming process, all practical onboard fuel reformers must use a catalyst. The catalyst is sensitive to sulfur contamination from the fuel. To avoid reduced performance over time, the reformer unit must incorporate a fuel cleaning process before the fuel enters the reformer. The gas exiting the reformer (commonly referred to as synthesis gas) has a high concentration of CO and H_2. A major difference between the different reforming processes is in the relative concentrations of CO and H_2 in the synthesis gas. For fuel cell applications, the synthesis gas must be further processed to convert the CO into CO_2 via the water-gas shift reaction

$$CO + H_2O \rightarrow CO_2 + H_2 \qquad (7.4)$$

This reaction is slightly exothermic, producing an energy of 1.46 kJ/gm CO. In practice, the conversion generally is performed in two stages: a high-temperature stage operating near 400°C, followed by a low-temperature reactor near 300°C [Docter *et al.* (1999)]. The water-gas shift reaction is important for fuel cell applications because CO will poison the fuel cell catalyst. An advantage of fuel reforming for SI engines is that they are not sensitive to CO within the reformate gas. Thus, fuel reformer units for SI engine applications can be potentially simplified relative to those developed for fuel cells.

Fuel reforming of both gaseous and liquid HC fuels has the potential of generating solid carbon within the reactor. Carbon formation is sensitive to the temperature and composition of the reactants entering the reformer. When the temperature of liquid HC fuels is increased in the absence of oxygen, the fuels can undergo cracking (breaking of C–C bonds) and dehydration reactions that produce solid carbon. Coking is a significant problem that must be avoided because it will foul the reformer catalyst. A major challenge of fuel reformer development is making the reformers robust to variations in operating conditions and fuel composition such that coking will always be avoided.

The basic principles of the different reformer types will be reviewed in the following sections. The experimental work that has been performed to develop fuel reformers to reduce cold-start HC emissions will be discussed. In addition, the use of fuel reformers to assist in the cold starting of alcohol-fueled vehicles also will be reviewed.

7.4.1 Steam Reforming

Steam reforming of natural gas (mainly composed of methane) is one of the standard processes used in the chemical engineering field. In steam reforming, the fuel is reacted with superheated steam according to the following general reaction of

$$CH_y + \beta H_2O \rightarrow \text{products} \quad (7.5)$$

where y is the hydrogen-to-carbon (H/C) ratio of the fuel, and β is the molar ratio of steam to carbon (S/C). The major products of the reaction are CO and H_2, with only trace amounts of CO_2 and H_2O. The chemically correct S/C ratio to convert all of the fuel into H_2 and CO is 1. Using a typical gasoline H/C ratio of 1.85 and an S/C ratio of 1, the resulting reformate is composed of 34% CO and 66% H_2. The heating value of the reformate is 23.5 kJ/gm, which is roughly half that of gasoline on a mass basis. Steam reforming typically is carried out using nickel-based catalysts, although cobalt and other noble metals also can be used [Docter et al. (1999)].

The S/C ratio is an important operating parameter for the steam reforming process. Thermodynamic equilibrium considerations can be used to provide an approximation for carbon formation boundaries [*Fuel Cell Handbook* (2000)]. At an S/C ratio near 1, the reaction temperature must be maintained higher than 800 to 850°C to avoid coke formation [Seo et al. (2002)]. When the S/C ratio is increased to 1.5, the critical reactor temperature for coking is decreased to the 650 to 700°C range. For most practical steam reforming processes, the

S/C ratio is larger than 2 because the excess steam promotes completion of the reaction and suppresses coke formation. The ability to use additional steam to lower the reactor temperature is good in terms of reformer catalyst durability, but the energy required to produce the steam represents yet another energy loss to the vehicle. When steam reforming a typical gasoline at an S/C ratio of 1, the mass flow rate of steam to the reformer is roughly 1.3 times larger than the fuel flow rate. Using a more typical S/C ratio of 2, the mass flow rate of steam is increased to 2.6 times the fuel flow rate. Thus, if steam reforming is carried out onboard, a significant source of water must be supplied to the reformer. Storing water onboard the vehicle is difficult, particularly in cold northern climates during winter months.

The reaction in Eq. 7.5 is highly endothermic, and an external heat source must be supplied to make the reaction take place. The energy required to steam reform iso-octane (H/C ratio of 2.25) at an S/C ratio of 1 is 11.5 kJ/gm fuel [Wieland *et al.* (2001)]. In steam reformers, the water-gas shift reaction typically takes place within the reformer without requiring post processing of the synthesis gas. However, even with the exothermic energy release from the water-gas shift reaction, the net energy requirement for steam reforming iso-octane is near 11 kJ/gm fuel. Considering that the heating value of a typical gasoline is ~44 kJ/gm, this means that the energy equivalent of nearly 0.25 grams of fuel is consumed per gram of fuel processed in the reformer. This potentially represents a 25% increase in fuel consumption during cold start. The engine starting strategy must take advantage of the increased dilution tolerance of the engine during cold start to minimize the impact of the increase in cold-start fuel consumption on FTP fuel economy.

7.4.2 *Partial Oxidation Reforming*

In partial oxidation reforming (POx), the fuel is reacted with air at a fuel/air equivalence ratio (Φ) less than stoichiometric. The overall reaction is written as

$$CH_y + 4.76\,\alpha(0.21\,O_2 + 0.79\,N_2) \rightarrow \text{products} \qquad (7.6)$$

where the molar ratio of air to fuel is equal to 4.76 α. The chemically correct value of α for converting all of the fuel into CO and H_2 is 0.5. Using an H/C ratio of 1.85 and $\alpha = 0.5$, the reformate is composed of 26% CO, 24% H_2, and 50% N_2. Significantly less hydrogen is generated from POx reforming than from steam reforming. The large amounts of nitrogen present in the reformate mixture from a POx reformer significantly reduce the heating value of the reformate on a total mass basis. The heating value of the CO and H_2 mixture is 17 kJ/gm. However, when the nitrogen in the reformate is included in the calculation, the heating value is decreased to a mere 6.2 kJ/gm. Note that this is really the correct heating value to use because the nitrogen in the reformate mixture cannot be removed. The reduced heating value of the reformate implies that there must be an increased fuel mass flow to the engine to obtain approximately the same total energy flow to the engine.

The A/F ratio at which the POx reactor is operated has a large effect on the fuel conversion efficiency and coking tendency of the reformer. Equilibrium calculations show that solid carbon formation reaches a maximum near an A/F equivalence ratio $\Phi = 7.3$ and is decreased to near zero at $\Phi = 2.4$ [Seo *et al.* (2002)]. At $\Phi = 2.4$, the reformate contains the highest CO levels. Unlike steam reforming, the carbon formation limits are relatively insensitive to the temperature of the reactants entering the reformer.

The partial oxidation reaction shown in Eq. 7.6 is exothermic. For iso-octane reformed in a POx reactor at $\Phi = 3.1$ (corresponding to $\alpha = 0.5$), the energy released is 5.5 kJ/gm fuel. Temperatures in excess of 1000°C are needed to get the reaction to proceed, so a catalyst is commonly used to speed the reaction rate [Ahmed *et al.* (1998)]. Typical catalysts that are used in POx reformers are palladium and ruthenium. However, to avoid damaging the catalyst, the reaction temperature must be maintained below roughly 800 to 850°C. A heat exchanger system must be used to vaporize the fuel and preheat the reactants to 200 to 300°C before entering the reactor. The composition of gasoline can have a large effect on the performance of POx reformers [Kopasz *et al.* (2002)]. Experiments performed with fuels with high aromatic and naphthenic content showed a lower conversion efficiency and were sensitive to the reactor temperature. In addition, fuels with high aromatic content are more susceptible to carbon formation.

An advantage of POx reforming that makes it attractive for automotive applications is that because the reaction is exothermic, it does not require an external heat source, and energy release from combustion of the fuel can be used to light off the reformer catalyst relatively quickly. The thermal efficiency of the fuel processor is key to all fuel reformer systems. From an energy efficiency basis, the most efficient way to reform gasoline is by POx reforming [Seo *et al.* (2002)].

7.4.3 Autothermal Reforming

A major disadvantage of steam reforming is that an external energy source is required for the reformer, whereas with POx reforming, a significant amount of waste heat must be dealt with. Autothermal reforming combines these two processes in such a way that the overall reaction can be carried out adiabatically. To accomplish this, the fuel is reformed in the presence of both water and air. The overall general reaction may be written as

$$CH_y + 4.76\,\alpha(0.21\,O_2 + 0.79\,N_2) + \beta H_2O \rightarrow \text{products} \quad (7.7)$$

where the molar A/F ratio is 4.76α, and β is the S/C ratio. It is possible to find combinations of the A/F and S/C ratios such that the overall reaction is energy neutral. Autothermal reactors can be operated at lower temperatures than a partial oxidation reformer. In terms of hydrogen generation, steam reforming produces the highest concentration of H_2 in the reformate (near 66%), and partial oxidation produces the lowest (24%). Because autothermal reforming is a combination of these two processes, the reformate composition from autothermal reforming lies between these two extremes. Using gasoline and an S/C ratio of 0.5, the reformate is composed of 30% CO, 42% H_2, and 28% N_2. The heating value of the reformate is equal to 11 kJ/gm based on the total mass of the mixture (which again contains a significant amount of nitrogen).

Autothermal reforming is not as efficient as POx reforming in terms of overall energy efficiency [Seo *et al.* (2002)]. The main reason for this is the large energy penalty that results from the need to provide the reformer with a source of high-temperature steam. The energy required to vaporize water and heat it to 200°C is roughly 2.5 kJ/gm water. Using an S/C ratio of 1, this energy requirement represents an approximate 7% increase in fuel flow. Thus, while the reforming process may be adiabatic, the energy requirement of having to provide a source of steam offsets these benefits. A variation of autothermal reforming is the recovery of the sensible energy and water vapor from the exhaust gas to reform the fuel. However, during a cold start, the exhaust gas temperature initially is low, and all of the enthalpy within the exhaust must be used to rapidly heat the catalyst. Although autothermal reforming using the exhaust gas is possible for fully warmed-up operation, it is unlikely that this approach will work for an engine cold start.

7.4.4 Cold-Start Performance Improvements

The benefits of cold-starting an engine with reformate have been extensively evaluated by Kirwan *et al.* using both synthetic reformate [Kirwan *et al.* (1999)] and reformate generated from a prototype POx fuel reformer [Kirwan *et al.* (2002)]. All of the experiments reported to date have been performed on a single-cylinder engine under a simulated cold-start transient. The control strategy developed was to start the engine on POx gas during cranking and the initial first idle, and then after (engine) catalyst light-off, the engine was transitioned to operate on gasoline. Figure 7.5 shows the HC emissions following a simulated cold start using the prototype POx fuel reformer. For these tests, the engine was operated at both stoichiometric and lean equivalence ratios. Due to the wider flammability limits of the reformer gas, the engine was able to be cold started at a significantly leaner fueling level than it could with gasoline. Engine-out HC emissions were reduced by 75% relative to those obtained with starting on gasoline.

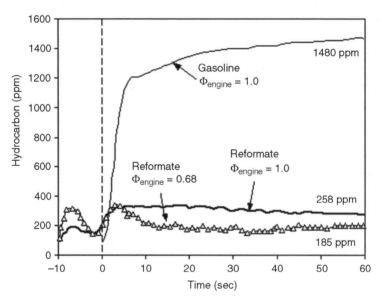

Figure 7.5 Comparison between HC emissions produced during simulated cold start using gasoline and POx gas [Kirwan et al. (2002)].

While starting the engine with POx gas has been shown to result in significant reductions in HC emissions, the challenge now becomes one of "cold starting" the fuel reformer. The reformer must be brought up to its operating temperature before it will efficiently reform the fuel. This is one of the most significant development issues that must be addressed for fuel reformers because there can be significant HC breakthrough before it is at its optimal operating temperature. To rapidly heat the reformer catalyst, Kirwan *et al.* (2002) developed a strategy where a stoichiometric A/F mixture was initially supplied to the reformer to preheat the catalyst. After the reformer temperature was increased, the equivalence ratio was adjusted to fuel-rich operation ($\Phi = 2.5$) for reformer gas production. As expected, the performance of the reformer is very sensitive to the preheat time. With the initial prototype reformer, a preheating time of 10 seconds was needed to obtain good performance and significant reductions of HC emissions. An important development issue is to shorten the preheat time requirements for the reformer. Depending on the operation of the reformer, the reformer gas contained 26 to 23% H_2 and roughly an equal concentration of CO.

The transition from operation with POx gas to gasoline must be handled correctly to avoid combustion instability and increased HC emissions. One of the issues with the transition is the effect of HC fuels on the flame speeds and combustion characteristics of the reformer gas, and in particular the effects on the lean combustion limit. Computational results on the effects of n-butane to reformer gas mixtures showed that the lean flammability limit is dramatically reduced by the addition of small amounts of the HC fuel [Yang *et al.* (2002)]. The reason for this is that the radicals produced during the reaction preferentially react with the HCs that are present instead of with the H_2 and CO in the reformer gas.

One advantage of operating with the hydrogen-rich POx gas is that the engine can be cold started at significantly leaner equivalence ratios than with gasoline. Kirwan *et al.* (1999) determined a lean misfire limit for a simulated cold start on a ULEV certified engine of $\Phi = 3.3$ using synthetic POx gas. The computational results indicate that the reformer gas mixture must have a very low HC concentration to obtain the maximum lean-limit extension during cold start.

7.4.5 *Improved Catalyst Light-Off*

The primary advantage of starting the engine on reformer gas is the potential to obtain lower engine-out HC emissions. Adding reformate to the exhaust gas during cold start offers the potential for obtaining faster catalyst light-off. Hydrogen will react with air over a standard three-way catalyst at room temperature [Kirwan et al. (1999)]. In addition, it is known that CO will react at lower temperatures than HCs over a catalyst when sufficient O_2 is available. Thus, it is expected that adding reformate to the exhaust will increase the rate of catalyst heating. An additional benefit is obtained if a POx reformer is used to produce the reformate because the exotherm from the reaction can be used to increase the temperature of the gas going into the converter. Kirwan et al. (1999) evaluated the effects of synthetic POx gas on catalyst heating using a single-cylinder engine. Tests were performed with fueling the engine during cold start with POx gas as well as supplying POx gas to the catalyst. In a simulated cold-start test, the time to obtain 50% CO conversion in the catalyst was reduced from 50 seconds with gasoline to 22 seconds with POx gas. When POx gas was supplied to the catalyst, significant reactions began in the catalyst at a temperature of approximately 125°C.

The effect of hydrogen addition to the catalyst was evaluated on a production vehicle during an FTP by Heimrich and Andrews (2000). In these tests, the hydrogen was generated using an onboard electrolyzer to generate hydrogen from water. A secondary air pump also was used in conjunction with the hydrogen addition, in order to supply additional O_2 with which the hydrogen can react over the catalyst. The system was tested on a 1996 Toyota Camry that was LEV certified. The hydrogen-heated catalyst system was able to reduce the HC emissions by 50% and CO emissions by 40% over the FTP cycle. The bulk of the emissions reduction was obtained from reduced cold-start emissions due to faster catalyst light-off. For the baseline vehicle tests, the catalyst reached a temperature of 300°C after 40 seconds. By comparison, with the hydrogen addition, the catalyst reached a temperature of 300°C after only 10 seconds of operation. Similar reductions should be obtained if the hydrogen was produced from a fuel reformer as opposed to the electrolysis of water.

7.4.6 Cold-Starting Alcohol-Fueled Vehicles

Another application of fuel reformers is to aid in the starting of alcohol-fueled vehicles at low ambient temperatures. Alcohol-fueled vehicles are difficult to start at low temperatures due to the low volatility of alcohol fuels. At atmospheric pressure, the boiling temperature of methanol is 65°C, and the boiling temperature of ethanol is 78°C. The latent heats of vaporization are 1.17 and 0.93 kJ/gm, respectively, which are a factor of 8 to 10 times higher than that of gasoline. The minimum starting temperature without external aids for neat methanol has been reported to be between 0 and 16°C, and for ethanol, it is approximately 43°C [Owen and Coley (1995)]. One potential means to address the startability problem of alcohol fuels is to blend a more volatile component into the fuel that will vaporize at low temperatures and can be used to start the engine. Gasoline generally is added to ethanol in small levels to aid engine starting. For example, E95 is a blend of 95% ethanol and 5% gasoline that is commonly available. As an alternative to using fuel blending to assist in cold starts, fuel reformers may be used to gasify the alcohol into reformate that can be used to start the vehicle.

The effect of a POx reformer on cold starting an E95-fueled vehicle was investigated by Drobot and Loftus (1998). A prototype onboard fuel reformer was developed for a demonstration vehicle. Similar to the strategy used by Kirwan *et al.* (2002), the POx reformer initially was fueled at near-stoichiometry to obtain fast light-off of the fuel reformer. Reliable light-off of the reformer was demonstrated down to temperatures of –20°C. The preheat delay time for the reformer was roughly 10 seconds. The system demonstrated that reliable vehicle starts could be achieved down to ambient temperatures of –20°C. Hydrocarbon emissions during a cold start at 50°C were reduced by 80% relative to starting without the reformer.

7.5 Onboard Fuel Distillation

One of the distinguishing characteristics of gasoline is that it vaporizes over a wide range of temperatures. A typical gasoline vaporizes over a range of temperatures from a low of 30 to 40°C to a high of 190°C. The model fuel composition shown in Table 7.2 for California Phase II RFG indicates that roughly 15 to 19% of the compounds in the fuel vaporize at temperatures below 30°C. These light compounds are butanes and pentanes, and their content in the fuel changes seasonally as the RVP is varied by the manufacturers. If these low-boiling-point compounds could be separated from the gasoline, it would be possible to start the engine at low ambient temperatures with these compounds and avoid having to over-fuel the engine to obtain a combustible mixture. Two such systems that have been developed for changing the vaporization characteristics of a fuel will be reviewed in this section.

Oakley *et al.* (2001) developed an onboard fractioning system to separate gasoline into two separate streams. The main objective of this work was to develop a high-octane fuel that could be used to obtain an increase in full power performance from the engine. The higher-boiling-point compounds in gasoline are largely aromatic species (e.g., toluene, alkylated aromatics) that have higher knock resistance than the lower-boiling-point straight-chain paraffinic species. The performance of the engine could be improved if the engine could be operated on these less volatile, high-octane species at high loads. This would provide the opportunity for either operating with more spark advance or potentially with higher compression ratios.

The system worked by flowing fuel into a temperature-controlled vaporization chamber. This system separated the fuel into two streams, consisting of a vapor fraction with lower-boiling-point compounds and a heavy fraction with relatively higher-boiling-point compounds. A copper coil immersed in the fuel was used to heat the liquid in the vaporization chamber. The system was designed to work at steady-state conditions to produce a steady flow of distilled fuel. Producing a stable vapor stream from the system at transient conditions was difficult, and the system temperatures had to be adjusted to maintain a stable flow. For cold-start operation, it will be necessary to develop a storage system to store the distillate onboard so that it is readily available for a start.

Table 7.3 shows the compositions of the two fuel streams taken from the prototype system. The temperature of the vaporization chamber was maintained at 65°C for these results. The light fraction is composed primarily of paraffinic compounds, whereas the heavy fraction is higher in aromatics. Vaporization of a fuel mixture with a large number of components such as gasoline is not as simple as it may first appear. The boiling point of a single component at

TABLE 7.3
FUEL COMPOSITION FROM ONBOARD EVAPORATION SYSTEM [OAKLEY ET AL. (2001)]

		Base Fuel	Light Fraction	Heavy Fraction
Percent of total fuel		100	48	52
Paraffins	(% volume)	58.3	72.1	42.6
Olefins	(% volume)	7.6	12.4	9.9
Aromatics	(% volume)	34.1	15.5	47.5
RON		94.7	93.5	98
MON		84.5	85	86.5
$\dfrac{R + M}{2}$		89.6	89.3	92.3

atmospheric pressure is not the same temperature at which it vaporizes in the mixture because of interactions with the other components in the fuel mixture. Thus, while the light fraction of the fuel contains a high percentage of lower-boiling-point compounds, it also has aromatic species whose single boiling points are higher than that of the vaporization chamber temperature. The heavy fraction of the fuel derived from batch vaporization also is a complex fuel with a wide range of compounds.

The octane rating $\left(\dfrac{R + M}{2}\right)$ of the heavy fuel fraction was increased from the base gasoline by almost three octane numbers. Dynamometer tests indicated a subsequent improvement in peak power from the engine due to the more advanced spark timings that could be used with the higher octane number fuel. Although no cold-start tests were performed, it seems clear that using the lighter fractions of the fuel will result in the ability to reduce the amount to over-fueling required to start the engine, thus reducing the cold-start HC emissions.

An onboard distillation system (OBDS) was developed by Ku *et al.* (2000) to aid in the cold starting of an E85-fueled vehicle. The distillation unit was designed to address both the cold-starting issues on the FTP as well as starting at low ambient temperatures and improving cold-weather driveability. Unlike the system developed by Oakley *et al.* using a vaporization chamber that was designed for steady-state operation, the OBDS unit was developed specifically for improving cold-starting engine behavior, and this system included an additional fuel tank onboard the vehicle to store the high-volatility distillate fuel for starting the engine.

The distillation unit consisted of a heat exchanger, distillation column, vapor condenser, and cold-start fuel tank in which to hold the distilled fuel. A heat exchanger was used to heat the fuel and vaporize the more volatile compounds out of the mixture. Engine coolant was used to heat the fuel to a temperature of 85 to 90°C. The distillation column was a vertical tube packed with steel wool to obtain a large surface area for fractionization. The vapor temperature was controlled with a cooling fan. The heated fuel entered the middle section of the column and flowed down to the bottom of the column. The most important requirements for the distillation system were to obtain intimate contact between the vapor and the liquid phases, and this typically dictated the column length [Ku *et al.* (2000)]. The important operating parameters were the temperature gradient along the column and the volatility differences among the different components in the mixture. The fuel flow to the OBDS unit was the return fuel from the engine. A sensor was used to detect the fuel level in the cold-start tank, and a control unit activated the distillation process when the fuel level in the cold-start tank dropped below a specified level. The unit shut off when the cold-start tank was full. For the system to utilize only one fuel injection system, the controller must be able to purge the fuel system of the base fuel and precharge the fuel system with the distilled fuel in preparation for a cold start.

Although the initial OBDS was designed to be used with alcohol fuels, the same approach will work equally well with gasoline. A third-generation fuel distillation system has been developed and tested using a fully blended gasoline [Ashford *et al.* (2003)]. Figure 7.6 shows ASTM distillation curves for both the baseline gasoline and the OBDS fuel. Also shown in the figure are the boiling-point temperatures for iso-pentane and n-pentane. As expected, the OBDS fuel is composed primarily of C_5 HCs.

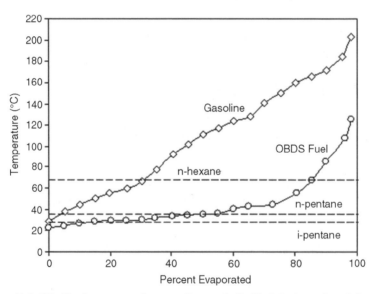

Figure 7.6 Distillation curves for gasoline and distilled fuel produced from an OBDS system [Ashford et al. (2003)].

The cold-start calibration of the vehicle was modified to take advantage of the improved distillation characteristics of the OBDS fuel. In particular, the fueling levels during the crank and the initial speed flare were reduced so that the engine operated closer to stoichiometric during cold start, then warmed up very lean ($\Phi \sim 0.7$) during the transition to closed-loop operation. A full FTP test was performed to determine the potential of the system to reduce HC emissions over the FTP test cycle. The calibration was modified so that the catalyst light-off times were reduced from 40 seconds to 20 to 30 seconds. Tailpipe HC emissions over the FTP were reduced by 40 to 45% using the OBDS system, and CO emissions were reduced by roughly 75%. A breakdown of the emissions over the FTP showed that, as expected, the same percent reductions in engine-out emissions were measured from cold start and the first cycle. This is the same overall level of change in HC emissions as obtained by reducing the fuel DI by 55°C [Kirwan *et al.* (1999)]. It also was found that the distillate could withstand more spark retard than gasoline. However, the calibration reported for stoichiometric crank, lean warm-up, and retarded spark did not show much improvement over the lean-fueling strategy alone [Ashford *et al.* (2003)]. In subsequent work [Ashford *et al.* (2005)], they improved the "lean, retarded" calibration and achieved a 55% reduction in light-off time, an 81% decrease in reactivity adjusted nonmethane organic gas (NMOG), and an 80% decrease in CO over the FTP [Matthews (2003)].

7.6 Summary

Onboard fuel reformers and fuel distillation systems are emerging technologies that can be used to obtain reduced cold-start HC emissions. At the current SULEV and PZEV emissions levels, these technologies have not been used in production vehicles because more cost-effective means of meeting the emissions standards have been developed. However, as the emissions standards are further reduced and the existing engine and vehicle technology is pushed to its limit, these systems may become cost-effective means of meeting future emissions regulations.

From an overall vehicle efficiency perspective, POx fuel reformers are the most attractive means to reform fuel. The main advantage of POx reforming over steam reforming and autothermal reforming is that it does not require a source of high-temperature steam for the reforming process. Although the development work that has been reported using POx reforming has shown promise in reducing cold-start HC emissions and obtaining fast catalyst light-off, this initial work has been performed on single-cylinder engines and simulated cold starts. Additional development work must be performed on actual vehicles and engine systems to demonstrate the real-world benefits of this technology.

A key element to successfully develop a fuel reformer system for vehicle cold starts is finding a means to rapidly light off the fuel reformer catalyst. It is not acceptable to increase engine start times by having to wait for the fuel reformer catalyst to light off. One approach that has been used to obtain fast light-off is to initially fuel the reformer stoichiometrically and use the energy release from combustion to rapidly heat the catalyst. Another approach is to use an electrically heated catalyst. Another potential means to solve this problem is to incorporate an onboard storage system for the reformer gas that can be used to cold start the vehicle while the reformer lights off. An important issue that must be addressed with fuel reformers is obtaining a smooth transition to operation on gasoline after the engine catalyst reaches its operating temperature. This transition must be done so that the engine does not misfire and is not perceived by the driver.

Fuel distillation systems have demonstrated impressive emissions benefits over both the FTP cycle and during real-world driving conditions. More development work must be done to reduce the package size of the distillation unit and to integrate it better with the vehicle. The disadvantage of these systems at their current state of development is that they are used only to improve performance during cold start. Automobile manufacturers are reluctant to add hardware and cost to a vehicle if the only benefit is obtained during the 20 seconds after a cold start. Thus, work should be performed to determine if fuel distillation can be used to improve the dilution tolerance of the engine at part-load conditions so that a fuel economy improvement can be gained. With increasing pressure on the automotive industry to improve vehicle fuel economy, cylinder deactivation and idle shut-off on hybrid powertrains are becoming more commonplace. For these applications, it is important to minimize the amount of fuel used to restart the engine to maintain the fuel economy benefit of shutting off the engine. Fuel distillation systems should be effective in reducing the amount of over-fueling that is required to obtain a smooth engine start.

7.7 References

1. Ahmed, S., Krumpelt, R., Kumar, S.H., Lee, D., Carter, J.D., Wilkenhoener, R., and Marshall, C. (1998), "Catalytic Partial Oxidation Reforming of Hydrocarbon Fuels," 1998 Fuel Cell Seminar, November 16–19, Palm Springs, CA.

2. Aquino, C., Plensdorf, W., Lavoie, G., and Curtis, E. (1998), "The Occurrence of Flash Boiling in a Port Injected Gasoline Engine," SAE Paper No. 982522, Society of Automotive Engineers, Warrendale, PA.

3. Ashford, M., Matthews, R., Hall, M., Kiehne, T., Dai, W., Curtis, E., and Davis, G. (2003), "An On-Board Distillation System to Reduce Cold-Start Hydrocarbon Emissions," SAE Paper No. 2003-01-3239, Society of Automotive Engineers, Warrendale, PA.

4. Ashford, M.D. and Matthews, R.D. (2005), "Further Development of an On-Board Distillation System for Generating a Highly Volatile Cold-Start Fuel," SAE Paper No. 2005-01-0233, Society of Automotive Engineers, Warrendale, PA.

5. Barker, D.A., Gibbs, L.M., and Steinke, E.D. (1988), "The Development and Proposed Implementation of the ASTM Driveability Index for Motor Gasoline," SAE Paper No. 881668, Society of Automotive Engineers, Warrendale, PA.

6. Bazzani, R., Brown, M., Kuck, K., Kwon, Y., and Schmidt, M. (2000), "The Effects of Driveability on Emissions in European Gasoline Vehicles," SAE Paper No. 2000-01-1884, Society of Automotive Engineers, Warrendale, PA.

7. Benson, J.D. and Dana, G.J. (2002), "The Impact of MMT Gasoline Additive on Exhaust Emissions and Fuel Economy of Low Emissions Vehicles (LEV)," SAE Paper No. 2002-01-2894, Society of Automotive Engineers, Warrendale, PA.

8. Boyle, R.J., Boam, D.J., and Finlay, I.C. (1993), "Cold-Start Performance of an Automotive Engine Using Prevaporized Gasoline," SAE Paper No. 930710, Society of Automotive Engineers, Warrendale, PA.

9. Burns, V.R., Rapp, L.A., Koehl, W.J., Benson, J.D., Hochhauser, A.M., Knepper, J.C., Leppard, W.R., Painter, L.J., Reuter, R.M., Rippon, B., and Rutherford, J.A. (1995), "Gasoline Reformulation and Vehicle Technology Effects on Emissions—Auto/Oil Air Quality Improvement Research Program," SAE Paper No. 952509, Society of Automotive Engineers, Warrendale, PA.

10. Chen, K.C., De Witte, K., and Cheng, W.K. (1994), "A Species-Based Multi-Component Volatility Model for Gasoline," SAE Paper No. 941877, Society of Automotive Engineers, Warrendale, PA.

11. Cheng, W.K. and Santoso, H. (2002), "Mixture Preparation and Hydrocarbon Emissions Behaviors in the First Cycle of SI Engine Cranking,"

SAE Paper No. 2002-01-2805, Society of Automotive Engineers, Warrendale, PA.

12. Chevron Products Company (1996), Motor Gasoline Technical Review FTR-1, San Ramon, CA; www.chevron.com/products/prodserv/Fuels.

13. Curtis, E.C., Russ, S., Aquino, C., Lavoie, G., and Trigui, N. (1998), "The Effects of Injector Targeting and Fuel Volatility on Fuel Dynamics in a PFI Engine During Warm-Up: Part II—Modeling Results," SAE Paper No. 982519, Society of Automotive Engineers, Warrendale, PA.

14. Docter, A. and Lamm, A. (1999), "Gasoline Fuel Cell Systems," *Journal of Power Sources*, Vol. 84, No. 2, pp. 194–200.

15. Drobot, K. and Loftus, P.J. (1998), "Using On-Board Fuel Reforming by Partial Oxidation to Improve SI Engine Cold-Start Performance and Emissions," SAE Paper No. 980939, Society of Automotive Engineers, Warrendale, PA.

16. Ferguson, T.J. and Griffin, J.R. (2000), "High DI Fuel Detection via Exhaust Gas Temperature Measurement for ULEV," SAE Paper No. 2000-01-0893, Society of Automotive Engineers, Warrendale, PA.

17. *Fuel Cell Handbook, 5th Edition* (2000), EG&G Services, U.S. Department of Energy, Morgantown, WV.

18. Furey, R.L. (1985), "Volatility Characteristics of Gasoline-Alcohol and Gasoline-Ether Fuel Blends," SAE Paper No. 852116, Society of Automotive Engineers, Warrendale, PA.

19. Gibbs, L.M. (1993), "How Gasoline Has Changed," SAE Paper No. 932828, Society of Automotive Engineers, Warrendale, PA.

20. Gibbs, L.M. (1996), "How Gasoline Has Changed II—The Impact of Air Pollution Regulations," SAE Paper No. 961950, Society of Automotive Engineers, Warrendale, PA.

21. Heiken, J.G., Darlington, T.L., Kahlbaum, D., and Herwick, G.A. (2001), "Vehicle Exhaust Emissions Benefit from a Regulatory Cap in Gasoline Distillation Index," SAE Paper No. 2001-01-1963, Society of Automotive Engineers, Warrendale, PA.

22. Heimrich, M.J. and Andrews, C.C. (2000), "On-Board Hydrogen Generation for Rapid Catalyst Light-Off," SAE Paper No. 2000-01-1841, Society of Automotive Engineers, Warrendale, PA.

23. Huang, Y., Sung, C.J., and Eng, J.A. (2002), "Effects of n-Butane Addition on Reformer Gas Combustion: Implications for the Potential of Using Reformer Gas for an Engine Cold-Start," 28th International Symposium on Combustion, The Combustion Institute, pp. 1320–1331.

24. Jorgenson, S.W., Musser, G.S., Uihlein, J.P., and Evans, B. (1996a), "A New CRC Cold-Start and Warm-Up Driveability Test and Associated Demerit Weighting Procedure for MPFI Vehicles," SAE Paper No. 962024, Society of Automotive Engineers, Warrendale, PA.

25. Jorgenson, S.W. and Benson, J.D. (1996b), "A Correlation Between Tailpipe Hydrocarbon Emissions and Driveability," SAE Paper No. 962023, Society of Automotive Engineers, Warrendale, PA.

26. Jorgensen, S.W., Eng, K.D., Evans, B., McNally, M.J., Whelan, D., Ziegel, E.R., and Musser, G.S. (1999), "Evaluation of New Volatility Indices for Modern Fuels," SAE Paper No. 1999-01-1549, Society of Automotive Engineers, Warrendale, PA.

27. Kaiser, E.W., Siegl, W.O., Lawson, G.P., Connolly, F.T., Cramer, C.F., Dobbins, K.L., Roth, P.W., and Smokovitz, M. (1996), "Effect of Fuel Preparation on Cold-Start Hydrocarbon Emissions from a Spark Ignition Engine," SAE Paper No. 961957, Society of Automotive Engineers, Warrendale, PA.

28. Kidokoro, T., Hoshi, K., Hiraku, K., Satoya, K., Watanabe, T., Fujiwara, T., and Suzuki, H. (2003), "Development of PZEV Exhaust Emission Control System," SAE Paper No. 2003-01-0817, Society of Automotive Engineers, Warrendale, PA.

29. Kirwan, J.E., Quader, A.A., and Grieve, J. (1999), "Advanced Engine Management Using On-Board Gasoline Partial Oxidation Reforming for Meeting Super-ULEV (SULEV) Emissions Standards," SAE Paper No. 1999-01-2927, Society of Automotive Engineers, Warrendale, PA.

30. Kirwan, J.E., Quader, A.A., and Grieve, M.J. (2002), "Fast Start-Up On-Board Gasoline Reformer for Near-Zero Emissions in Spark-Ignition Engines," SAE Paper No. 2002-01-1011, Society of Automotive Engineers, Warrendale, PA.

31. Kishi, N., Kikuchi, S., Seki, Y., Kato, A., and Fujimori, K. (1998), "Development of the High-Performance L4 Engine ULEV System," SAE Paper No. 980415, Society of Automotive Engineers, Warrendale, PA.

32. Kopasz, J.P., Ahmed, S., and Devlin, P. (2001), "Challenges in Reforming Gasoline: All Components Are Not Created Equal," SAE Paper No. 2001-01-1915, Society of Automotive Engineers, Warrendale, PA.

33. Kopasz, J.P., Miller, L.E., Ahmed, S., Devlin, P., and Pacheco, M. (2002), "Reforming Petroleum-Based Fuels for Fuel Cell Vehicles: Composition–Performance Relationships," SAE Paper No. 2002-01-1885, Society of Automotive Engineers, Warrendale, PA.

34. Ku, J., Huang, Y., Matthews, R.D., and Hall, M.J. (2000), "Conversion of a 1999 Silverado to Dedicated E85 with Emphasis on Cold Start and Cold Driveability," SAE Paper No. 2000-01-0590, Society of Automotive Engineers, Warrendale, PA.

35. Leppard, W.R., Rapp, L.A., Burns, V.R., Gorse, R.A. Jr., Knepper, J.C., and Koehl, W.J. (1992), "Effects of Gasoline Composition on Vehicle Engine-Out and Tailpipe Hydrocarbon Emissions—The Auto/Oil Air Quality Improvement Research Program," SAE Paper No. 920329, Society of Automotive Engineers, Warrendale, PA.

36. Leppard, W.R., Koehl, W.J., Benson, J.D., Burns, V.R., Hochhauser, A.M., Knepper, J.C., Painter, L.J., Rapp, L.A., Rippon, B.H., Reuter, R.M., and Rutherford, J.A. (1995), "Effects of Gasoline Properties (T_{50}, T_{90} and Sulfur) on Exhaust Hydrocarbon Emissions of Current and Future Vehicles: Model Analysis—The Auto/Oil Air Quality Improvement Research Program," SAE Paper No. 952504, Society of Automotive Engineers, Warrendale, PA.

37. McCabe, R.W., DiCicco, D., Guo, G., and Hubbard, C.P. (2004), "Effects of MMT Fuel Additive on Emission System Components: Comparison of Clear- and MMT-Fueled Escort Vehicles from the Alliance Study," SAE Paper No. 2004-01-1084, Society of Automotive Engineers, Warrendale, PA.

38. Oakley, A., Zhao, H., Ladommatos, N., and Ma, T. (2001), "Feasibility Study of an Online Gasoline Fractionating System for Use in Spark-Ignition Engines," SAE Paper No. 2001-01-1193, Society of Automotive Engineers, Warrendale, PA.

39. Oguma, H., Koga, M., Nishizawa, K., Momoshima, S., and Yamamoto, S. (2003), "Development of Third Generation of Gasoline PZEV Technology," SAE Paper No. 2003-01-0816, Society of Automotive Engineers, Warrendale, PA.

40. Owen, K. and Coley, T. (1995), *Automotive Fuels Reference Book, 2nd Edition*, Society of Automotive Engineers, Warrendale, PA.

41. Porse, E.C. (2002), "MTBE: Examining the Oxygenate Requirement and Remediation Costs: A Study in Science and Technology Policy Implementation," SAE Paper No. 2002-01-1268, Society of Automotive Engineers, Warrendale, PA.

42. Quader, A.A., Kirwan, J.E., and Grieve, M.J. (2003), "Engine Performance and Emissions Near the Dilute Limit with Hydrogen Enrichment Using an On-Board Reforming Strategy," SAE Paper No. 2003-01-1356, Society of Automotive Engineers, Warrendale, PA.

43. Seo, Y.S., Shirley, A., and Kolaczkowski, S.T. (2002), "Evaluation of Thermodynamically Favorable Operating Conditions for Production of Hydrogen in Three Different Reforming Technologies," *Journal of Power Sources*, Vol. 108, No. 1–2, pp. 213–225.

44. Shayler, P.J., Davies, M.T., Colechin, M.J.F., and Scarisbrick, A. (1996), "Intake Port Fuel Transport and Emissions: The Influence of Injector Type and Fuel Composition," SAE Paper No. 961996, Society of Automotive Engineers, Warrendale, PA.

45. Takeda, K., Yaegashi, T., Sekiguchi, K., Saito, K., and Imatake, N. (1995), "Mixture Preparation and HC Emissions of a Four-Valve Engine During Cold Starting and Warm-Up," SAE Paper No. 950074, Society of Automotive Engineers, Warrendale, PA.

46. Wieland, S., Baumann, F., and Starz, K.A. (2001), "New Powerful Catalysts for Autothermal Reforming of Hydrocarbons and Water-Gas Shift Reaction for On-Board Hydrogen Generation in Automotive PEMFC Applications," SAE Paper No. 2001-01-0234, Society of Automotive Engineers, Warrendale, PA.

47. Yang, Y., Sung, C.J., and Eng, J.A. (2002), "Effects of n-Butane Addition on Reformer Gas Combustion," Technical Meeting of the Central States Section of The Combustion Institute, Knoxville, TN.

48. Zimmermann, F., Ren, W.-M., Bright, J., and Imoehl, B. (1999), "An Internally Heated Tip Injector to Reduce HC Emissions During Cold-Start," SAE Paper No. 1999-01-0792, Society of Automotive Engineers, Warrendale, PA.

CHAPTER 8

Advanced Catalyst Design

Paul J. Andersen, Todd H. Ballinger, and David S. Lafyatis
Johnson Matthey

8.1 Introduction

Regulated emissions levels from automobiles continue to be lowered around the world to improve air quality. The most stringent regulations for gasoline-powered vehicles are now approaching near-zero levels for tailpipe emissions. The first examples of this were the super ultra-low emissions vehicle/partial zero emissions vehicle (SULEV/PZEV) categories in California's Low Emissions Vehicle II (LEV II) program and the Bin 3/Bin 2 categories in the U.S. Environmental Protection Agency (EPA) Tier 2 emissions regulations. Between 2003 and 2009, automotive manufacturers will phase in the production of light-duty vehicles (passenger cars), medium-duty passenger vehicles (minivans, sport utility vehicles [SUVs], and full-size vans with gross vehicle weights up to 10,000 lb), and light-duty trucks (gross vehicle weights up to 8500 lb) that meet these near-zero emissions standards.

The central exhaust component necessary to meet the near-zero tailpipe emissions standards is the advanced three-way catalyst (TWC), which contains individual or a combination of the platinum group metals (PGM) palladium, platinum, and rhodium (Pd, Pt, and Rh, respectively) to oxidize the hydrocarbons (HCs) and carbon monoxide (CO) to carbon dioxide [CO_2] and water [H_2O] and to reduce the nitrogen oxides (NOx) to nitrogen. Modern advanced TWCs can achieve virtually 100% conversion of the three pollutants (hence near-zero emissions) when two important criteria are met: (1) the catalyst must be at operating temperatures (temperature greater than 300°C, ideal temperature greater than 450°C), and (2) the exhaust gas mixture is at the optimum stoichiometric air/fuel (A/F) ratio. Under most driving conditions, the exhaust temperature is high enough, and an onboard computer (using oxygen sensor feedback) maintains engine conditions such that the exhaust gas is near the stoichiometric point, and near-zero emissions can be realized. However, if near-zero emissions are to be achieved under all driving conditions, two important problems must be overcome: (1) brief transients away from the stoichiometric A/F ratio during hard accelerations and decelerations, and (2) rapid warm-up of the catalyst to reach operational temperatures when a vehicle is started with a cold engine (so-called cold starts).

The first problem has been significantly minimized by tremendous improvements in engine control strategies, as described in Chapter 5. The first SULEV/PZEV vehicles sold in California used multiple oxygen sensors located before and after the catalysts in the exhaust to provide feedback to the onboard computer, which utilized sophisticated algorithms for precise control of the exhaust A/F ratio around the optimum stoichiometric point [Kishi *et al.* (1999); Inoue *et al.* (2000); Kitagawa (2000); Nishizawa *et al.* (2000); Webster (2000)]. Subsequent improvements on these vehicles further refined the algorithms and the exhaust A/F ratio control [Nishizawa *et al.* (2001); Kidokoro *et al.* (2003); Oguma *et al.* (2003)]. Indeed, all near-zero emissions vehicles currently produced use multiple oxygen sensors and have excellent warmed-up A/F ratio control such that emissions from the warmed-up portions of the regulated test procedures have essentially zero emissions. An example of the effect of improved warmed-up A/F ratio control to improve TWC efficiency will be shown later in this chapter.

The second problem, rapid catalyst warm-up during engine cold starts, has proven to be the most challenging problem to overcome for achieving near-zero emissions. Because TWCs are not active below 300°C, pollutants emitted from the engine (mainly HCs because little NOx is produced at the low engine temperature and load conditions present during a cold idle) while the catalyst warms up to this temperature pass over the catalyst and are emitted into the atmosphere. Near-zero emissions vehicles must solve this problem to meet the near-zero emissions standards. From a catalyst development standpoint, much research has gone into the development of low light-off catalysts, which are catalysts that have catalytic activity below 300°C. Although low light-off catalysts have been developed, which overcame some of the kinetic and thermodynamic limitations, such catalysts were hindered by two fundamental problems. The first problem was the adsorption of various species on the PGM used in the catalysts, which blocks these active sites from reacting with pollutants in the exhaust. The adsorbed species include exhaust components (mainly H_2O and HCs) that adsorb on the catalyst after the engine is turned off, and poisons (mainly sulfur and phosphorus) that come from gasoline and oil. Water and HCs will desorb from the PGM sites beginning at temperatures above 100°C (thus limiting reactivity below this temperature), but even small amounts of adsorbed sulfur and phosphorus poison the low-temperature activity of low light-off catalysts. The second problem was the lack of thermal durability of low light-off catalysts. Platinum group metals and washcoat component sintering at the upper temperature range of normal driving conditions degrade the low-temperature activity of these catalysts. Thus, because TWC activity is limited below 300°C, other engineering solutions are needed to quickly heat the TWC to operational temperatures.

One of the most obvious solutions for rapid catalyst heat-up is to move the catalyst as close as possible to the engine. (Catalysts located close to the engine have been named close-coupled [CC] catalysts.) This requires the catalysts to have very high thermal durability, but, as will be discussed later, advanced TWCs with high thermal durability have been developed. Close-coupled catalysts were used to help meet the ultra-low emissions vehicle (ULEV) emissions standards in the mid- to late 1990s. However, CC catalysts on a typical ULEV application reach operating temperatures approximately 20 seconds after an engine cold start. For the near-zero emissions standards, the amount of HCs emitted from a conventional engine in this time frame will exceed the SULEV/PZEV HC emissions requirements. Therefore, other engineering solutions were needed to further assist rapid catalyst heat-up.

Some of the more common engineering solutions include the use of high-cell-density, thin-wall substrate; dual-wall exhaust manifolds; secondary air injection; and engine cold-start strategies such as spark retard, variable valve timing (VVT), enhanced swirl combustion, and improved fuel injection enabling lean, hotter cold starts. These engine cold-start strategies also reduced the HCs emitted from the engine, which further alleviated cold-start emissions problems. Although the impact of each of the solutions on catalyst warm-up will be discussed briefly, the first SULEV/PZEV vehicles used combinations of these solutions to achieve rapid catalyst warm-up [Kishi *et al.* (1999); Inoue *et al.* (2000); Kitagawa (2000); Nishizawa *et al.* (2000); Webster *et al.* (2000)]. The success of these strategies is shown by their continued use (in various forms) on all near-zero emissions vehicles.

The effect of high-cell-density, thin-wall substrate (both ceramic and metallic) on catalyst HC light-off has been well documented in recent work and references

[Kikuchi *et al.* (1999); Ball *et al.* (2000); Holy *et al.* (2000); Nishizawa *et al.* (2000); Will *et al.* (2000); Kubsh (2001); Brueck *et al.* (2002); Mueller *et al.* (2002); Schaper *et al.* (2002); Kidokoro *et al.* (2003)]. The basic impact of high-cell-density substrate on HC light-off is improved mass transport of pollutant molecules to the catalyst coating on the walls of the substrate (i.e., smaller cell dimensions and higher geometric area of substrate), while the thinner walls resulted in lower thermal mass and faster substrate heat-up. Both effects have a positive impact on reducing the HC light-off/rapid heat-up of the catalyst, as will be shown later in this chapter. Furthermore, the improved mass transport improves the warmed-up HC and NOx activity of catalysts coated on high-cell-density substrate. However, one study [Will *et al.* (2000)] showed that the use of high-cell-density substrate can result in more severe catalyst temperatures under some engine conditions, emphasizing the need for thermally durable catalysts when high-cell-density substrates are used. Several studies conclude that when ceramic substrate is used, 900-cpsi substrate is optimal [Mueller *et al.* (2002); Kubsh (2001); Will *et al.* (2000)].

Engine cold-start strategies such as spark retard (Chapter 5), VVT (Chapter 1), and enhanced swirl combustion and improved fuel injection (Chapter 2) have been discussed in detail in previous chapters. These strategies have a positive impact on the catalyst HC light-off by producing lean, hotter cold starts with fewer engine-out HCs—all of which improve catalyst warm-up. An example of the effect of an engine cold-start strategy will be shown later in this chapter. Other demonstrations can be found in recent work [Kidokoro *et al.* (2003); Oguma *et al.* (2003); Karwa and Biel (2001); Acke (2001); Nishizawa *et al.* (2001); Inoue *et al.* (2000); Ueno (2000); Kitagawa (2000)].

Usually coupled with engine cold-start strategies are dual-wall exhaust manifolds that also have been optimized for improved exhaust flow. Such strategies maximize the transfer of the heat generated in the engine to the CC catalysts and minimize the A/F ratio fluctuations from individual cylinders through better exhaust mixing before the catalyst. These improvements again benefit tailpipe emissions by reducing the catalyst warm-up time. Almost all of the references listed in the preceding paragraphs utilize dual-wall manifolds for near-zero emissions vehicles.

Secondary air addition utilizes a slightly different strategy to achieve rapid catalyst light-off than some of the strategies already mentioned and has been used in vehicles where engine cold-start strategies are not practical. Further details of secondary air injection were discussed in Chapter 6. Secondary air can be used with rich A/F ratio cold starts to either combust exhaust (particularly CO and H_2) before the catalyst [Borland and Zhao (2002)] to increase the exhaust temperature, or to combust the exhaust in the catalyst [Baumgarten *et al.* (2001); Marsh *et al.* (2000)] to generate an exotherm in the catalyst. Both methods are effective ways to assist rapid warm-up of the catalyst but require an air pump to be added to the vehicle.

Another cold-start technology that has received a lot of attention is the HC trap. The basic concept of an HC trap is to use a material, most commonly a zeolite, that adsorbs HCs at low temperatures and then desorbs them at higher temperatures to be oxidized by a catalyst. In principle, this would be an excellent solution to the cold-start HC problem; however, two issues are associated with HC traps. The first problem is that the zeolite HC desorption temperature (for most HC species) often is below the light-off temperature of a catalyst; thus, although the HCs are trapped, they are released below the catalyst HC light-off temperature, pass through the exhaust, and are emitted into the atmosphere. The second problem is that zeolites do not have high thermal durability. Although this problem can be solved by placing the HC trap in an underfloor location in the exhaust, this further delays the downstream catalyst light-off. Various engineering approaches have been used to overcome this problem in near-zero emissions vehicles, including the use of a two-stage HC trap system [Nishizawa *et al.* (2000)], a complex design underbody catalyst that diverts exhaust flow through an HC trap surrounding the underbody TWC during cold starts [Inoue *et al.* (2000)], an HC trap bypass exhaust leg [Tayama *et al.* (1998)], or an HC trap combined with a downstream electrically heated catalyst (EHC) [Baumgarten *et al.* (2001); Morris *et al.* (1999); Brück *et al.* (1999)]. Other studies have investigated the optimization of engine control for use with optimized HC trap exhaust configurations [Nakagawa *et al.* (2003); Yamamoto *et al.* (2000)], including running the engine lean during HC desorption to assist HC oxidation over the catalyst. Another investigation showed the impact of engine cold-start strategies [Ballinger and Andersen (2002)] on the trapping efficiency of HC traps. In some cases, these strategies can reduce the effectiveness of the HC trap. Despite the complexity and increased cost associated with these engineering solutions, two SULEV/PZEV vehicles continue to use HC traps [Kidokoro *et al.* (2003); Oguma *et al.* (2003)]. In addition, recent work has investigated metal impregnation into zeolites to raise the HC desorption temperature of HC traps [Higashiyama *et al.* (2003)].

Other less-common cold-start strategies have been studied for near-zero emissions vehicles. Such technologies include EHCs used without [Holy *et al.* (2000); Tayama *et al.* (1998)] or with HC traps [Baumgarten *et al.* (2001); Morris *et al.* (1999); Brück *et al.* (1999)], chemically heated catalyst systems

[Marsh et al. (2000); Morris et al. (1999)], exhaust gas ignition (EGI) systems [Morris et al. (1999)], vacuum insulated catalysts to retain heat between cold starts [Karwa and Biel (2001)], onboard fuel reforming systems [Kirwan et al. (2002)], and a canister bypass system to adsorb cold-start HCs similar to evaporative emissions systems [Marsh et al. (2000)]. Although these strategies have proven effective to reduce cold-start HCs to near-zero emissions levels, the cost and complexity (including long-term durability) inhibit these strategies from practical use.

As already discussed, although the durability of low light-off catalysts is a major challenge, other catalyst formulation effects have a beneficial impact on HC light-off and are being used in near-zero emissions vehicles. One of the most common effects is to increase the PGM loading of the catalyst, as will be shown in this chapter. Palladium (Pd) is the best PGM for HC oxidation and typically is used in high loadings in CC catalysts. Several studies have concluded that the highest optimal Pd loading is 200 g/ft^3 [Mueller et al. (2002); Lindner et al. (2000)]. Indeed, in most investigations of near-zero emissions vehicles, high Pd-loaded Pd-only, Pd:Rh, or Pt:Pd:Rh CC catalysts are used for rapid HC light-off [Mueller et al. (2002); Kubsh (2001); Karwa and Biel (2001); Ball et al. (2000); Lindner et al. (2000)]. In addition to increasing the PGM loading, there are also washcoat formulation effects, such as optimized PGM-support interactions, layered catalyst structures, and improved catalyst manufacturing techniques that improve HC light-off. These will be discussed in more detail in this chapter and have been reported in studies by others [Kubsh (2001); Kidokoro et al. (2003)].

Finally, TWC improvements also significantly impact warmed-up activity. Stabilized alumina and stabilized ceria-zirconia (oxygen storage components) have greatly improved the thermal durability of TWCs, allowing the TWC to not only survive higher exhaust temperatures, but also to maintain high pollutant conversions throughout the 150,000-mile durability requirement. As will be discussed in this chapter, optimized PGM-support interactions and layered catalyst structures also are used to improve catalyst activity and provide thermal durability. Consideration also should be given to the types of PGM used in near-zero emissions systems. As stated, Pd:Rh and tri-metal catalysts are used in close-coupled locations for rapid HC light-off and warmed-up HC and CO oxidation. Rhodium (Rh) has been demonstrated to be the most effective PGM for NOx reduction; hence, near-zero emissions systems use Pd:Rh or Pt:Rh underfloor catalysts that have been optimized for high NOx conversion activity, as will be shown here and has been shown by others [Kubsh (2001); Karwa and Biel (2001); Ball et al. (2000); Lindner et al. (2000)].

This chapter will describe the factors involved in advanced catalyst design for near-zero emissions vehicles. The first section will demonstrate the effect of PGM, washcoat components, and catalyst formulations on improving TWC activity. The second section will show how high-activity TWCs are incorporated into an exhaust system and how the combination of other exhaust strategies is used to produce near-zero emissions systems.

8.2 Advanced Three-Way Catalyst Concepts and Design

Three-way catalysts for meeting PZEV emissions must have high HC and NOx conversion activities. Thoss and Rieck (1997) describe a study in which the three PGM employed in TWCs were examined to determine their utility for converting these pollutants. Three-way catalysts were prepared with a wide range of PGM loads on a washcoat typical of formulations employed in the mid-1990s. The same washcoat was used for all PGM combinations. The catalysts were dynamometer aged for 45 hours on a two-mode cycle that reached maximum catalyst temperatures of 1050°C. After aging, the catalysts (1.42L) were tested on a 1.9L vehicle.

The HC and NOx Federal Test Procedure (FTP) emissions for three types of catalyst systems were measured as a function of Pd or Pt load for Pd only, Pd:Rh, and Pt:Rh catalysts. The HC emissions (Figure 8.1) were clearly a strong function of Pd or Pt load. Additionally, Pd was more effective for converting HC emissions than Pt, whether or not Rh was present. However, one interesting observation is that Pd-only catalysts had higher HC conversion activity than Pd:Rh catalysts at comparable Pd loads. The Rh-containing systems had substantially lower NOx emissions than the Pd-only system (Figure 8.2). Additional Pd or Pt in the Rh-containing catalyst had only a secondary impact on NOx emissions, with the Pt:Rh catalysts having somewhat lower NOx emissions than the Pd:Rh catalysts.

These data indicated that the Pd component was critical to achieving low HC emissions, and the Rh component was critical to achieving low NOx emissions. However, these data also suggested that improvement of the NOx conversion characteristics over the Pd-only catalyst or the Pd component in the bimetallic TWCs is a potential area for TWC improvement. Palladium was deemed to be a more appropriate focus because it has significant TWC activity on its own. Platinum (Pt), while being a good catalyst for HC and CO oxidation, has little or no capacity to catalyze NOx reduction (i.e., a Pt-only catalyst) after typical automotive aging conditions.

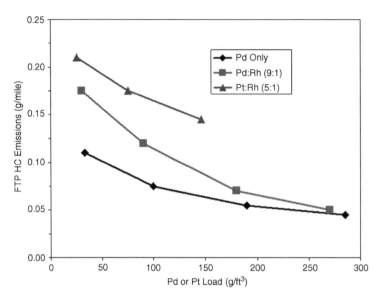

Figure 8.1 Federal Test Procedure HC emissions as a function of Pd or Pt load [Thoss and Rieck (1997)].

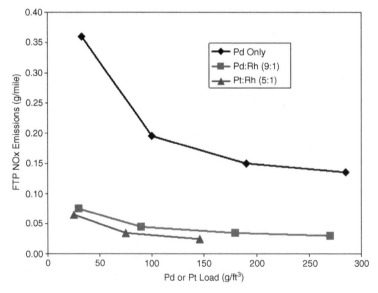

Figure 8.2 Federal Test Procedure NOx emissions as a function of Pd or Pt load [Thoss and Rieck (1997)].

Because the main weakness of Pd-only catalysts is NOx conversion activity, much of the catalyst development work has focused on improving this aspect of the catalyst function. Sulfur can have a significant poisoning effect on Pd-only catalysts (and Pd catalyst components). The data in Figure 8.3 [Andersen (1997)] demonstrate the effect of sulfur on the TWC activity of Pd-based catalysts. A typical Pd-only catalyst (110 g/ft^3 Pd) was hydrothermally aged in a laboratory furnace at 1050°C under conditions that cycled between lean and rich stoichiometry (redox aging). The catalyst then was tested in a laboratory reactor at 500°C under simulated exhaust conditions except for the presence of sulfur. The reactant mix was adjusted to the point where the CO and NOx conversion were equal. Then, at a specific time (t = 0), 20 ppm sulfur dioxide (SO_2) was introduced into the exhaust mixture, and the NOx conversion dropped significantly. After an extended poisoning period (240 minutes), the SO_2 was removed, and most of the NOx conversion activity was recovered.

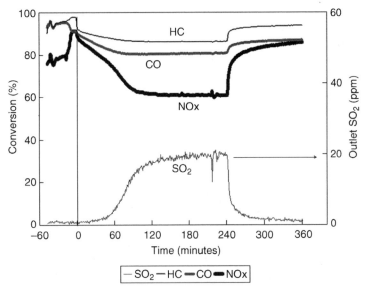

Figure 8.3 *Response of Pd-only catalyst activity to the addition and removal of SO_2; reaction temperature = 500°C, aging temperature = 1050°C [Andersen (1997)].*

The response to sulfur can be affected by altering the composition and preparation of the catalyst washcoat. The washcoat consisted primarily of alumina, a ceria-zirconia mixed oxide (oxygen storage material component [OSC]), and a basic promoter such as lanthanum (La) or alkaline earth materials. Careful selection of the basic promoter significantly improved activity under both low- and high-sulfur conditions (Figure 8.4).

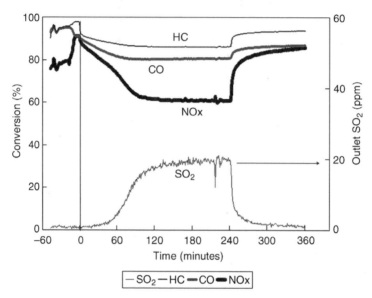

Figure 8.4 SO_2 *poisoning profile for a Pd-only catalyst with optimized basic promoter; reaction temperature = 500°C, aging temperature = 1050°C [Andersen (1997)].*

Modifying the distribution of Pd between the alumina and CeZrOx supports in the Pd-only catalyst also can have a strong impact on activity and the response of the catalyst to sulfur poisoning. Figure 8.5 shows the sulfur poisoning response of Pd-only catalysts with varying extents of Pd/CeZrOx interaction. The catalyst with the highest amount of Pd/OSC interaction showed the highest activity under zero sulfur conditions, but the catalyst with no Pd/OSC interaction showed the highest activity after sulfur poisoning. This Pd/OSC interaction can be adjusted to give catalysts with the optimum balance of low-sulfur and high-sulfur performance for a given application. Adjustments in the Pd/OSC balance can give improved vehicle performance, as will be demonstrated in the later discussion of PZEV systems.

Advanced Catalyst Design

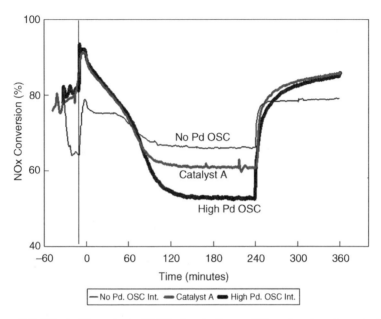

Figure 8.5 *Effect of the extent of Pd/Ce interaction on SO_2 poisoning characteristics of Pd-only catalysts; reaction temperature = 500°C, aging temperature = 1050°C [Andersen (1997)].*

As discussed in the Introduction to this chapter, for near-zero emissions vehicles, Pd-based catalysts usually are located in close-coupled positions and are exposed to high temperatures during operation. Consequently, the thermal durability of the Pd-containing catalysts also is critical to performance. To examine thermal durability, separate 1.24L pieces of an identical Pd-only catalyst were dynamometer aged for 120 hours with four-mode aging cycles with a maximum temperature of either 1000 or 1150°C. The catalysts then were fitted to a single exhaust bank of a 1995 vehicle with a 4.6L engine and FTP tested. The results (Figure 8.6) indicated that following the higher-temperature aging, CO and NOx conversions decreased significantly more than HC conversions. For example, the percent of unconverted NOx increased from 4.5 to 11.2%, but the percent of unconverted HC increased from only 5.7 to 6.9%.

Due to the larger sensitivity of CO/NOx activity to thermal degradation, studies have focused primarily on improving activity for converting those components after high-temperature aging. Much of the focus has been on improving the

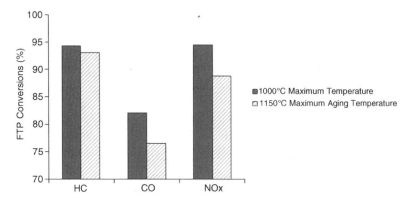

Figure 8.6 Effect of aging temperature on FTP conversions of Pd-only catalysts.

stability of the OSC component in the catalyst. In general, it is desired to have a homogeneously mixed CeZrOx phase that does not separate into separate CeO_2 and ZrO_2 phases after exposure to high temperatures. Significant efforts are spent on modifying compositions and preparation methods to achieve this goal. The data in Table 8.1 show that improvements in the CeZrOx properties can substantially affect thermal durability. Carbon monoxide and NOx conversions were measured in a sulfur-free synthetic exhaust mixture, as described previously for catalysts subjected to redox aging at either 1050 or 1100°C. The catalyst prepared with OSC i or ii dropped significantly when the aging temperature was increased from 1050 to 1100°C; however, the catalyst prepared with OSC iii was not affected by the increased aging temperatures.

TABLE 8.1
EFFECT OF OSC TYPE ON CO/NOx ACTIVITY
AFTER VARIOUS AGING TREATMENTS

OSC Type	1050°C Aging, 50 Hours	1100°C Aging, 50 Hours
i	91	76
ii	94	80
iii	95	97

Rhodium can be added to the Pd-only compositions to make highly active Pd:Rh catalysts as well. Indeed, the value of Rh to promote NOx reduction is well known and is demonstrated in Figure 8.2 on a previous generation of catalysts. Figure 8.7 shows the A/F sweep activity of Pd-only, Pd:Rh, and Rh-only catalysts when tested on an engine dynamometer after aging for 120 hours on a 1000°C four-mode aging, where the washcoat has been optimized for high-activity Pd-only catalysts as previously described. The sweep performance is represented by the CO/NOx conversions at the A/F ratio at which those two conversions are equal (i.e., the CO/NOx crossover point, or COP). The HC conversion at the COP A/F ratio also is shown. Figure 8.7 shows that Pd removal markedly decreased catalyst activity, suggesting that Rh was making a relatively small contribution to catalyst performance. The same catalysts were FTP tested on the 1995 4.6L vehicle as described earlier. The FTP data shown in Figure 8.8 (same conditions as described for Figure 8.6) indicated the same trend for CO and NOx activity and that the Pd-free catalyst had quite low HC conversion activity. Thus, it was felt that significant improvements in catalyst activity may be possible by improving the TWC activity of the Rh component.

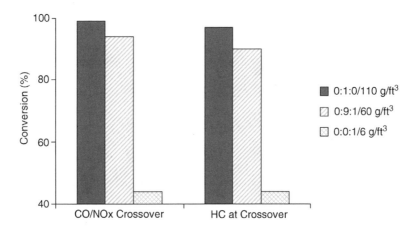

Figure 8.7 *Air/fuel sweep conversions for Pd-only, Pd:Rh, and Rh-only catalysts; inlet temperature = 450°C; space velocity = 85,000 hr^{-1}.*

The main focus of development work to improve the Rh component has been on optimizing the composition and the preparation method of the CeZrOx OSC component used to support Rh in the catalyst. As with Pd-only catalysts, the preparation method for the OSC Rh support and the manner in which Rh was deposited on that support were critical to achieving high activity and thermal durability. The sweep data shown in Figure 8.9 indicate the magnitude of the

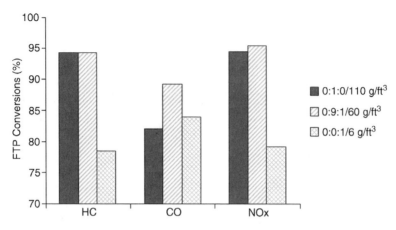

Figure 8.8 Federal Test Procedure conversions for Pd-only, Pd:Rh, and Rh-only catalysts.

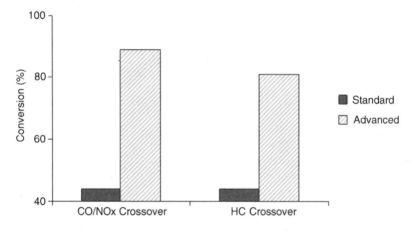

Figure 8.9 Air/fuel sweep conversions for an improved Rh-based catalyst.

improvements that can be made through optimization of these parameters with COP values increasing from 44 to 89% through washcoat modifications.

However, the HC activity of even these improved catalysts is still relatively low compared to Pt- or Pd-containing catalysts. For that reason, Pt or Pd (or both) generally are combined with these Rh components to make TWCs with

Advanced Catalyst Design

high activity for HC, CO, and NOx. Frequently, the Pt and Pd are placed in separate washcoat layers to avoid negative interactions that may arise between catalyst components. Figure 8.10 demonstrates the FTP performance of these advanced, layered Pd:Rh and Pt:Rh catalysts. Platinum and Pd addition significantly boosts HC activity while slightly improving NOx activity.

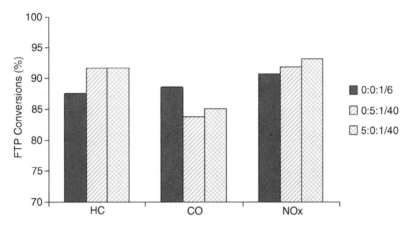

Figure 8.10 *Federal Test Procedure conversions for advanced Rh-only, Pd:Rh, and Pt:Rh catalysts.*

The next section describes the application of these advanced Pd-only, Pd:Rh, and Pt:Rh catalysts to vehicle systems to achieve PZEV emissions standards.

8.3 Catalyst System Design Principles for Meeting Partial Zero Emissions Vehicle Emissions Standards

Initial system concepts to reach the PZEV emissions standards included strategies such as electrically heated catalysts and underfloor HC traps. However, these strategies may add significant cost and complexity to the exhaust system design and make them undesirable. As already discussed, significant advances have been made in the activity and durability of catalyst washcoats. Implementation of these improved washcoats into an optimal system design gives the opportunity to approach the PZEV target with passive means. The following discussion centers on the system design philosophies that, when combined, may allow PZEV to be achieved. Strategies for both the close-coupled and underfloor converters to greatly reduce both NOx and nonmethane hydrocarbon (NMHC) emissions over the FTP cycle will be demonstrated.

The first area of challenge is light-off. A conventional vehicle will fail the PZEV standard for HCs in the first 20 seconds of the FTP test, while the vehicle is still idling after cold start. However, several improvements may be made in the design of the CC catalyst to significantly reduce light-off emissions. One well-known method of improving cold-start emissions of HC and NOx is by the use of increased PGM content on the CC catalyst. Palladium is known to be an excellent PGM for light-off purposes following high-temperature aging. Figure 8.11 shows the effect of increasing Pd loading in the close-coupled position of an LEV-calibrated, V-engine equipped vehicle on FTP emissions following an aging that represents 50,000 miles. This chart clearly shows that increasing the Pd loading from 54 to 200 g/ft^3 leads to a significant decrease in the Bag 1 emissions for NMHC, NOx, and CO.

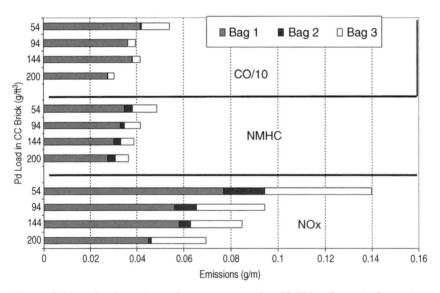

Figure 8.11 Federal Test Procedure emissions after 50,000-mile equivalent aging.

Another important strategy for reducing cold-start emissions is the use of high-cell-density, thin-wall substrates. The 400/6 ceramic substrate that has been common in the automotive industry usually will account for 70 to 80% of the thermal mass of the washcoated catalyst. A washcoated 900/2 ceramic substrate typically will have a 30% reduced thermal mass and a 60% higher geometric surface area compared to a coated 400/6 catalyst. The combination of increased geometric surface area (which leads to improved heat and mass transfer within the substrate) and lower thermal mass (which leads to more rapid heat-up) that these materials offer leads to substantial advantages during the cold-start portion of an FTP test. Figure 8.12 shows the temperature measured in the catalyst bed for two catalysts with identical coatings and PGM loadings but on different substrates. The more rapid heat-up of the 900/2 material translates directly into reduced tailpipe emissions from early in the FTP.

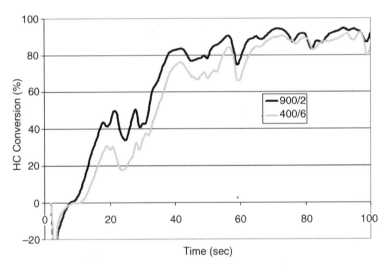

Figure 8.12 Hydrocarbon light-off comparison on a 1.9L vehicle, 400/6 versus 900/2 substrate.

Close-coupled catalyst formulations designed specifically for light-off also can lead to emissions improvements during cold start. Figure 8.13 compares the HC light-off performance of two different Pd-Rh catalysts at an identical PGM loading, following an aging to represent 50,000 miles. The formulation B leads to significantly improved HC performance during the first 60 seconds of the FTP drive cycle. This improved performance is particularly evident during the transient conditions of Hill 1.

The CC catalyst is responsible for more than light-off in the emissions system. It also is a critical component for controlling NMHC and NOx during the

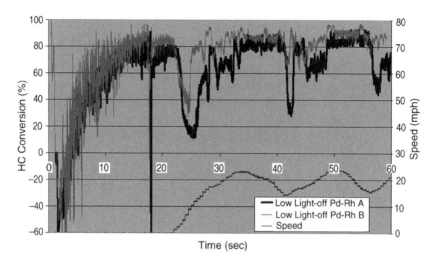

Figure 8.13 Effect of washcoat type on HC light-off for a PZEV-type vehicle.

warmed-up sections of the driving cycle. Figure 8.11 demonstrates that the strategy of increased Pd loading in the CC catalyst leads to improved Bag 2 and Bag 3 emissions, in addition to the improvement in Bag 1 performance that was already noted. Figure 8.14 demonstrates another PGM strategy for improving FTP performance. This figure compares the emissions results over the close-coupled portion of the exhaust system from a V6-equipped passenger car from CC catalyst systems that are equipped with 200 g/ft^3 Pd in both cases. However, in one case, the CC catalyst contains only Pd; in the other case, the CC catalysts also contain an additional 6 g/ft^3 Rh. This demonstrates that in these advanced washcoats, the addition of only this small amount of Rh to the CC catalyst reduces NOx emissions by roughly 20%, while leaving HC emissions unchanged.

To meet the low emissions levels required for PZEV certification, exhaust systems are likely to have both a CC catalyst mounted close to the engine and additional catalysts farther downstream in the exhaust system, commonly referred to as underbody catalysts. These catalysts play a complementary role to the CC catalysts for exhaust emissions control. Although they generally play little role in light-off performance, they are critical for cleaning emissions that pass through the CC catalyst during warmed-up operation due to either high-space velocity mass transfer limited operation or A/F excursions during

Advanced Catalyst Design

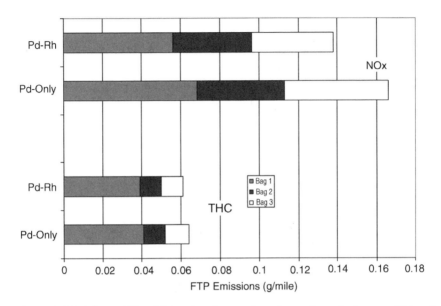

Figure 8.14 *Effect of Rh addition in advanced washcoat formulations on HC and NOx emissions.*

transient driving events. Their requirements for thermal durability may be reduced compared to CC catalysts because of their greater distance from the engine. It is important in system design to recognize that the function of these catalysts is significantly different from that of CC catalysts and to optimize the system configuration to account for this.

Figure 8.15 shows an example of the dramatic effect on overall emissions that the design of the underbody catalyst can have. This study was conducted on a V8-equipped SUV, using a common set of CC catalysts but varying the underbody catalysts. Each of the systems was aged to 50,000 miles. Three different underbody catalyst systems were tested: a standard Pd-Rh catalyst (0:9:1/60), an advanced Pd-Rh catalyst (0:9:1/60), and an advanced Pt-Rh catalyst (9:0:1/60). Each of the underbody catalysts also had been aged to represent 50,000 miles. The data in Figure 8.15 indicate that the change of the underbody catalyst formulation from the standard Pd-Rh to the advanced Pt-Rh has reduced NMHC tailpipe emissions by approximately 10%, but more impressively, NOx emissions have been reduced by 75%!

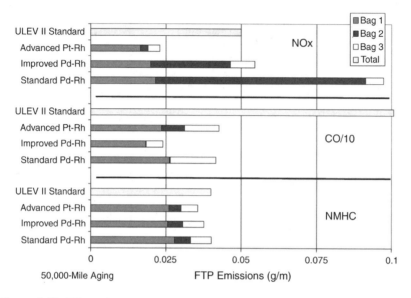

Figure 8.15 Effect of rear catalyst on FTP emissions.

All of the topics discussed in the previous paragraphs have focused on changes that can be made to components in the exhaust stream to reduce emissions while using a fixed vehicle architecture and calibration strategy. However, it is important to recognize that parallel to the development of better catalyst components and system design strategies, vehicle manufacturers have revolutionized the vehicle control algorithms during both cold start and warmed-up conditions. The results of these changes when combined with the catalyst component optimization described above can be truly remarkable.

Figure 8.16 is a schematic diagram of experiments performed on a V8 SUV. Initially, a 1999 Model Year LEV was tested using its OEM calibration after aging its original emissions system to the equivalent of 50,000 miles. That same vehicle then was upgraded with an advanced calibration. This calibration incorporated algorithms both for a more rapid warm-up and for more responsive A/F control during transient driving conditions. In addition, the exhaust system on the vehicle was upgraded to incorporate changes such as the use of high-cell-density, thin-wall substrate, replacing the Pd-only catalysts in the close-coupled position with Pd-Rh catalysts and replacing the Pd-Rh catalysts in the underbody location with the advanced Pt-Rh washcoat.

Advanced Catalyst Design

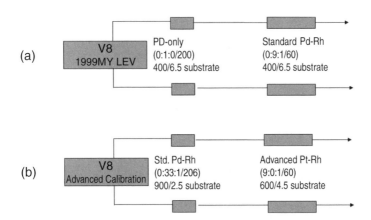

Figure 8.16 *Schematic diagram comparing OEM and advanced technology systems. (a) OEM catalyst system/OEM calibration. (b) Advanced catalyst system/advanced calibration.*

Figure 8.17 shows the improvement in FTP emissions realized because of these changes. The combination of the advanced vehicle calibration with the improved exhaust system design led to a large reduction in all three legislated emissions, including an 89% reduction in NOx emissions. Significant improvements in all three phases of the FTP cycle were seen for NOx, while the improvements for both NMHC and CO emissions were mainly in Bag 1. One key to the Phase 1 improvements demonstrated for all three components is the rapid catalyst warm-up demonstrated by the advanced system, which is a combination of the changes to both the vehicle calibration and the close-coupled substrate. Figure 8.18 compares the warm-up of the two systems, using a thermocouple positioned in the catalyst bed 1 inch from the front face. The critical time for warming the close-coupled converter to light-off temperature has been reduced by approximately 50%, resulting in a major benefit in emissions during the critical first 50 seconds of the FTP cycle.

The preceding example shows the value in combining advanced calibrations with advanced catalyst technology on a large V8-powered vehicle to reduce emissions. In practice, most manufacturers opt to generally release smaller, I4-equipped vehicles for their initial PZEVs because the smaller engine size and lower inertial weight lead to easier implementation for meeting these radically reduced emissions standards. The same principles in system design described above apply to the smaller vehicles. Figure 8.19 shows a schematic

261

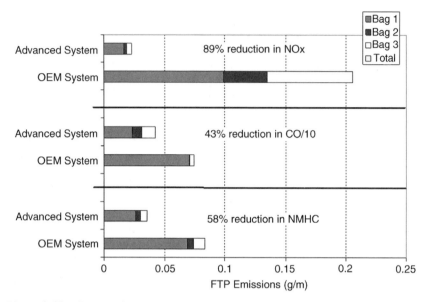

Figure 8.17 Effect of advanced engine control and catalyst strategies on V8-equipped SUV FTP emissions following 50,000-mile aging.

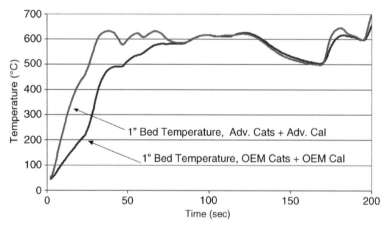

Figure 8.18 Improved catalyst temperature ramp during startup for an advanced system.

Advanced Catalyst Design

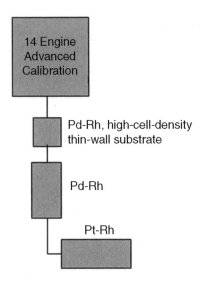

Figure 8.19 Schematic of a PZEV-type emissions system.

system on an I4 vehicle with PZEV-style calibration and hardware. The exhaust system includes in the close-coupled location a small, high-PGM Pd-Rh, high-cell-density, thin-wall metallic converter optimized for light-off performance, followed by a Pd-Rh catalyst and a Pt-Rh catalyst. The catalyst system was aged for using an accelerated aging protocol period, meant to represent 50,000 to 100,000 miles. As can be seen in Figure 8.20, the vehicle is able to meet the PZEV emissions standards for NMHC and NOx (CO standards also are easily met) with the aged exhaust system.

8.4 Summary

Tailpipe emissions levels from mobile sources continue to be dramatically reduced around the world. Legislation in the United States, both from the California Air Resources Board (CARB) and the EPA, has introduced new emissions targets that require extremely low emissions for both NOx and NMHC. These new emissions limits have required the entire industry to focus on cost-effective emissions control technologies. Improved catalyst washcoat materials that have dramatically increased the activity and durability of TWCs containing Pt, Pd, and Rh have played an important role in this process. By combining these advanced catalyst washcoats with other exhaust

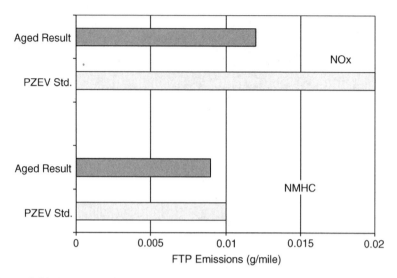

Figure 8.20 *Emissions from an I4 vehicle equipped with a PZEV-type calibration strategy and catalyst system.*

system improvements and revolutionary advances in the calibration and control strategies, standards such as PZEV have been achieved at a minimal increase in system cost and complexity.

8.5 References

1. Acke, F. (2001), "LEV II Applications Based on a Lean Start Calibration," SAE Paper No. 2001-01-1311, Society of Automotive Engineers, Warrendale, PA.

2. Andersen, P.J. (1997), "Advances in Pd Containing Three Way Catalyst Activity," SAE Paper No. 970739, Society of Automotive Engineers, Warrendale, PA.

3. Ball, D., Richmond, R.P., Kirby, C.W., Tripp, G.E., and Williamson, B. (2000), "Ultrathin Wall Catalyst Solutions at Similar Restriction and Precious Metal Loading," SAE Paper No. 2000-01-1844 Society of Automotive Engineers, Warrendale, PA.

4. Ballinger, T. and Andersen, P.J. (2002), "Vehicle Comparison of Advanced Three-Way Catalysts and Hydrocarbon Trap Catalysts," SAE Paper No. 2002-01-0730, Society of Automotive Engineers, Warrendale, PA.

5. Baumgarten, H., Weinowski, R., Habermann, K., Breuer, M., and Goetz, W. (2001), "Low Emission Concept for SULEV," SAE Paper No. 2001-01-1313, Society of Automotive Engineers, Warrendale, PA.

6. Borland, M. and Zhao, F. (2002), "Application of Secondary Air Injection for Simultaneously Reducing Converter-In Emissions and Improving Catalyst Light-Off Performance," SAE Paper No. 2002-01-2803, Society of Automotive Engineers, Warrendale, PA.

7. Brück, R., Maus, W., and Hirth, P. (1999), "The Necessity of Optimizing the Interactions of Advanced Post-Treatment Components in Order to Obtain Compliance with SULEV-Legislation," SAE Paper No. 1999-01-0770, Society of Automotive Engineers, Warrendale, PA.

8. Brueck, R., Konieczny, R., Schaper, K., Zinecker, R., and Dietsche, A. (2002), "New Design of Ultra-High Cell Density Metal Substrates," SAE Paper No. 2002-01-0353, Society of Automotive Engineers, Warrendale, PA.

9. Higashiyama, K., Nagayama, T., Nagano, M., Nakagawa, S., Tominaga, S., Murakami, K., and Hamada, I. (2003), "A Catalyzed Hydrocarbon Trap Using Metal-Impregnated Zeolite for SULEV Systems," SAE Paper No. 2003-01-0815, Society of Automotive Engineers, Warrendale, PA.

10. Holy, G., Brück, R., and Hirth, P. (2000), "Improved Catalyst Systems for SULEV Legislation: First Practical Experience," SAE Paper No. 2000-01-0500, Society of Automotive Engineers, Warrendale, PA.

11. Inoue, T., Kusada, M., Kanai, H., Hino, S., and Hyodo, Y. (2000), "Improvement of a Highly Efficient Hybrid Vehicle and Integrating Super Low Emissions," SAE Paper No. 2000-01-2930, Society of Automotive Engineers, Warrendale, PA.

12. Karwa, M.K. and Biel, J.P. (2001), "Development of a SULEV Capable Technology for a Full-Size Gasoline PFI V8 Passenger Car," SAE Paper No. 2001-01-1314, Society of Automotive Engineers, Warrendale, PA.

13. Kidokoro, T., Hoshi, K., Hiraku, K., Satoya, K., Watanabe, T., Fujiwara, T., and Suzuki, H. (2003), "Development of PZEV Exhaust Control System," SAE Paper No. 2003-01-0817, Society of Automotive Engineers, Warrendale, PA.

14. Kikuchi, S., Hatcho, S., Inose, S., and Ikeshima, K. (1999), "High Cell Density and Thin Wall Substrate for Higher Conversion Ratio Catalyst," SAE Paper No. 1999-01-0268, Society of Automotive Engineers, Warrendale, PA.

15. Kirwan, J.E., Quader, A.A., and Grieve, M.J. (2002), "Fast Start-Up On-Board Gasoline Reformer for Near-Zero Emissions in Spark-Ignition Engines," SAE Paper No. 2002-01-1011, Society of Automotive Engineers, Warrendale, PA.

16. Kishi, N., Kikuchi, S., Suzuki, N., and Hayashi, T. (1999), "Technology for Reducing Exhaust Gas Emissions in Zero-Level Emission Vehicles (ZLEV)," SAE Paper No. 1999-01-0772, Society of Automotive Engineers, Warrendale, PA.

17. Kitagawa, H. (2000), "L4-Engine Development for a Super-Ultra-Low Emissions Vehicle," SAE Paper No. 2000-01-0887, Society of Automotive Engineers, Warrendale, PA.

18. Kubsh, J.E. (2001), "Pushing the Envelope to Near-Zero Emissions on Light-Duty Gasoline Vehicles," SAE Paper No. 2001-01-3840, Society of Automotive Engineers, Warrendale, PA.

19. Marsh, P., Gottberg, I., Thorn, K.B., Lundgren, M., Acke, F., and Wirmark, G. (2000), "SULEV Emission Technologies for a Five-Cylinder N/A Engine," SAE Paper No. 2000-01-0894, Society of Automotive Engineers, Warrendale, PA.

20. Morris, D., Twigg, M.V., Collins, N.R., Brisley, R.J., Lafyatis, D.S., and Ballinger, T.H. (1999), "Catalyst Strategies for Meeting Super-Ultra-Low-Emissions-Vehicle Standards," SAE Paper No. 1999-01-3067, Society of Automotive Engineers, Warrendale, PA.

21. Mueller, W., Schmidt, J., Busch, M., Enderle, C., Heil, B., Merdes, N., Franz, J., Mowll, D., Brady, M.J., Kreuzer, T., Lindner, D., Lox, E., Bog, T., Clark, D., Henninger, R., Steopler, W., Buckel, T., Ermer, H., Kunz, A., Vogt, C.-D., Abe, F., and Makino, M. (2002), "Utilization of

Advanced Three-Way Catalyst Formulations on Ceramic Ultra-Thin-Wall Substrates for Future Legislation," SAE Paper No. 2002-01-0349, Society of Automotive Engineers, Warrendale, PA.

22. Nakagawa, S., Minowa, T., Nagano, M., Katogi, K., Higashiyama, K., and Hamada, I. (2003), "A New Catalyzed Hydrocarbon Trap Control System for ULEV/SULEV Standard," SAE Paper No. 2003-01-0567, Society of Automotive Engineers, Warrendale, PA.

23. Nishizawa, K., Momosima, S., Koga, M., Tsuchda, H., and Yamamoto, S. (2000), "Development of New Technologies Targeting Zero Emissions for Gasoline Engines," SAE Paper No. 2000-01-0890, Society of Automotive Engineers, Warrendale, PA.

24. Nishizawa, K., Mori, K., Mitsuishi, S., and Yamamoto, S. (2001), "Development of Second Generation of Gasoline PZEV Technology," SAE Paper No. 2001-01-1310, Society of Automotive Engineers, Warrendale, PA.

25. Oguma, H., Koga, M., Nishizawa, K., Momoshima, S., and Yamamoto, S. (2003), "Development of Third Generation of Gasoline PZEV Technology," SAE Paper No. 2003-01-0816, Society of Automotive Engineers, Warrendale, PA.

26. Roshan, A., Garr, G., Lindner, D.H., Mussmann, L., van den Tillaart, J.A., Lox, E., and Beason, R. (2000), "Comparison of Pd-Only, Pd/Rh, and Pt/Rh Catalysts in TLEV, LEV Vehicle Applications—Real Vehicle Data Versus Computer Modeling Results," SAE Paper No. 2000-01-0501, Society of Automotive Engineers, Warrendale, PA.

27. Tayama, A., Kanetoshi, K., Tsuchida, H., and Morita, H. (1998), "A Study of a Gasoline-Fueled Near-Zero-Emission Vehicle Using an Improved Emission Measurement System," SAE Paper No. 982555, Society of Automotive Engineers, Warrendale, PA.

28. Thoss, J.E. and Rieck, J. (1997), "The Impact of Fuel Sulfur Level on FTP Emissions—Effect of PGM Catalyst Type," SAE Paper No. 970737, Society of Automotive Engineers, Warrendale, PA.

29. Ueno, M. (2000), "A Quick Warm-Up System During Engine Start-Up Period Using Adaptive Control of Intake Air and Ignition Timing," SAE Paper No. 2000-01-0551, Society of Automotive Engineers, Warrendale, PA.

30. Webster, L., Nishizawa, K., Momoshima, S., and Koga, M. (2000), "Nissan's Gasoline SULEV Technology," SAE Paper No. 2000-01-1583, Society of Automotive Engineers, Warrendale, PA.

31. Will, N., Martin, A., Lafyatis, D., Rieck, J., and Cox, J.P. (2000), "Use of High Cell Density Substrates and High Technology Catalysts to Significantly Reduce Vehicle Emissions," SAE Paper No. 2000-01-0502, Society of Automotive Engineers, Warrendale, PA.

32. Yamamoto, S., Matsushita, K., Etoh, S., and Takaya, M. (2000), "In-Line Hydrocarbon (HC) Adsorber System for Reducing Cold-Start Emissions," SAE Paper No. 2000-01-0892, Society of Automotive Engineers, Warrendale, PA.

CHAPTER 9

The Hydrocarbon Trap

Kimiyoshi Nishizawa
Nissan Motor Co., Ltd.

9.1 Introduction

The hydrocarbon (HC) trap is the key, and the most recent, technology to reduce cold-start emissions. The 2000 Nissan Sentra CA was the world's first partial zero emissions vehicle (PZEV). This car used an HC trap system as the main technology to reduce emissions to the level required to meet PZEV standards. An HC trap catalyst consists of an HC trapping material, such as zeolite, and a catalyst coating. Before the catalyst is warmed up, the trapping material captures the HC. After the catalyst is warmed up, the trapping material releases the HC, and the catalyst coating converts it.

This chapter describes the functions of HC traps and factors to control their efficiency, and it provides some application examples. An HC trap sometimes is called an HC adsorber; however, the phrase "HC trap" will be used throughout this chapter.

9.2 Functions of the Hydrocarbon Trap

9.2.1 Hydrocarbon Trap System

The following is an example of the HC trap system that is used in the 2000 Nissan Sentra CA (Figure 9.1). This system consists of a close-coupled three-way catalyst (TWC), followed by an HC trap catalyst located under the oil pan, and a second HC trap catalyst located under the floor. Following a cold start, the following sequence of events occurs [Nishizawa *et al.* (2000)]:

- The first HC trap catalyst traps the HC until the close-coupled catalyst is warmed up.

- After some time, the first HC trap catalyst begins releasing trapped HC and converting some of it.

- The HC released from the first trap that is not converted is then trapped in the second HC trap catalyst.

- After some time, the HC is released from the second HC trap catalyst with some conversion.

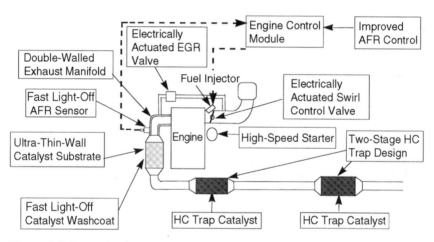

Figure 9.1 Example of an HC trap system. (AFR = Air/fuel ratio)

9.2.2 Materials

The material used for trapping HC is a type of zeolite that is coated on a substrate. Figure 9.2 shows a schematic of the coating of an HC trap catalyst. The bottom coat contains an HC trapping material, such as zeolite (Figure 9.3). The top coat contains the TWC to convert the HC released from the trapping material.

The Hydrocarbon Trap

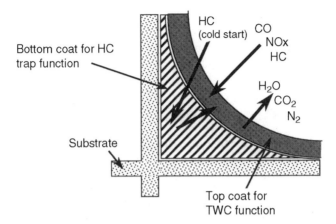

Figure 9.2 Washcoat structure.

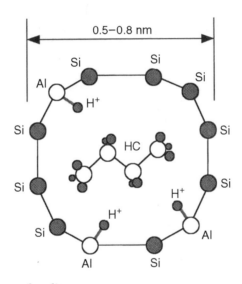

Figure 9.3 Structure of zeolite.

During a cold start, the zeolite traps the HC molecules in the holes in its porous, crystalline structure. The holes in the zeolite structure are 0.5 to 0.8 nm, while the HC molecules range in size from 0.4 to 0.7 nm. After the zeolite temperature reaches 150 to 250°C, the trapped HC is released from the material. The zeolite has an aluminosilicate crystal structure. SiO_4 and aluminum tetraoxide (AlO_4) tetrahedrons with oxygen in common form a three-dimensional network. The shape of the structure and the size of the pores are affected by the way the network is connected [Mitsuishi et al. (1999)].

9.3 Factors to Control Efficiency

Figure 9.4 shows the HC emissions during the first 40 seconds after engine start. The thin line represents the HC concentration at the inlet to the HC trap catalyst. The bold line represents the HC concentration at the outlet. The

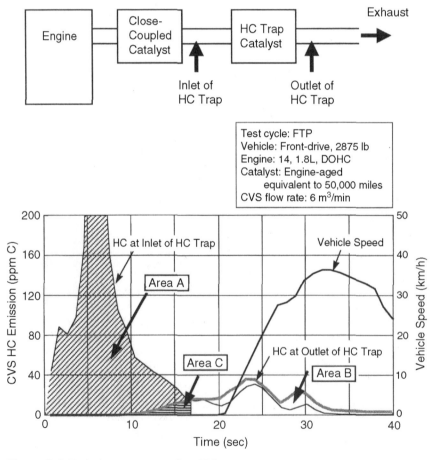

Figure 9.4 Emissions patterns of an HC trap system.

shaded area A shows the amount of HC trapped in the HC trap catalyst. The shaded area B is the unconverted HC emitted from the HC trap catalyst. The larger the area A is, the more HC is trapped, representing better trapping performance. The smaller the area B is, the more HC is converted upon release from the trap, representing better conversion performance.

Even after aging, it is not difficult to make a trap catalyst with good trapping performance, but it is difficult to maintain good conversion performance. Catalyst aging, especially at high temperatures, typically shifts the light-off temperature of the catalyst higher. If shifted high enough, the HC trap will release before the catalyst is lit off. The most important, and the most difficult, thing in the design of an HC trap system is to maximize the conversion of the trapped HC, minimizing area B in Figure 9.4. The metrics for measuring HC trap catalyst performance are trapping efficiency (TE) and conversion efficiency (CE). These two terms are defined as [Nishizawa *et al.* (2000 and 2001); Oguma *et al.* (2003)]

$$TE = \frac{A}{A+C} \times 100(\%)$$

$$CE = \left(1 - \frac{B}{A}\right) \times 100(\%)$$

9.3.1 Selecting and Developing Trapping Material

One way to improve the conversion efficiency of an HC trap system is to select and develop a trapping material that can release the trapped HC slowly. Figure 9.5 shows a comparison of the adsorption performance of various materials. Although each material had a suitable pore size for the adsorption of toluene, the adsorption efficiency varied between 50 and 70%. This variation was due to the differences in the zeolite pore size, the pore size distribution, and the frame structure. Among the traps tested, material J showed the best adsorption performance (80% or more) because of its suitable pore size distribution.

Hydrocarbons trapped by the material during the cold-start phase are desorbed as a result of a rise in the HC trap temperature due to the influence of the exhaust gas. The two most desirable properties of desorption are a high temperature before beginning to release the trapped HC and a slow rate of desorption as the trap is heated. Figure 9.5 also shows a comparison of the desorption rates for the trap materials tested. Although a large difference in the temperature at

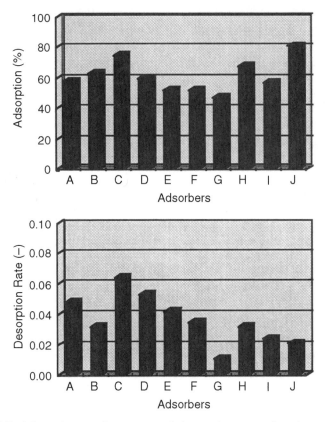

Figure 9.5 Adsorption performance and desorption rate of various adsorbers using model gas [Yamamoto et al. (2000)].

which the traps begin to release HC was not found, desorption rates did vary widely. Although material G showed the lowest rate of release, its inferior trapping performance precludes its consideration. Material J shows only a slightly higher release rate, along with superior trapping performance.

These results confirmed that the adsorption and desorption of cold-start HCs are influenced by the pore size and frame structure of the zeolite. Therefore, optimization of the pore size and distribution is needed to optimize the trapping performance of the zeolite.

Figure 9.6 shows the results obtained for several candidate adsorbers that were being tested for structural integrity. The samples were aged by steaming them at temperatures as high as 900°C. Material J exhibited the best structural stability of the samples tested and is most likely to meet high-temperature durability requirements. Based on all of the results shown, material J shows the highest potential for successful use in an HC trap catalyst [Yamamoto et al. (2000)].

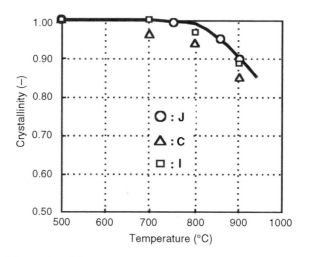

Figure 9.6 *Thermal stability of HC adsorbers in H_2O 10% air [Yamamoto* et al. *(2000)].*

9.3.2 Selecting and Developing the Catalyst Coating

In addition to selecting and developing the trapping material, another way to improve the efficiency of the HC trap catalyst system is to improve the performance of the catalyst coating on the HC trap catalyst. This can be done chiefly through a reduction in the light-off temperature of the catalytic coating. There are two ways to lower the light-off temperature of the TWC. The first way is to increase the precious metal loading; the other is to apply improved materials in the catalyst coating. Figure 9.7 shows the effect of increasing the loading of the precious metals in the TWC coating. By increasing the loading from 40 to 120 g/cf, the light-off temperature was reduced by 50°C. This

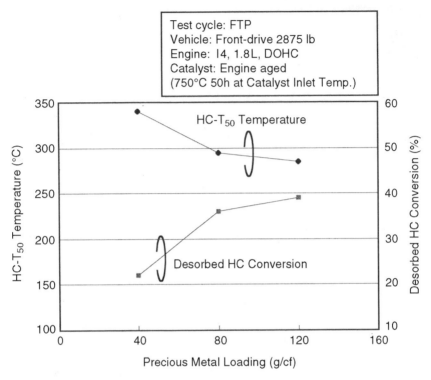

Figure 9.7 Effect on lowering light-off temperature.

reduction in light-off temperature increased the conversion efficiency by 20% in this case.

9.3.3 Selecting the Shape of the Catalyst Substrate

The substrate on which the HC trap catalyst coating is applied affects the desorbed HC conversion performance. It is known that increasing the geometric surface area (GSA) is effective in improving the light-off and conversion performance of TWCs. Figure 9.8 shows the desorption rate and desorbed HC conversion rate as a function of GSA. Increasing the GSA improves the light-off slightly, but the HC desorption rate increases, causing lower conversion rates. Reducing the GSA has the effect of delaying desorption, resulting in higher conversion rates. It is thought that the effect of GSA change is due to changes in the gas diffusion state in the overcoat layer. These results show

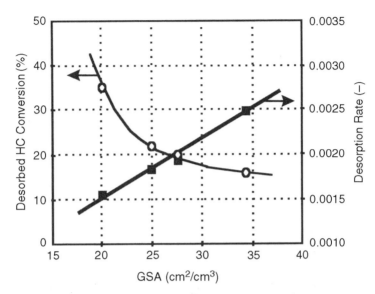

Figure 9.8 *Conversion and desorption characteristics of GSA in an HC trap system [Yamamoto et al. (2000)].*

that by changing the substrate of the HC trap catalyst, HC conversion was improved to between 45 and 50% [Yamamoto *et al.* (2000)].

9.4 Measures for Improving System Efficiency

Many measures to delay the release of HC and/or to achieve faster catalyst light-off were considered. One of the methods tried was active control of the exhaust gas path. Examples of these measures are shown in this section.

9.4.1 Actively Controlled Systems

9.4.1.1 Bypass Systems

Figures 9.9 and 9.10 show examples of the bypass type of HC trap exhaust systems. Both systems operate on the same principal but differ in the number of TWC bricks required. This bypass trap system consists of a conventional TWC, a bypass loop containing the HC trap, and a second TWC downstream. The bypass loop can be isolated from the exhaust stream by means of valves.

Near-Zero-Emission Gasoline-Powered Vehicles

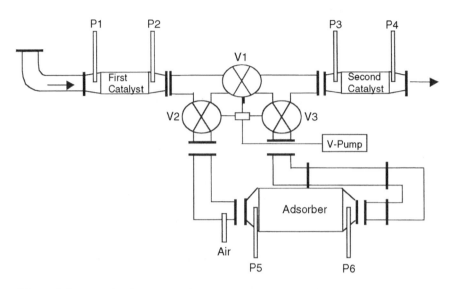

Figure 9.9 Example of one type of bypass HC trap system [Williams et al. (1996)].

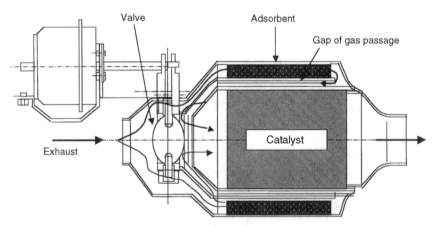

Figure 9.10 Example of a different type of bypass HC trap system [Kanazawa et al. (2001)].

The operation of the bypass type of system shown in Figure 9.9 is as follows [Heimrich *et al.* (1992); Williams *et al.* (1996); Kanazawa *et al.* (2001)]:

- From 0 to 70 seconds after engine start, valves V2 and V3 open, and valve V1 closes. This exhaust path allows the gases to heat the front catalyst toward its light-off temperature and allows any unconverted HC to flow over the HC trap.

- At 70 seconds, valve V1 opens, and valves V2 and V3 close. The first catalyst is now active, and the goal is to heat the second catalyst to its light-off temperature.

- When both catalysts reach their full operating temperatures, valves V2 and V3 are opened, and valve V1 is partially closed. This diverts part of the exhaust flow through the HC trap to heat it and release its trapped HCs. The rate of this HC release can be controlled by the amount of exhaust diverted through the bybass loop.

9.4.1.2 Substrate Hole

Figure 9.11 shows a second type of actively controlled system. It consists of a first catalyst, an adsorber with a hole in the center, and a second catalyst downstream. The hole in the adsorber is intended to allow some fraction of the exhaust gases to pass directly to the second catalyst to heat it to its light-off temperature more quickly. This design requires a tradeoff between trapping efficiency and the speed with which the second catalyst lights off. A small hole will increase the trapping efficiency but slow the light-off of the second catalyst. A large hole decreases trapping efficiency but speeds light-off of the second catalyst. It was even proposed that the hole could be "blocked" using injected air aimed at the hole to influence the exhaust gases to pass through the HC trap, thus increasing trapping efficiency. Due to the complication of the design tradeoffs for this design, it has never seen a production application [Hertl *et al.* (1996); Patil *et al.* (1996); Noda *et al.* (1997 and 1998)].

9.4.2 Improved Passive Systems

One step taken to improve system efficiency without the addition of active components is to implement a two-stage trap system. This system consists of two HC trap catalysts in sequence in the same exhaust flow. Figure 9.12 shows the improvements resulting from the two-stage trap. The conversion efficiency of desorbed HC was between 30 and 40% with a single-trap system. The adoption of a two-stage system improved conversion performance to 60%. As a

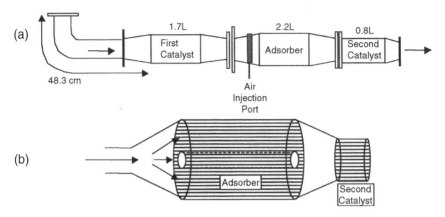

Figure 9.11 (a) Schematic diagram of an HC adsorber system, and (b) an enlarged view of the adsorber and the second catalyst [Hertl et al. (1996)].

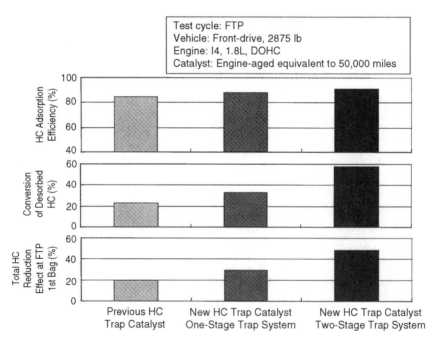

Figure 9.12 Effect of an improved HC trap system [Nishizawa et al. (2000)].

result, the total quantity of HCs in the first bag of the Federal Test Procedure (FTP) test was reduced by 50%.

In the two-stage trap system, a temperature differential exists between the two HC trap catalysts due to the heat capacity of the first unit. Hydrocarbons initially are trapped and released from the first HC trap catalyst. Although the first HC trap catalyst is hot enough to release the HCs, the second is still cool enough to adsorb the HCs. This mechanism improves system performance significantly [Nishizawa *et al.* (2000 and 2001); Oguma *et al.* (2003)].

9.5 Summary

The application of HC trap catalyst systems has only started, and the number of automotive manufacturers using such systems remains small at present. Although many improvements still must be made to HC trap catalyst systems and their application is not easy to accomplish, they have a large potential for reducing HC emissions. Hydrocarbon trap catalyst systems are the most promising way to eliminate HC emissions ultimately. It is expected that the performance of HC trap catalysts will be continually improved and that their application will be expanded widely in the near future.

9.6 References

1. Heimrich, Martin J., Smith, Lawrence R., and Kitowski, Jack (1992), "Cold-Start HC Collection for Advanced Exhaust Emission Control," SAE Paper No. 920847, Society of Automotive Engineers, Warrendale, PA.

2. Hertl, W., Patil, M.D., and Williams, J.L. (1996), "Hydrocarbon Adsorber System for Cold Start Emissions," SAE Paper No. 960347, Society of Automotive Engineers, Warrendale, PA.

3. Kanazawa, Takaaki and Sakurai, Kazuhiro (2001), "Development of the Automotive Exhaust Hydrocarbon Adsorbent," SAE Paper No. 2001-01-0660, Society of Automotive Engineers, Warrendale, PA.

4. Mitsuishi, Shunichi, Mori, Kouichi, Nishizawa, Kimiyoshi, and Yamamoto, Shinji (1999), "Emission Reduction Technologies for Turbocharged Engine," SAE Paper No. 1999-01-3629, Society of Automotive Engineers, Warrendale, PA.

5. Nishizawa, Kimiyoshi, Momoshima, Sukenori, Koga, Masaki, and Tsuchida, Hirofumi (2000), "Development of New Technology Targeting Zero Emissions for Gasoline Engine," SAE Paper No. 2000-01-0890, Society of Automotive Engineers, Warrendale, PA.

6. Nishizawa, Kimiyoshi, Mitsuishi, Shunichi, Mori, Kouichi, and Yamamoto, Shinji (2001), "Development of Second Generation of Gasoline P-ZEV Technology," SAE Paper No. 2001-01-1310, Society of Automotive Engineers, Warrendale, PA.

7. Noda, Naomi, Takahashi, Akira, and Mizuno, Hiroshige (1997), "In-Line Hydrocarbon Adsorber System for Cold Start Emission," SAE Paper No. 970266, Society of Automotive Engineers, Warrendale, PA.

8. Noda, Naomi, Takahashi, Akira, and Mizuno, Hiroshige (1998), "In-Line Hydrocarbon Adsorber System for Cold Start Emission—Part II," SAE Paper No. 980423, Society of Automotive Engineers, Warrendale, PA.

9. Oguma, Hajime, Koga, Masaki, Momoshima, Sukenori, Nishizawa, Kimiyoshi, and Yamamoto, Shinji (2003), "Development of Third Generation Gasoline P-ZEV Technology," SAE Paper No. 2003-01-0816, Society of Automotive Engineers, Warrendale, PA.

10. Patil, M.D., Hertl, W., Williams, J.L., and Nagel, J.N. (1996), "In-Line HC Adsorber System for ULEV," SAE Paper No. 960348, Society of Automotive Engineers, Warrendale, PA.

11. Williams, J.L., Patil, M.D., and Hertl, W. (1996) "By-Pass Hydrocarbon Adsorber System for ULEV," SAE Paper No. 960343, Society of Automotive Engineers, Warrendale, PA.

12. Yamamoto, Shinji, Matsushita, Kenzirou, Etoh, Satomi, and Takayo, Masahiro (2000), "In-Line Hydrocarbon Adsorber System for Reducing Cold-Start Emissions," SAE Paper No. 2000-01-0892, Society of Automotive Engineers, Warrendale, PA.

CHAPTER 10

Three-Way Catalytic Converter System Modeling

Tariq Shamim
The University of Michigan—Dearborn

10.1 Introduction

Due to progressively stricter emissions regulations, catalytic converter design and performance must undergo continuous modification. In the past, many of the design and engineering processes to optimize various components of engine and emissions systems have involved prototype testing. The complexity of modern systems and the resulting flow dynamics and thermal and chemical mechanisms have increased the difficulty in assessing and optimizing system operation. Due to the overall complexity and increased costs associated with these factors, modeling continues to be pursued as a method of obtaining valuable information supporting the design and development processes associated with exhaust emissions system optimization.

Mathematical models of catalytic converters have been employed for nearly 30 years, and different approaches simulate the phenomena of exhaust systems. In some studies, models of transient heat transfer have been used to assess the temperature distribution along the exhaust systems as dimensions, materials, and insulating properties are modified to obtain a workable system [Moore and Mondt (1993); Chen (1993); Wendland (1993)]. These models are used to define the state of the inlet gases into the catalyst as their positions change along the exhaust system. In other studies, models of flow distribution are applied to predict the non-uniformity of flow at the catalyst entrance and to predict pressure losses in the emissions system [Lai *et al.* (1992); Baxendale (1993); Lloyd-Thomas *et al.* (1993)]. These parameters have been found to influence converter performance. However, these studies emphasize only the flow patterns in the diffusion section of the catalytic converters; they do not include the full flow and kinetic processes in the total length of the converter. Detailed models of catalyst processes have been used to assess catalyst functions [Young and Finlayson (1976a and 1976b); Otto and LeGray (1980); Oh (1988); Montreuil *et al.* (1992); Pattas *et al.* (1994); Siemund *et al.* (1996); Koltsakis *et al.* (1997 and 1998); Shamim *et al.* (2002)]. In principle, these models can be used to support the understanding of the catalyst process. Some studies also focused on investigating the light-off behavior of converters [Oh and Cavendish (1985a and 1985b)] and the catalyst performance during different driving test cycles [Pattas *et al.* (1994); Shamim *et al.* (2002)]. Modeling studies also have generated new insight into the catalyst dynamic behavior during transient driving conditions [Koltsakis and Stamatelos (1999); Shamim and Medisetty (2003); Shamim (2005a and 2005b)]. Due to its various advantages, the use of modeling in designing and optimizing the catalytic converter system is now becoming common practice.

10.2 Modeling Approaches

Figure 10.1 shows a schematic of a catalytic converter. The various approaches for modeling catalytic converter systems may be classified into two main categories: (1) single-channel-based one-dimensional modeling [Pattas *et al.* (1994); Shamim *et al.* (2002)], and (2) multidimensional modeling of the entire catalytic monolith [Braun *et al.* (2000); Windmann *et al.* (2003)]. The former approach has been the most common and the most popular. It offers simplified, less computationally intensive one-dimensional handling and practically equivalent accuracy levels [Koltsakis and Stamatelos (1997)]. These approaches are explained briefly in the following sections.

10.2.1 Single-Channel-Based One-Dimensional Modeling

The governing equations are developed by considering the conservation of mass, energy, and chemical species in a single representative channel. The following phenomena are explicitly included in the model: (1) convective heat and mass transfer from the exhaust gas to the catalytic surface, (2) heterogeneous chemical reactions taking place on the catalytic surface, (3) the capacity of the catalyst to store extra oxygen under lean conditions and release it under rich

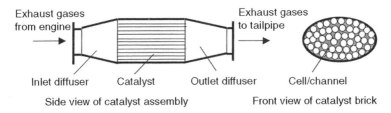

Figure 10.1 Schematic of a three-way catalytic converter.

conditions, (4) heat losses to the surroundings, and (5) heat conduction along the catalytic converter.

To simplify the mathematical model and for computational expediency, the non-uniform flow distribution at the face of the monolith generally is neglected. This simplification results in a one-dimensional model, with the resulting governing conservation equations as listed here.

The gas phase energy equation:

$$\rho_g C_{P_g} \left(\varepsilon \frac{\partial T_g}{\partial t} + v_g \frac{\partial T_g}{\partial z} \right) = -h_g G_a \left(T_g - T_s \right) \quad (10.1)$$

The gas phase species equation for gas species j:

$$\left(\varepsilon \frac{\partial C_g^j}{\partial t} + v_g \frac{\partial C_g^j}{\partial z} \right) = -km^j G_a \left(C_g^j - C_s^j \right) \quad (10.2)$$

The number and type of species considered in different modeling studies vary. However, most studies consider carbon monoxide (CO), oxides of nitrogen (NOx), and hydrocarbons (HC). Various HC species in engine exhaust generally are classified into one of two categories: (1) fast burning (represented by propylene, C_3H_6), or (2) slow burning (represented by propane, C_3H_8). Some studies also consider an additional category of inert HCs (represented by methane, CH_4). The effect of homogeneous reactions in gas phase is very small and generally is neglected.

The surface energy equation:

$$(1-\varepsilon)\rho_s C_{Ps}\frac{\partial T_s}{\partial t} =$$

$$(1-\varepsilon)\lambda_s \frac{\partial^2 T_s}{\partial z^2} + h_g G_a (T_g - T_s) - h_\infty S_{ext}(T_s - T_\infty)$$

$$+ G_a \sum_{k=1}^{n_{reaction}} R^k\left(T_s, C_s^1, \cdots, C_s^{n_{species}}\right) \cdot \Delta H^k \qquad (10.3)$$

The surface species equation for surface species j:

$$(1-\varepsilon)\frac{\partial C_s^j}{\partial t} = km^j G_a\left(C_g^j - C_s^j\right) - G_a R^j\left(T_s, C_s^1, \cdots, C_c^{N_{species}}\right) \qquad (10.4)$$

The number and type of surface species are similar to those of gas species. Equation 10.4 excluding the convective mass transport term also is used to represent the surface oxygen storage capacity (OSC) of the catalyst.

10.2.2 Multidimensional Modeling

The adequacy of single-channel-based one-dimensional models in the design and optimization of full-scale catalytic converters has recently been questioned [(Mazumder and Sengupta (2002)]. This is because of the strong coupling of the individual channels of the monolith through heat transfer and the inherent non-uniformities in flow distribution within real-life catalytic converters. For example, the conversion efficiency distribution across the monolith can be significantly non-uniform for flow maldistribution, as shown in Figure 10.2. Similarly, the temperature profiles in different channels may vary, as shown in Figure 10.3. These examples clearly indicate the importance of including multidimensional effects in monolith modeling.

In contrast to the variety of studies on reactive flows in single channels, there have been a relatively small number of sophisticated models for simulation of the entire catalytic monolith using simultaneously detailed models of transport and chemistry in the single channels [Windmann et al. (2003)]. A major difficulty in three-dimensional modeling of a full-scale catalytic converter is the

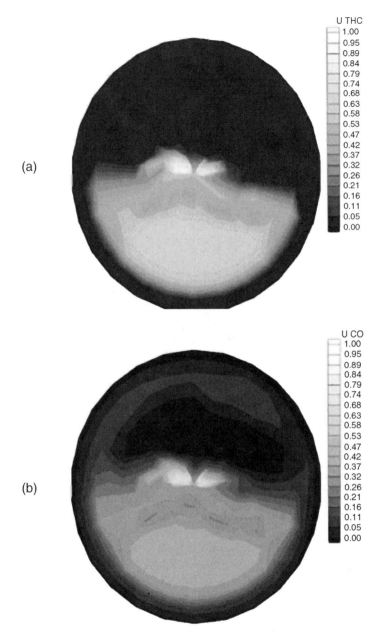

Figure 10.2 Distribution of the conversion efficiencies across the monolith 40 seconds after startup. (a) Total hydrocarbon (THC); (b) CO [Windmann et al. (2003)].

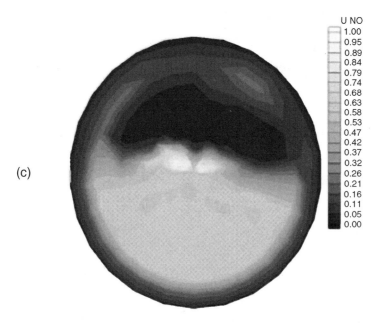

Figure 10.2 (Cont.) (c) NO [Windmann et al. (2003)].

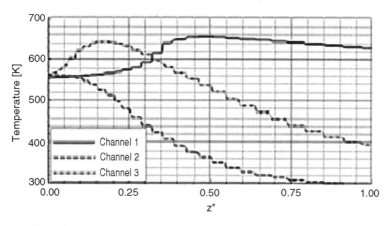

Figure 10.3 Distribution of the wall temperatures of the channels along the monolith. The relative axial coordinate is z^* [Windmann et al. (2003)].

large amount of computation, which may take several weeks on a high-end processor. To circumvent this difficulty, many studies employ a mixed dimensional approach. This approach exploits the fact that monolith channels are coupled to each other only through heat transfer; thus, it is possible to link one-dimensional transport-reaction models within the monolith channels with three-dimensional heat conduction models for the entire converter using prescribed heat and mass transfer coefficients to attain the desired coupling [Mazumder and Sengupta (2002)]. This approach has been successful in modeling the global three-dimensional effects and in predicting the effect of temperature non-uniformity on reaction kinetics and conversion efficiency [Chen *et al.* (1988); Jahn *et al.* (1997); Braun *et al.* (2000); Windmann *et al.* (2003)].

Recently, a new approach for modeling of full-scale catalytic converters with low computational cost is presented by Mazumder and Sengupta (2002). In this approach, the entire catalytic monolith is treated as an anisotropic porous medium, and sub-grid scale models are employed to represent the heterogeneous chemical reactions occurring at the solid-fluid interfaces within the monolith. Full coupling between fluid flow, heat transfer, species transport, and heterogeneous chemical reactions is achieved through flux balance of species and energy at the solid-fluid interfaces. The model allows for an unlimited number of finite-rate reaction steps and species, including surface-adsorbed species and site coverage effects.

10.3 Chemical Reaction Mechanisms

A number of heterogeneous reactions occur on the catalyst surface as the exhaust gases pass through the catalyst. Adequate kinetics data for these reactions generally are lacking, which poses a major difficulty in the modeling of catalytic converters. Furthermore, the effect of thermal degradation, which greatly influences the catalyst operation, is largely unknown and is not incorporated into the kinetic data. Most modeling studies rely on the classical work of Voltz *et al.* (1973) to simulate chemical reactions and to calculate the reaction rates. This work involved the measurement of pellet-type platinum (Pt) catalyst performances and derived kinetic reaction rate expressions for the oxidation of CO and C_3H_6 under lean conditions. The rate expressions are of the Langmuir-Hinshelwood type and take into account the inhibition due to NO. These rate expressions with some modifications have been used for non-Pt catalysts due to lack of data for other types of catalysts. A methodology for updating steady-state kinetic data for Pt/Rh and Pd/Rh catalysts has been presented by Montreuil *et al.* (1992).

The catalyst heterogeneous reactions have been represented by several chemical reaction mechanisms in literature. These mechanisms vary in details, accuracy, and convenience of use. The following sections briefly describe these mechanisms.

10.3.1 Three-Step Chemical Reaction Mechanism

The reaction mechanism listed below was proposed by Oh and coworkers [Oh and Cavendish (1982); Chen *et al.* (1988)]. This model considers only the oxidation of CO, HC, and H_2. The HC oxidation is represented by the reaction of propylene ("fast-oxidizing HC"); the effects of other HCs are neglected.

$$CO + \frac{1}{2}O_2 \rightarrow CO_2 \qquad \text{Reaction heat} = -2.832*10^5 \text{ (J/mol)} \qquad (10.5)$$

$$C_3H_6 + \frac{9}{2}O_2 \rightarrow 3CO_2 + 3H_2O \quad \text{Reaction heat} = -1.928*10^6 \text{ (J/mol)} \quad (10.6)$$

$$H_2 + \frac{1}{2}O_2 \rightarrow H_2O \qquad \text{Reaction heat} = -2.42*10^5 \text{ (J/mol)} \qquad (10.7)$$

The expressions for reaction rates are

$$R_1 = k_1 C_{CO} C_{O_2}/G \qquad (10.8)$$

$$R_2 = k_2 C_{C_3H_6} C_{O_2}/G \qquad (10.9)$$

$$R_3 = k_1 C_{H_2} C_{O_2}/G \qquad (10.10)$$

where

$$k_1 = 6.699 \times 10^9 \exp(-12{,}556/T_S) \qquad \text{mol/cm}^2 \cdot \text{s} \qquad (10.11)$$

$$k_2 = 1.392 \times 10^{11} \exp(-14{,}556/T_S) \qquad \text{mol/cm}^2 \cdot \text{s} \qquad (10.12)$$

$$G = T_S\left(1 + K_1 C_{CO} + K_2 C_{C_3H_6}\right)^2 \left(1 + K_3 C_{CO}^2 C_{C_3H_6}^2\right)\left(1 + K_4 C_{NO}^{0.7}\right) \quad (10.13)$$

$K_1 = 65.5 \, \text{EXP}(961/T_S)$ Dimensionless   (10.14)

$K_2 = 2.08 \times 10^3 \, \text{EXP}(361/T_S)$ Dimensionless   (10.15)

$K_3 = 3.98 \, \text{EXP}(11{,}611/T_S)$ Dimensionless   (10.16)

$K_4 = 4.79 \times 10^5 \, \text{EXP}(-3733/T_S)$ Dimensionless   (10.17)

10.3.2 Four-Step Chemical Reaction Mechanism

This widely used reaction scheme considers the reactions described in Eqs. 10.5 through 10.7 with the nitric oxide (NO) reduction by CO. This mechanism has been shown to simulate the catalyst reactions with reasonable accuracy [Siemund et al. (1996)]. The NO reduction used in this mechanism is

$$CO + NO = CO_2 + \frac{1}{2}N_2 \quad \text{Reaction heat} = -3.73 * 10^5 \, (\text{J/mol}) \quad (10.18)$$

The rate expression for NO reduction by CO is generally taken from Subramanian and Verma (1984 and 1985) as

$$R_4 = \frac{k_4 C_{CO}^{1.4} C_{O_2}^{0.3} C_{NO}^{0.13}}{T_S^{-0.17}(T + k_5 C_{CO})^2} \quad (10.19)$$

where

$k_4 = 3.067 \times 10^8 \exp(-8771/T_S)$ mol/cm²·s   (10.20)

$k_5 = 1.2028 \times 10^5 \exp(653.5/T_S)$ K   (10.21)

All unburned HCs are represented by CH_y (y is the ratio of hydrogen to carbon in the fuel). The heat of formation of CH_y is assumed to be one-third of C_3H_6. Due to its simplicity and accuracy, the four-step mechanism is widely used in simulating catalyst performance. However, this mechanism ignores the variation in the reaction rates of several HC species by lumping them into one category. It also ignores the effect of water-gas shift and steam reforming effects, which become especially important during severe transient driving conditions.

10.3.3 Modified Four-Step Chemical Reaction Mechanism

This reaction mechanism is essentially similar to the four-step mechanism described previously. The only difference is the modification in the rate expression of NO reduction. Based on comparison with the experimental measurement, Shamim et al. (2000) modified the exponent of CO concentration in the rate expression R_4 from 1.4 to 1.9. The modified reaction rate of NO reduction is

$$R_4 = \frac{k_4 C_{CO}^{1.9} C_{O_2}^{0.3} C_{NO}^{0.13}}{T_S^{-0.17}(T + k_5 C_{CO})^2} \quad (10.22)$$

Shamim et al. (2000) showed that the modification resulted in much better agreement with the measurements, particularly the predictions of CO and HC conversions (Figure 10.4).

10.3.4 Five-Step Chemical Reaction Mechanism

This mechanism is obtained by adding a steam reforming reaction in the four-step reaction scheme (described in Eqs. 10.5 through 10.7 and Eq. 10.18). Consideration of the steam reforming reaction is especially important for catalysts with rhodium. The reaction can be expressed as

$$C_3H_6 + 3H_2O = 3CO + 6H_2 \quad \text{Reaction heat} = 3.7346*10^5 \text{ (J/mol)} \quad (10.23)$$

The reaction rate expression for the steam reforming also includes the effects of the chemical equilibrium of the reaction [Koltsakis et al. (1997)]. In the

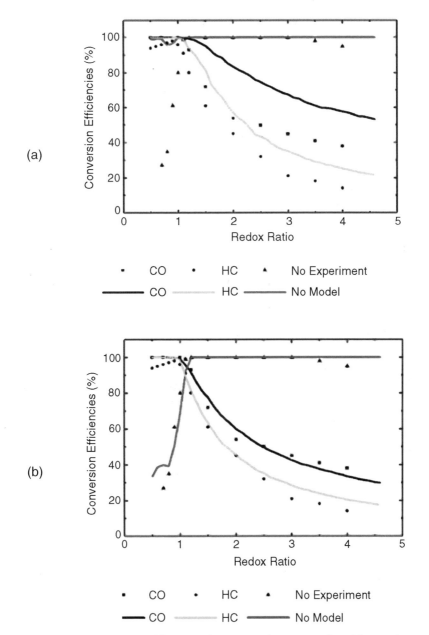

Figure 10.4 Comparison of four-step kinetic mechanism results with experimental measurements under steady-state conditions. (a) Standard four-step mechanism; (b) modified four-step mechanism [Shamim et al. (2000)].

absence of precise kinetic measurements, Koltsakis et al. incorporated this effect by multiplying the reaction rate expression by a factor as

$$R_5 = \left(k_5 C_{C_3H_6} C_{H_2O}/G\right) \cdot \Theta \tag{10.24}$$

where

$$k_5 = 1.7 \times 10^{12} \exp(-12{,}630/T_S) \tag{10.25}$$

and

$$\Theta = 1 - \frac{C_{CO}^3 \, C_{H_2}^6}{C_{C_3H_6}^3 \, C_{H_2O}^3 \, K_p(T)} \tag{10.26}$$

The value of G is as defined in Eq. 10.13.

10.3.5 Six-Step Chemical Reaction Mechanism

This mechanism, which has been used by Koltsakis et al. (1997), is similar to the five-step mechanism with the exception that HC is divided into two categories: (1) fast burning, and (2) slow burning. Different oxidation rates are used for the two HC categories. Only fast-burning HC is considered to participate in the steam reforming reaction. The reaction scheme and the kinetic rate expressions used by Kolstakis et al. (1997) are as follows:

$$CO + \frac{1}{2}O_2 \rightarrow CO_2 \tag{10.27}$$

$$H_2 + \frac{1}{2}O_2 \rightarrow H_2O \tag{10.28}$$

$$C_xH_y + \left(x + \frac{y}{4}\right)O_2 \rightarrow xCO_2 + \frac{y}{2}H_2O \quad (HC_F) \tag{10.29}$$

$$C_xH_y + \left(x + \frac{y}{4}\right)O_2 \rightarrow xCO_2 + \frac{y}{2}H_2O \quad (HC_S) \tag{10.30}$$

$$C_xH_y + xH_2O \rightarrow xCO + \left(x + \frac{y}{2}\right)H_2 \qquad (HC_F) \qquad (10.31)$$

$$CO + NO \rightarrow CO_2 + \frac{1}{2}N_2 \qquad (10.32)$$

The corresponding reaction rate expressions are as follows:

$$R_1 = k_1 C_{CO} C_{O_2}/G \qquad (10.33)$$

$$R_2 = k_2 C_{H_2} C_{O_2}/G \qquad (10.34)$$

$$R_3 = k_3 C_{C_xH_y} C_{O_2}/G \qquad (10.35)$$

$$R_4 = k_4 C_{C_xH_y} C_{O_2}/G \qquad (10.36)$$

$$R_5 = \left(k_5 C_{C_xH_y} C_{H_2O}/G\right) \cdot \Theta \qquad (10.37)$$

$$R_6 = k_6 C_{CO}^m C_{NO}^{0.5} C_{NO}/G \qquad (10.38)$$

where

$$k_1 = 5 \times 10^{16} \exp(-11{,}426/T_S) \qquad \text{mol·K/m}^3\text{·s} \qquad (10.39)$$

$$k_2 = 5 \times 10^{16} \exp(-11{,}426/T_S) \qquad \text{mol·K/m}^3\text{·s} \qquad (10.40)$$

$$k_3 = 1 \times 10^{18} \exp(-12{,}630/T_S) \qquad \text{mol·K/m}^3\text{·s} \qquad (10.41)$$

$$k_4 = 1.2 \times 10^{18} \exp(-15{,}035/T_S) \qquad \text{mol·K/m}^3\text{·s} \qquad (10.42)$$

$$k_5 = 1.7 \times 10^{12} \exp(-12{,}630/T_S) \qquad \text{mol·K/m}^3\text{·s} \qquad (10.43)$$

$$k_6 = 1.5 \times 10^6 \exp(-8420/T_S) \qquad \text{mol·K/m}^3\text{·s} \qquad (10.44)$$

$$m = -0.19\left(1 - 6.266 \cdot \exp(-m_1 C_{CO})\right) \quad (10.45)$$

Here m_1 is a tunable factor. HC_F and HC_S represent fast- and slow-burning HCs, respectively. G, K, and Θ are as defined.

10.3.6 Thirteen-Step Chemical Reaction Mechanism

This mechanism consists of 13 independent forward pathways for oxidation of CO, H_2, C_3H_6, C_3H_8, and NH_3 with O_2 and NO as oxidizing agents, and their corresponding rich and lean kinetic rate expressions. This chemical reaction scheme and kinetic data originally were presented by Otto and LeGray (1980) and later were presented by Montreuil et al. (1992) with the modified kinetic data. The reaction scheme is as follows:

$$CO + \frac{1}{2}O_2 \rightarrow CO_2 \quad (10.46)$$

$$CH_\alpha + M_1 \cdot O_2 \rightarrow CO_2 + \frac{\alpha}{2} \cdot H_2O \quad (HC_F) \quad (10.47)$$

$$CH_\alpha + M_2 \cdot O_2 \rightarrow CO + \frac{\alpha}{2} \cdot H_2O \quad (HC_F) \quad (10.48)$$

$$H_2 + \frac{1}{2}O_2 \rightarrow H_2O \quad (10.49)$$

$$NH_3 + \frac{3}{4}O_2 \rightarrow 1.5H_2O + \frac{1}{2}N_2 \quad (10.50)$$

$$CO + NO \rightarrow CO_2 + \frac{1}{2}N_2 \quad (10.51)$$

$$2.5CO + NO + 1.5H_2O \rightarrow NH_3 + 2.5CO_2 + \frac{1}{2}N_2 \quad (10.52)$$

$$H_2 + NO \rightarrow H_2O + \frac{1}{2}N_2 \quad (10.53)$$

$$2.5H_2 + NO \rightarrow NH_3 + H_2O \qquad (10.54)$$

$$CH_\alpha + M_1 \cdot O_2 \rightarrow CO_2 + \frac{\alpha}{2} \cdot H_2O \qquad (HC_S) \qquad (10.55)$$

$$\frac{1}{2M_2}CH_\alpha + NO \rightarrow \frac{1}{2M_2}CO + \frac{\alpha}{4M_2} \cdot H_2O + \frac{1}{2}N_2 \qquad (HC_F) \qquad (10.56)$$

$$CH_\alpha + M_3 \cdot NO \rightarrow CO_2 + \frac{\alpha}{2} \cdot H_2O + \frac{M_3}{2} \cdot N_2 \qquad (HC_F) \qquad (10.57)$$

$$\frac{2.5}{2M_2}CH_\alpha + NO + \frac{(3-\alpha)}{4M_2} \cdot H_2O \rightarrow NH_3 + \frac{2.5}{2M_2}CO \qquad (HC_F) \qquad (10.58)$$

where

$$M_1 = \left[1 + \frac{\alpha}{4}\right]; \quad M_2 = \left[\frac{1}{2} + \frac{\alpha}{4}\right]; \quad M_3 = \left[2 + \frac{\alpha}{2}\right] \qquad (10.59)$$

Here, α is the hydrogen-to-carbon ratio. HC_F and HC_S represent fast- and slow-burning HCs, respectively, which generally are represented by propylene and propane, respectively. The corresponding reaction rate expressions are listed in Section 10.11 Appendix at the end of this chapter. For different catalyst formulations, the coefficients in the reaction rate expressions are different and vary with the aging of the catalytic converters. The mechanism has been found to simulate the performance of catalytic converters under both steady-state and transient conditions with good accuracy [Montreuil et al. (1992); Shamim et al. (2000 and 2002)]. However, the use of these detailed kinetic rate expressions requires the knowledge of 97 constants, which requires detailed experimentation for different types of catalysts.

10.3.7 Multistep Chemical Reaction Mechanism with Elementary Reaction

The global model used in most studies neglects the various single reactions that occur on the surface. An alternate approach is the description of the chemical reactions by a set of elementary reaction steps. The reaction equations of the elementary steps describe the reactions on a molecular level, so the approach is much more accurate than a globally fitted kinetic. The main advantage of these detailed reaction mechanisms is their potential to predict the behavior of the chemical system at different external conditions. The disadvantage of using elementary chemical reactions is the large number of reaction equations, which demands a large computational capacity. Furthermore, the rate coefficients of all the single steps must be known [Braun *et al.* (2000)].

A detailed surface reaction mechanism in a three-way catalyst (TWC) is proposed by Chatterjee *et al.* (2001)]. The mechanism considers the sample exhaust gas mixture is composed of C_3H_6, CH_4, CO_2, H_2O, CO, NO, O_2, and N_2. The surface reaction scheme consists of 62 reaction steps, among these eight gas-phase and further 29 adsorbed chemical species. It is assumed that all species are adsorbed competitively. The model also considers the different adsorption sites (platinum or rhodium) of the surface. However, on rhodium, surface reactions are considered only among NO, CO, and O_2. The kinetic data of the mechanism were taken either from literature or fits to experimental data [Windmann *et al.* (2003)]. The mechanism is described in more detail by Chatterjee *et al.* (2001).

10.3.8 Influence of Catalyst Deactivation on Reaction Mechanism

Over time and usage, the catalyst performance (activity and/or selectivity) deteriorates for a variety of reasons. For example, a catalytic converter may be poisoned or fouled by fuel or lubricant additives and/or engine corrosion products [Franz *et al.* (2005)]. Exposure to high temperature also may result in thermal degradation, which may result in the formation of active phase crystallite growth, collapse of the carrier (support) pore structure, and/or solid-state reactions of the active phase with the carrier or promoters [Bartholomew (2001)]. A good review of the state of knowledge regarding catalyst deactivation mechanisms is provided by Bartholomew (2001).

Catalyst deactivation can influence the reaction rate in three ways: (1) decrease the number of active sites, (2) decrease the quality of the active sites, and (3) degrade the accessibility of the pore space [Moulijn *et al.* (2001)]. The catalyst activity as a function of temperature and time may be described by an empirical relation

$$\frac{da}{dt} = k_d a^m \tag{10.60}$$

where

$$k_d = k_{d0} \exp\left(-\frac{E_a}{RT}\right) \tag{10.61}$$

Here, a is the catalyst activity, and k_d is the deactivation rate constant. The exponent m in Eq. 10.60 often has a value of 2. However, the models for predicting the deactivation of automotive catalytic converter based on first principle are not available in open literature.

10.4 Oxygen Storage Mechanism

The efficient operation of a TWC requires that an engine burn a stoichiometric mixture of air and fuel. This is ensured by employing a feedback control system comprising a lambda sensor and a control unit. However, the response lag of the system (due to the exhaust gas travel time and the response delay of the sensor) causes the air/fuel (A/F) ratio to fluctuate rapidly around the control set point. The negative impact of such modulation on the catalyst performance can be avoided by storing the extra oxygen under fuel lean conditions and releasing it under rich conditions [Gandhi et al. (1976)]. The released oxygen may participate in the reactions with the reducing agents, thereby increasing the conversion of CO and HC in a rich exhaust-gas environment [Gandhi et al. (1976); Koltsakis and Stamatelos (1997)]. Such an OSC is developed in the modern catalyst by coating its substrate with a washcoat material containing ceria. The OSC is recognized as an important mechanism affecting catalyst behavior during vehicle acceleration and deceleration.

In addition to the pathways that specify the kinetics over the noble metal sites, another kinetic mechanism is required to represent the OSC. Two of the various OSC mechanisms presented in literature are described in the following sections.

10.4.1 Simple Single-Step Oxygen Storage Capacity Mechanism

In this simple model, presented by Koltsakis et al. (1997), the oxygen storage mechanism is described by the oxidation and reduction of the cerium (Ce) oxides present in the washcoat, according to the reaction

$$2CeO_2 \Leftrightarrow Ce_2O_3 + \frac{1}{2}CO_2 \qquad (10.62)$$

The reaction to the right denotes the release of an oxygen atom, which is made available to react with a reducing species of the exhaust gas (e.g., CO). The left direction of the reaction represents the storage of an oxygen atom by increasing the oxidation state of Ce_2O_3. An auxiliary number ψ is defined to represent the fractional extent of oxidation of the oxygen storage component as

$$\psi = \frac{2 \times \text{moles of } CeO_2}{2 \times \text{moles of } CeO_2 + \text{moles of } Ce_2O_3} \qquad (10.63)$$

The oxidation reaction rate is expressed as

$$R_{ox} = k_{ox} C_{s,O_2} \Psi_{cap} (1-\psi) \qquad (10.64)$$

where $\Psi_{cap}(1-\psi)$ represents the available active sites of "reduced-state" cerium oxide, and

$$k_{ox} = 9 \times 10^8 \exp(-10{,}825/T_S) \qquad \text{mol/m}^3 \cdot \text{s} \qquad (10.65)$$

The reduction reaction rate is expressed in a similar manner as

$$R_{red} = k_{red} C_{CO} \Psi_{cap} \psi \qquad (10.66)$$

where

$$k_{ox} = 3 \times 10^8 \exp(-10{,}825/T_S) \qquad \text{mol/m}^3 \cdot \text{s} \qquad (10.67)$$

The variation of the oxidation state can be found at each location by

$$\frac{d\psi}{dt} = -\frac{R_{red}}{\Psi_{cap}} + \frac{R_{ox}}{\Psi_{cap}} \qquad (10.68)$$

10.4.2 Detailed Nine-Step Oxygen Storage Capacity Mechanism

In this more detailed reaction scheme, postulated by Otto (1984), the OSC is simulated by designating two kinds of sites that can be oxidized and reduced through a nine-step site reaction mechanism. The metal reduction site on the surface is defined as <S>, and the oxidized site is defined as <OS>. The accuracy of this mechanism in simulating the OSC phenomenon during the transient driving conditions is shown by Shamim et al. (2000 and 2002). The mechanism is described by the following nine reactions:

$$<S> + \frac{1}{2}O_2 \rightarrow <OS> \qquad \text{Site oxidation} \qquad (10.69)$$

$$<OS> + CO \rightarrow <S> + CO_2 \qquad \text{Site reduction by CO} \qquad (10.70)$$

$$H_2O + CO \rightarrow H_2 + CO_2 \qquad \text{Water-gas shift} \qquad (10.71)$$

$$<OS> + H_2 \rightarrow <S> + H_2O \qquad \text{Site reduction by } H_2 \qquad (10.72)$$

$$CH_\alpha + H_2O \rightarrow CO + \left(1 + \frac{\alpha}{2}\right) \cdot H_2 \qquad \text{Steam reforming} \qquad (10.73)$$

$$\frac{3}{2}CH_\alpha + <OS> \rightarrow <S> + \frac{1}{2}C + CO + \left(\frac{3}{4}\alpha\right)H \qquad \text{Reduction by HC} \qquad (10.74)$$

$$C + O_2 \rightarrow CO_2 \qquad \text{Coke burn-off} \qquad (10.75)$$

$$<S> + NO \rightarrow <OS> + \frac{1}{2}N_2 \qquad \text{NO storage} \qquad (10.76)$$

$$<OS> + \frac{2}{5}NH_3 \rightarrow <S> + \frac{2}{5}NO + \frac{3}{5}H_2O \qquad NH_3 \text{ site reduction} \qquad (10.77)$$

where α is the hydrogen-to-carbon ratio of the HCs.

Each reaction has two rate expressions, one being the fast site rate expression, and the other being the slow site rate expression. Each expression is associated with the materials in the catalyst washcoat. The total sites are conserved for both fast site and slow sites. Therefore,

$$S_{total,\,f} = <S>_f + <OS>_f \tag{10.78}$$

$$S_{total,\,s} = <S>_s + <OS>_s \tag{10.79}$$

The rates of the transient reactions are of the form

$$R_{transient} = \frac{(OXSW) \cdot CTR \cdot e^{-E/RT_s} \cdot \prod_{i=1}^{N_{specie}} \left(X_s^i\right)^{ex(i)}}{1 + \sum_{n=1}^{N_{specie}} K_n \cdot X_s^n} \tag{10.80}$$

where OXSW is an oxidation switch and equals 1 if the redox ratio is less than 1, or 0 if the redox ratio is greater than 1. The redox ratio is defined as

$$\frac{[CO] + [H_2] + 6 \cdot \left(1 + \frac{\alpha}{4}\right)[HC]}{[NO] + 2 \cdot [O_2]} \tag{10.81}$$

The corresponding coefficients for calculating the transient reaction rates are based on experimental data and are listed by Li *et al.* (1996).

10.5 Heat and Mass Transfer Phenomena

In addition to chemical reactions, the catalyst performance is strongly influenced by the transport mechanisms of mass and heat transfer from the exhaust gases to the catalyst surface. Especially at high temperatures, the mass transfer process becomes rate limiting to an increasing extent [Holmgren and Andersson (1998)]. Hence, a correct description of the transport processes is important in catalyst modeling [Shamim (2003)].

The convective heat transfer between the exhaust gas and the catalyst substrate, and the heat generated during exothermic reactions, are the main heat transfer modes in the catalytic converter. Due to low catalyst operation temperatures, the radiation heat exchange between the substrate and the surrounding walls generally is neglected. Most studies also include the effect of convective heat loss from the converter surface to the ambient.

The mass transfer in the catalyst channels is due to the concentration gradients between the exhaust gas and the reactive washcoat [Koltsakis and Stamatelos (1997)]. Due to the low concentrations involved, the mass transfer is governed by the laws of diffusion in dilute mixtures, and the analogies between heat and mass transfer are fully applicable [Mondt (1987)].

Most studies modeled the convective heat and mass transfer processes by using the simplified one-dimensional film model. In the film model, the dimensionless Nusselt (Nu) and Sherwood (Sh) numbers describe the heat and mass transfer rates. The heat and mass transfer coefficients (h_g and k_m^j) in the governing equations are calculated from the following:

$$h_g = \frac{Nu \, \lambda_g}{D_h} \qquad (10.82)$$

$$k_m^j = \frac{Sh \, D_j}{D_h} \qquad (10.83)$$

Without being too sophisticated, the film model is a good compromise to obtain a reasonable prediction of the converter efficiency [Massing *et al.* (2000)]. However, many uncertainties remain about the estimate of Nu and Sh numbers [Ryan *et al.* (1991)]. Some studies [e.g., Young and Finlayson (1976a and 1976b)] compute Nu and Sh numbers assuming the flow in monolith channels is fully developed laminar. Other studies [e.g., Hawthorns (1974)] include the effect of developing laminar flow at the channel inlet in the calculation of Nu and Sh numbers. The effect of initial turbulence in the channel flow also has been proposed to be included in the analysis of the transport mechanism [Holmgren and Andersson (1998)]. Also, the ignition of reactions inside the monolith has been reported to influence the transfer processes [Hayes and Kolaczkowski (1994)]. In addition, the monolith cell density has been found to influence the mass transfer coefficient [Ullah *et al.* (1992)].

Hayes and Kolaczkowski (1994) reported that the correct value of Nu and Sh numbers in the fully developed region for an adiabatic cylindrical system at steady-state conditions is in the range 3.0 to 4.5. For other geometry, such as square or triangular ducts, the value may be lower. For sharp transients, they found that the values of Nu and Sh numbers might exhibit discontinuities. Holmgren and Andersson (1998) predicted a much higher range for Sh numbers (3.7 to 7) for square monolith channels. They ascribed the higher values of Sh numbers to the turbulence effects. Although the Reynolds (Re) numbers in the monolith channel generally are below 600, a certain degree of turbulence is present in the channel. The channel turbulence is created partly by the presence of turbulence in the entering flow, turbulence generated by monolith walls at the entrance, and the surface roughness of the channels. Enhancement of the heat and mass transfer processes as a result of the presence of surface roughness has been documented by several other studies [e.g., Mottahed and Molki (1996)]. Hatton *et al.* (1999) suggested the inclusion of a temperature-dependent term into the Sh relationship. In a numerical heat transfer study, Day *et al.* (2000) reported that the Nu number varies in the range 2.8 to 8 for a triangular channel. The value was found to remain close to 3 for most of the channel length.

Some modeling studies incorporate the inlet and other effects on the transport processes by determining the values of Nu and Sh numbers using the following type of empirical correlations with Re, Pr, and Sc numbers [Shamim (2003)]:

$$Nu = c\left(Re\, Pr\, \frac{L}{z}\right)^n \tag{10.84}$$

$$Sh = c\left(Re\, Sc\, \frac{L}{z}\right)^n \tag{10.85}$$

Here, c and n are empirical constants.

10.6 Inlet Flow Distribution

Converter performance is strongly influenced by the flow field in its channels. Non-uniformities in the velocity field at the inlet of the converter, as shown in Figure 10.5, may cause both poor converter performance due to localized high space velocities and increased aging due to poison accumulation in high-mass-flow-rate areas of the monolith. The flow maldistribution is due to the fact that the core area of the frontal face of a brick (approximately 35% of the total frontal area) receives more than 50% of the emerging flow with higher velocity, and the edge of the brick (approximately 20% of the total area) receives only

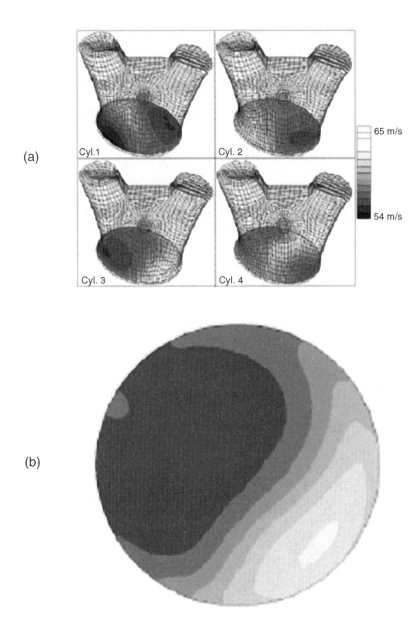

Figure 10.5 Non-uniformities in the catalyst flow. (a) Flow distribution at the catalyst inlet for a four-cylinder engine. (b) Velocity distribution at steady-state cold flow for one of the cylinders. The results were obtained using CFD calculations [Breuer et al. (2000)].

5 to 10% of the emerging flow [Kim and Son (1999)]. The effect is influenced by various parameters including inlet pipe bend, flow, Reynolds number, brick resistance, and inlet velocity profile [Lai *et al.* (1992)]. This effect is even more significant over a driving cycle with aged catalysts [Martin *et al.* (1999)]. Due to its practical importance, the catalyst flow maldistribution has been a subject of active research [Howitt and Sekella (1974); Bella *et al.* (1991); Lai *et al.* (1992); Martin *et al.* (1999); Kim and Son (1999); Breuer *et al.* (2000); Braun *et al.* (2000); Windmann *et al.* (2003)].

The effect of flow non-uniformity is incorporated into the catalyst modeling by describing the exhaust flow through the catalyst using conservation of mass and momentum equations. Using the general assumptions of a stationary, incompressible, three-dimensional turbulent flow, the time-averaged equations of motion can be written as listed here [Lai *et al.* (1992); Kim *et al.* (1995); Kim and Son (1999)].

Conservation of mass (continuity):

$$\frac{\partial U_j}{\partial X_j} = 0 \qquad (10.86)$$

Conservation of momentum:

$$U_j \frac{\partial U_i}{\partial X_j} = -2 \frac{\partial P}{\partial X_i} + \frac{\partial}{\partial X_j}\left(\frac{1}{Re} + v_t\right)\left(\frac{\partial U_i}{\partial X_j} + \frac{\partial U_j}{\partial X_i}\right) + BR \qquad (10.87)$$

where BR is the brick resistance formulated by Eq. 10.94, which is turned on only inside the brick along the streamwise direction. For turbulent flow calculation, the standard $k - \varepsilon$ model is used to calculate the turbulent eddy viscosity as

$$v_t = C_\mu \frac{k^2}{\varepsilon} \qquad (10.88)$$

The model equations for turbulent kinetic energy k and turbulent dissipation rate ε are as follows:

$$\frac{\partial U_j k}{\partial X_j} = \frac{\partial}{\partial X_j}\left(\frac{v_t}{\sigma_k} \frac{\partial k}{\partial X_j}\right) + G_k - \varepsilon \qquad (10.89)$$

$$\frac{\partial U_j \varepsilon}{\partial X_j} = \frac{\partial}{\partial X_j}\left(\frac{v_t}{\sigma_\varepsilon}\frac{\partial \varepsilon}{\partial X_j}\right) + C_1 \frac{\varepsilon}{k} G_k - C_2 \frac{\varepsilon^2}{k} \tag{10.90}$$

where G_k is the generation term of k,

$$G_k = v_t \left(\frac{\partial U_i}{\partial X_j} + \frac{\partial U_j}{\partial X_i}\right)\frac{\partial U_i}{\partial X_j} \tag{10.91}$$

The dimensionless empirical constants recommended for plane shear layers and plane jet [Launder and Spalding (1974)] are generally used by many researchers without alteration,

$$C_\mu = 0.09, \ C_1 = 1.44, \ C_2 = 1.92, \ \sigma_k = 1, \ \sigma_\varepsilon = 1.3 \tag{10.92}$$

The brick resistance formulation of the monolithic square-cell flow passage for steady, incompressible, fully developed laminar flow in a constant-area duct [White (1999)] is expressed as

$$f_r \, Re_{cell} = 56.91 \tag{10.93}$$

By choosing the diameter and mean velocity at the inlet pipe as the reference length and velocity scales, the formulation for the nondimensional pressure gradient within the duct is then

$$BR = E \frac{28.488}{Re_i} \frac{V_c}{D_{hc}^2} \tag{10.94}$$

where Re_i is the Reynolds number at the inlet pipe, V_c is the flow velocity inside the cell of the brick, and D_{hc} is the hydraulic diameter of the cell. Also, E is a correction factor for the entrance effect, and its values can by found by using the approach of Johnson and Chang (1974),

$$E = \left(1 + \frac{0.0445}{x^*}\right)^{0.5} \tag{10.95}$$

where $x^* = \dfrac{L}{D_{hc}}$. Re_c is the nondimensional brick length, and $Re = \dfrac{V_c DH_c}{\nu}$ is the Reynolds number inside the monolith cell. Simulated local flow velocity with high accuracy inside the monolith is required to prescribe the pressure gradient indicated in Eq. 10.94. To achieve this aim, the flow inside the monolith is treated as one-dimensional. Convection and diffusion transports in the other two directions are disabled. The effect of the porosity of the brick is included by partially blocking the streamwise convective and diffusive fluxes across the discretized cell surface. Local flow velocity inside the monolith V_c is solved together with the remainder of the converter in the multidimensional simulation of momentum transport. This formulation has been validated with experimental pressure drop data across bricks of different cell densities and at different flow rates.

10.6.1 Flow Distribution Index

The degree of flow non-uniformity is generally quantified by devising an index of flow distribution. There are different approaches to define such an index. Wendland and Matthes (1986) used the maximum variance for each axial cross section by utilizing the ensemble-averaged velocity. Lai *et al.* (1992) determined the flow distribution by using the ensemble-averaged standard deviation of the velocity field

$$M_{sd} = \dfrac{\left[\left(U_i - U_{area-mean}\right)^2\right]^{\frac{1}{2}}}{U_{area-mean}} \qquad (10.96)$$

In this approach, they used the area-weighted average to represent the mean velocity. In another approach, Weltens *et al.* (1993) proposed the uniformity index (γ) defined as

$$\gamma = 1 - \dfrac{\omega}{2} \qquad (10.97)$$

where ω is the non-uniformity index

$$\omega = \dfrac{\Sigma \omega_i}{n} \qquad (10.98)$$

and ω_i is the local non-uniformity index

$$\omega_i = \frac{\Sigma\left(w_i - w^*\right)^2}{w^*} \qquad (10.99)$$

The average velocity w* is defined as

$$w^* = \frac{\Sigma w_i}{n} \qquad (10.100)$$

This approach also was used recently by Park *et al.* (1999) to analyze the flow and catalytic characteristics for various catalyst cell shapes.

10.6.2 *Improvement of Flow Uniformity*

The concept of radially variable cell density, presented by Kim and Son (1999), is one of the new approaches to improve flow uniformity. This concept is based on the consideration that the resistance from the brick porosity is a function of the velocity inside the channel and the channel diameter. If the center area (high flow area) has the higher cell density, the upstream flow immediately feels higher resistance along the center and adjusts itself to send less flow into the center. In the same way, if the edge of the frontal face of a brick has lower cell density, the upstream flow knows lower resistance and sends more flow to the edge of the brick (Figure 10.6). This concept offers several benefits: a relatively constant space velocity (volume flow rate/brick volume) throughout the entire frontal surface, a reduction in the magnitude of the highest velocity, a reduction in the possibility of the thermal degradation caused by high velocity and temperature, and lower pressure drop along the entire brick. All these benefits can be realized with no significant cost increase and have a positive effect on the catalyst conversion efficiency and engine performance [Kim and Son (1999)].

10.7 Modeling of Catalyst Dynamic Behavior

The efficient operation of TWCs requires that they be operated at or near a stoichiometric A/F ratio. This is ensured by employing a closed-loop control fuel supply system. However, the response lag of the control system causes the A/F ratio to oscillate around the stoichiometric value. In addition to these fast

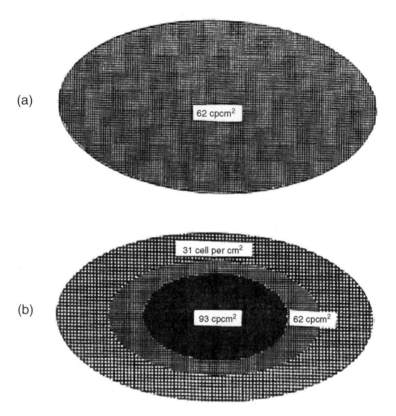

Figure 10.6 A brick with the (a) uniform and (b) radially variable cell density [Kim and Son (1999)].

oscillations, which have frequencies of 1 to 4 Hz, the catalysts are subjected to several slow fluctuations during a typical driving cycle [Herz (1987)]. These slower changes, which have frequencies of 1 Hz and less, occur as a result of acceleration and deceleration, and cause fluctuations in gas flow rates, temperatures, and compositions. The influence of these dynamic conditions makes the catalyst behavior differ significantly from that under steady-state conditions [Herz (1981 and 1987); Silveston (1995 and 1996); Koltsakis and Stamatelos (1999); Shamim and Medisetty (2003); Shamim (2005a and 2005b)].

The catalyst behavior under such dynamic conditions has a great deal of practical interest and has been investigated by many researchers [Herz (1981 and 1987); Schlatter *et al.* (1983); Taylor and Sinkevitch (1983); Matsunga *et al.* (1987); Silveston (1995 and 1996); Koltsakis and Stamatelos (1999); Shamim and Medisetty (2003)]. Several of these studies report enhancement of the activity or selectivity of the catalyst. Such a conversion enhancement is attributed to the washcoat oxygen storage capacity. The enhancement of CO conversion also is explained on the basis of temporarily increased activity of the water-gas shift reaction by some studies [Schlatter and Mitchell (1980)]. The catalyst temperature and operating conditions have been found to strongly influence the catalyst response to A/F ratio modulations [Silveston (1996); Koltsakis and Stamatelos (1999); Schlatter *et al.* (1983)]. The modulation is reported to have favorable effects on the CO conversion in the fuel-rich zone and on NO conversion in the lean zone [Schlatter *et al.* (1983)]. However, at the stoichiometric value, the modulation is reported to decrease the CO and NO conversions. An increase in temperature also is found to reduce the conversion enhancement [Silveston (1996)].

However, most past studies are limited to experimental work conducted under laboratory conditions. The use of mathematical modeling and numerical simulations, which can provide a cost-effective tool in analyzing the catalyst dynamic behavior, has been limited [Koltsakis and Stamatelos (1999); Shamim and Medisetty (2003); Shamim (2005a and 2005b)].

Modeling of catalyst dynamic behavior requires the inclusion of all highly transient phenomena occurring in a catalytic converter, in that it differs from converter models that are developed for predicting cumulative emissions during legislated driving cycles and are generally quasi-steady. The following are some important transient phenomena that should be included in dynamic modeling:

- The oxygen storage and release phenomena in the washcoat
- Accumulation effects in the conservation of mass, species, and energy
- Water-gas shift and steam reforming effects
- Transient effects on reaction kinetics, adsorption, and desorption

Among these, the first three can easily be incorporated into the model, as has been done by Shamim and Medisetty (2003) and Shamim (2005a and 2005b). However, the inclusion of the last item is more challenging because it requires the development of transient kinetic, adsorption, and desorption rates. The existing kinetic rates, which typically include the lumping effects of chemical kinetics, adsorption, and desorption, are generally derived from steady or quasi-steady measurements. This limits their range of application in truly transient conditions. However, the existing lumped kinetic rates may be used with caution to gain qualitative understanding of catalyst dynamic behavior.

Figure 10.7 is an example of such a modeling study [Shamim and Medisetty (2003)]. These results were obtained by considering the catalysts, which were initially at steady-state conditions and suddenly were subjected to sinusoidal modulations in A/F ratio. During these modulations, other inlet conditions and the concentrations of CO, HC, and NO remained unchanged. The A/F ratio was varied by changing the oxygen concentration. The figure shows the CO, HC, and NO conversion efficiencies as a function of imposed fluctuation time period for different frequencies. All these results are for catalysts, initially operating at an A/F ratio of 14.8 (stoichiometric value of the A/F ratio is 14.51) and subjected to sinusoidal oscillation in the A/F ratio of 5% amplitude. The A/F ratio ranges between 14.06 and 15.54, and the catalyst undergoes a transition between rich and lean operating conditions during each fluctuation time period. The catalyst responds to A/F ratio modulations with a time delay.

Figure 10.7 depicts that the catalyst response to imposed oscillation is maximum at low frequencies, and its amplitude decreases with an increase of the imposed frequency. With an increase in frequency, the catalyst response becomes relatively more sinusoidal. Higher frequencies also increase the initial phase lag in the catalyst response to imposed modulations. The catalyst becomes insensitive to imposed fluctuations at high frequencies. The "insensitivity" of the catalyst is due to effective neutralization of high-frequency fluctuations by diffusion processes over the time period required to convect them to the reaction sites. The cutoff frequency corresponding to the insensitivity of the catalyst is different for CO, HC, and NO. The effect of frequency on catalyst response also is found to be different for different pollutants in different frequency ranges, as shown in Figure 10.8, where the catalyst average response amplitude is plotted as a function of the imposed frequency. The results show that HC response is much more sensitive to the imposed modulation frequency at very low frequencies (below 1 Hz) than the CO and NO responses. These results clearly indicate the role of modeling in gaining an understanding of the catalyst dynamic behavior.

Three-Way Catalytic Converter System Modeling

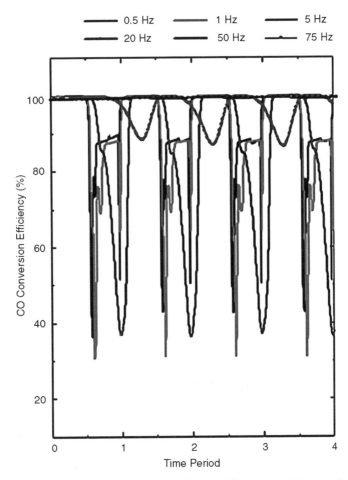

Figure 10.7 *Catalyst response to sinusoidal modulations in A/F ratio for different frequencies: (a) CO conversion efficiency response [Shamim and Medisetty (2003)].*

10.8 Summary

Modeling of catalytic converters has generated new insight into catalyst performance during light-off, steady-state, and transient driving conditions. Due to its various advantages, the use of modeling in designing and optimizing the catalytic converter system is becoming common practice. There are different

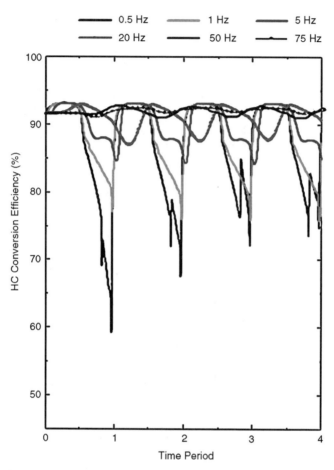

Figure 10.7 (Cont.) (b) HC conversion efficiency response [Shamim and Medisetty (2003)].

approaches to simulate the phenomena of exhaust systems. These modeling approaches may be classified into two main categories: (1) single-channel-based one-dimensional modeling, and (2) multidimensional modeling of the entire catalytic monolith.

The single-channel-based one-dimensional modeling approach has been the most common and popular. It offers simplified, less computationally intensive one-dimensional handling and practically equivalent accuracy levels. In

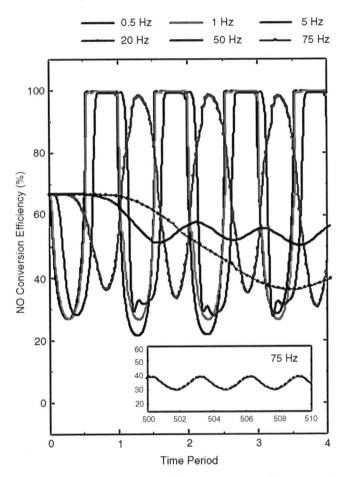

Figure 10.7 (Cont.) *(c) NO conversion efficiency response. The mean A/F ratio for the simulations was set at 14.8 (near stoichiometric conditions), and the modulation amplitude was 5% [Shamim and Medisetty (2003)].*

this approach, the non-uniform flow distribution at the face of the monolith is generally neglected. The multidimensional modeling approach accounts for the strong coupling of the individual channels of the monolith through heat transfer and the inherent non-uniformity in flow distribution. However, due to complexity and the large amount of computation, there are a small number of sophisticated models for the simulation of the entire catalytic monolith

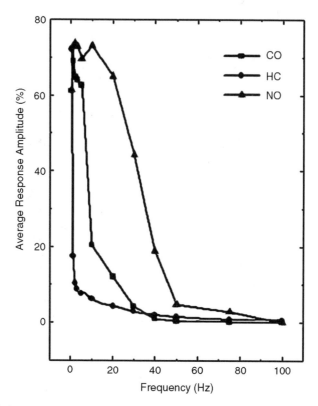

Figure 10.8 Catalyst response to sinusoidal modulations in A/F ratio for different oscillation frequencies. The mean A/F ratio for the simulations was set at 14.8 (near stoichiometric conditions), and the modulation amplitude was 5% [Shamim and Medisetty (2003)].

using simultaneously detailed models of transport and chemistry in each single channel.

There is also a quasi-multidimensional modeling approach. This approach exploits the fact that monolith channels are coupled to each other only through heat transfer; thus, it is possible to link one-dimensional transport-reaction models within the monolith channels with three-dimensional heat conduction models for the entire converter, using prescribed heat and mass transfer coefficients to attain the desired coupling.

An important component of all modeling studies is the simulation of catalyst chemical reactions. Due to low temperatures, the effect of homogeneous reactions in the gas phase is small and generally is neglected. The catalyst reactions, which mainly determine the catalyst performance, are the heterogeneous chemical reactions occurring on the monolith surface. Several reaction mechanisms appear in the literature, which vary in details, accuracy, and convenience of use. Most mechanisms are based on global reaction rates in which adsorption, desorption, and kinetic effects are lumped together. An alternate approach is the description of the chemical reactions by a set of elementary reaction steps. The catalyst oxygen storage capacity is modeled by incorporating an additional kinetic mechanism. The accurate submodels for the transport mechanisms of heat and mass transfer also are essential for predicting catalyst performance. Most studies model the convective heat and mass transfer processes by using the simplified one-dimensional film model.

In addition to the use of more detailed chemical kinetics, the prediction capabilities of catalyst models can be greatly improved by incorporating the transient effect. This will allow the prediction of catalyst performance under transient conditions, which is an area that needs future modeling efforts.

10.9 Mathematical Nomenclature

a	Catalyst activity
BR	One-dimensional pressure gradient due to monolith brick resistance
c	Empirical constant
C	Concentration, mol/m^3
C_1	Model constant
C_2	Model constant
C_g^j	Gas phase concentration of species j, mol/m^3
C_s^j	Surface concentration of species j, mol/m^3
C_μ	Model constant
Cp_g	Specific heat of gas, J/kg·K

Cp_s	Specific heat of substrate, J/kg·K
CTR	Empirical constant for oxygen storage reaction rates
D_h	Hydraulic diameter, m
D^j	Diffusion coefficient of species j, m²/s
E	Correction factor for the entrance effect
E_a, EA, E	Activation energy, Pa-m³/g-mol
ex	Empirical constant for reaction rates
f_r	Friction factor
G	Inhibition factor, K
G_a	Geometric surface area
G_k	Generation term of turbulent kinetic energy
h_g	Heat transfer coefficient between flow and substrate, J/m²·s·K
ΔH^k	Heat of reaction of species k, J/mol
h_∞	Heat transfer coefficient between substrate and atmosphere, J/m²·s·K
k	Turbulent kinetic energy
K	Adsorption equilibrium constant
k_d, k_{d0}	Catalyst deactivation rate constant
km^j	Mass transfer coefficient for species j, m/s
k_{ox}	Oxygen storage component oxidation rate constant, mol/m³·s
k_{red}	Oxygen storage component reduction rate constant, mol/m³·s
m, m_1	Tunable factor

M_{sd}	Ensemble-averaged standard deviation of the velocity field
n	Empirical constant, number of channels
Nu	Nusselt number
Re	Reynolds number
R^k	Reaction rate of kth reaction, mol/m²·s
Sc	Schmidt number
S_{ext}	External surface-to-volume area ratio, m²/m³
Sh	Sherwood number
t	Time, s
T_g	Gas temperature, K
T_s	Substrate temperature, K
T_∞	Ambient temperature, K
U_{area_mean}	Area-weighted average gas flow velocity, m/s
U_i	Gas flow velocity, m/s
V_c	Flow velocity inside the brick cell, m/s
v_g	Gas flow velocity, m/s
w^*	Average velocity, m/s
w_i	Local velocity, m/s
x^*	Nondimensional brick length
X_s^i	Mole fraction of species i in substrate
X_j	Cartesian coordinates
z	Coordinate along catalyst axis, m
z^*	Relative axial coordinate
α	Hydrogen-to-carbon ratio in the fuel

γ	Flow uniformity index
ε	Void volume fraction (value ranging from zero [for no void] to one)
ε	Turbulent dissipation rate
Θ	Reaction rate factor accounting for chemical equilibrium
λ_s	Thermal conductivity of substrate, J/m·s·K
ν_t	Turbulent eddy viscosity
ρ_g	Gas density, kg/m^3
ρ_s	Substrate density, kg/m^3
σ_k	Model constant
σ_ε	Model constant
ω	Flow non-uniformity index
ω_i	Local flow non-uniformity index
ψ	Oxidation fraction

10.10 References

1. Bartholomew, C.H. (2001), "Mechanisms of Catalyst Deactivation," *Applied Catalysis A: General*, Vol. 212, pp. 17–60.

2. Baxendale, A.J. (1993), "The Role of Computational Fluid Dynamics in Exhaust System Design and Development," SAE Paper No. 931072, Society of Automotive Engineers, Warrendale, PA.

3. Bella, G., Rocco, V., and Maggiore, M. (1991), "A Study of Inlet Flow Distortion Effects on Automotive Catalytic Converters," *ASME J. of Engineering for Gas Turbines and Power*, Vol. 113, pp. 419–426.

4. Braun, J., Hauber, T., Többen, H., Zacke, P., Chatterjee, D., Deutschmann, O., and Warnatz, J. (2000), "Influence of Physical and Chemical Parameters on the Conversion Rate of a Catalytic Converter: A Numerical Simulation

Study," SAE Paper No. 2000-01-0211, Society of Automotive Engineers, Warrendale, PA.

5. Breuer, M., Schernus, C., Bowing, R., Kuphal, A., and Lieske, S. (2000), "Experimental Approach to Optimize Catalyst Flow Uniformity," SAE Paper No. 2000-01-0865, Society of Automotive Engineers, Warrendale, PA.

6. Chatterjee, D., Deutschmann, O., and Warnatz, O. (2001), "Detailed Surface Reaction Mechanism in a Three-Way Catalyst," *Faraday Discussions*, Vol. 119, pp. 371–384.

7. Chen, D.K.S., Bisselt, E.J., Oh, S.H., and Van Ostrom, D.L. (1988), "A Three-Dimensional Model for the Analysis of Transient Thermal and Conversion Characteristics of Monolithic Catalytic Converters," SAE Paper No. 880282, Society of Automotive Engineers, Warrendale, PA.

8. Chen, D.K.S. (1993), "A Numerical Model for Thermal Problem in Exhaust Systems," SAE Paper No. 931070, Society of Automotive Engineers, Warrendale, PA.

9. Day, E.G.W., Benjamin, S.F., and Roberts, C.A. (2000), "Simulating Heat Transfer in Catalyst Substrates with Triangular and Sinusoidal Channels and the Effect of Oblique Inlet Flow," SAE Paper No. 2000-01-0206, Society of Automotive Engineers, Warrendale, PA.

10. Franz, J., Schmidt, J., Schoen, C., Harperscheid, M., Eckhoff, S., Roesch, M., and Leyrer, J. (2005), "Deactivation of TWC as a Function of Oil Ash Accumulation—A Parameter Study," SAE Paper No. 2005-01-1097, Society of Automotive Engineers, Warrendale, PA.

11. Gandhi, H.S., Piken, A.G., Shelef, M., and Delosh, R.G. (1976), "Laboratory Evaluation of Three-Way Catalysts," *SAE Transactions*, Vol. 85, Society of Automotive Engineers, Warrendale, PA, pp. 201–212.

12. Hatton, A., Birkby, N., and Hartick, J. (1999), "Theoretical and Experimental Study of Mass Transfer Effects in Automotive Catalysts," SAE Paper No. 1999-01-3474, Society of Automotive Engineers, Warrendale, PA.

13. Hawthorns, R.D. (1974), "Afterburner Catalysts—Effect of Heat and Mass Transfer Between Gas and Catalyst Surface," AIChE Symposium Ser., Vol. 70, American Institute of Chemical Engineers, New York, pp. 428–438.

14. Hayes, R.E. and Kolaczkowski, S.T. (1994), "Mass and Heat Transfer Effects in Catalytic Monolith Reactors," *Chemical Engineering Science*, Vol. 49, pp. 3587–3599.

15. Herz, R.K. (1981), "Dynamic Behavior of Automotive Catalysts: Part 1. Catalyst Oxidation and Reduction," *Industrial and Engineering Chemistry Product Research and Development*, Vol. 20, pp. 451–457.

16. Herz, R.K. (1987), "Dynamic Behavior of Automotive Three-Way Emission Control System," *Catalysis and Automotive Pollution Control*, Elsevier, Amsterdam, pp. 427–444.

17. Holmgren, A. and Andersson, B. (1998), "Mass Transfer in Monolith Catalyst—CO Oxidation Experiments and Simulation," *Chemical Engineering Science*, Vol. 53, pp. 2285–2298.

18. Howitt, J.S. and Sekella, T.C. (1974), "Flow Effect in Monolithic Honeycomb Automotive Catalytic Converters," SAE Paper No. 740244, Society of Automotive Engineers, Warrendale, PA.

19. Jahn, R., Snita, D., Kubicek, M., and Marek, M. (1997), "3-D Modelling of Monolith Reactors," *Catalysis Today*, Vol. 38, pp. 39–46.

20. Johnson, W.C. and Chang, J.C. (1974), "Analytical Investigation of the Performance of Catalytic Monoliths of Varying Channel Geometries Based on Mass Transfer Controlling Conditions," SAE Paper No. 740196, Society of Automotive Engineers, Warrendale, PA.

21. Kim, J.-Y., Lai, M.-C., Li, P., and Cuhi, G. (1995), "Flow Distribution and Pressure-Drop in Diffuser-Monolith Flows," *Journal of Fluids Engineering*, Vol. 117, No. 3, pp. 362–368.

22. Kim, J.Y. and Son, S. (1999), "Improving Flow Efficiency of a Catalytic Converter Using the Concept of Radially Variable Cell Density—Part I," SAE Paper No. 1999-01-0769, Society of Automotive Engineers, Warrendale, PA.

23. Koltsakis, G.C., Kandylas, I.P., and Stamatelos, A.M. (1998), "Three-Way Catalytic Converter Modeling and Applications," *Chemical Engineering Communications*, Vol. 164, pp. 153–189.

24. Koltsakis, G.C., Konstantinidis, P.A., and Stamatelos, A.M. (1997), "Development and Application Range of Mathematical Models for 3-Way

Catalytic Converters," *Applied Catalysis B: Environmental*, Vol. 12, pp. 161–191.

25. Koltsakis, G.C. and Stamatelos, A.M. (1997), "Catalytic Automotive Exhaust Aftertreatment," *Progress in Energy and Combustion Science*, Vol. 23, pp. 1–39.

26. Koltsakis, G.C. and Stamatelos, A.M. (1999), "Dynamic Behavior Issues in Three-Way Catalyst Modeling," *AIChE Journal*, Vol. 45, pp. 603–614.

27. Lai, M.C., Lee, T., Kim, J.Y., Li, P., Chui, G., and Pakko, J.D. (1992), "Numerical and Experimental Characterization of Automotive Catalytic Converter Internal Flows," *Journal of Fluids & Structures*, Vol. 6, pp. 451–470.

28. Launder, B.E. and Spalding, D.B. (1974), "The Numerical Computation of Turbulent Flows," *Computer Methods in Applied Mechanics and Engineering*, Vol. 3, pp. 269–289.

29. Li, P., Adamczyk, A.A., and Pakko, J.D. (1996), "Thermal Management of Automotive Emission Systems: Reducing the Environmental Impact," in Sengupta, S. and Sano, T. (eds.), *Thermal Engineering for Global Environmental Protection*, Begell House, New York, pp. 55–77.

30. Lloyd-Thomas, D.G., Ashworth, R., and Qiao, J. (1993), "Meeting Heat Flow Challenges in Automotive Catalyst Design with CFD," SAE Paper No. 931079, Society of Automotive Engineers, Warrendale, PA.

31. Martin, A.P., Massey, A.E., Twigg, M.V., Will, N.S., and Davidson, J.M. (1999), "A Chemical Method for the Visualisation of Flow Maldistribution in a Catalytic Converter," SAE Paper No. 1999-01-3076, Society of Automotive Engineers, Warrendale, PA.

32. Massing E., Brilhac, J.F., Brillard, A., Gilot, P., and Prado, G. (2000), "Modelling of the Behaviour of a Three Way Catalytic Converter at Steady State: Influence of the Propene Diffusion Inside the Catalytic Layer," *Chemical Engineering Science*, Vol. 55, pp. 1707–1716.

33. Matsunga, S.-I., Yokota, K., Muraki, H., and Fujitani, Y. (1987), "Improvement of Engine Emissions Over Three-Way Catalyst by the Periodic Operation," SAE Paper No. 872098, Society of Automotive Engineers, Warrendale, PA.

34. Mazumder, S. and Sengupta, D. (2002), "Sub-Grid Scale Modeling of Heterogeneous Chemical Reactions and Transport in Full-Scale Catalytic Converters," *Combustion and Flame*, Vol. 131, pp. 85–97.

35. Mondt, J.R. (1987), "Adapting the Heat and Mass Transfer Analogy to Model Performance of Automotive Catalytic Converters," *ASME J. of Engineering for Gas Turbines and Power*, Vol. 109, pp. 200–206.

36. Montreuil, C.N., Williams, S.C., and Adamczyk, A.A. (1992), "Modeling Current Generation Catalytic Converters: Laboratory Experiments and Kinetic Parameter Optimization—Steady State Kinetics," SAE Paper No. 920096, Society of Automotive Engineers, Warrendale, PA.

37. Moore, W.R. and Mondt, J.R. (1993), "Predicted Cold Start Emission Reductions Resulting from Exhaust Thermal Energy Conservation to Quicken Catalytic Converter Lightoff," SAE Paper No. 931087, Society of Automotive Engineers, Warrendale, PA.

38. Mottahed, B. and Molki, M. (1996), "Artificial Roughness Effects on Turbulent Transfer Coefficients in the Entrance Region of a Circular Tube," *International Journal of Heat and Mass Transfer*, Vol. 39, pp. 2515–2523.

39. Moulijn, J.A., van Diepen, A.E., and Kapteijn, F. (2001), "Catalyst Deactivation: Is It Predictable? What to Do?" *Applied Catalysis A: General*, Vol. 212, pp. 3–16.

40. Oh, S. (1988), "Thermal Response of Monolithic Catalytic Converters During Sustained Engine Misfiring: A Computational Study," SAE Paper No. 881591, Society of Automotive Engineers, Warrendale, PA.

41. Oh, S.H. and Cavendish, J.C. (1982), "Transients of Monolithic Catalytic Converters: Response to a Step Change in Free-Stream Temperature as Related to Controlling Automobile Emissions," *Ind. Eng. Chem. Prod. Res. Dev.*, Vol. 21, p. 29.

42. Oh, S.H. and Cavendish, J.C. (1985a), "Mathematical Modeling of Catalytic Converter Lightoff—Part 2: Model Verification by Engine-Dynamometer Experiments," *AIChE Journal*, Vol. 31, pp. 935–942.

43. Oh, S.H. and Cavendish, J.C. (1985b), "Mathematical Modeling of Catalytic Converter Lightoff—Part 3: Prediction of Vehicle Exhaust Emissions and Parametric Analysis," *AIChE Journal*, Vol. 31, pp. 943–947.

44. Otto, N.C. (1984), private communication.

45. Otto, N.C. and LeGray, W.J. (1980), "Mathematical Models for Catalytic Converter Performance," SAE Paper No. 800841, Society of Automotive Engineers, Warrendale, PA.

46. Park, S., Kim, H.Y., Cho, Y.J., Lee, S.Y., and Yoon, K.J. (1999), "Flow Analysis and Catalytic Characteristics for the Various Catalyst Cell Shapes," SAE Paper No. 1999-01-1541, Society of Automotive Engineers, Warrendale, PA.

47. Pattas, K.N., Stamatelos, A.M., Pistikopoulos, P.K., Koltsakis, G.C., Konstantinidis, P.A., Volpi, E., and Leveroni, E. (1994), "Transient Modeling of 3-Way Catalytic Converters," *SAE Transactions*, SAE Paper No. 940934, Society of Automotive Engineers, Warrendale, PA.

48. Ryan, M.J., Becke, E.R., and Zygourakis, K. (1991), "Light-Off Performance of Catalytic Converters: The Effect of Heat and Mass Transfer Characteristics," SAE Paper No. 910610, Society of Automotive Engineers, Warrendale, PA.

49. Schlatter, J.C. and Mitchell, P.J. (1980), "Three-Way Catalyst Response to Transients," *Industrial and Engineering Chemistry Product Research and Development*, Vol. 19, pp. 288–293.

50. Schlatter, J.C., Sinkevitch, R.M., and Mitchell, P.J. (1983), "Laboratory Reactor System for Three-Way Automotive Catalyst Evaluation," *Industrial and Engineering Chemistry Product Research and Development*, Vol. 22, pp. 51–56.

51. Shamim, T. (2003), "Effect of Heat and Mass Transfer Coefficients on the Performance of Automotive Catalytic Converters," *International Journal of Engine Research*, Vol. 4, No. 2, pp. 129–141.

52. Shamim, T. (2005a), "The Effect of Space Velocity on the Dynamic Characteristics of an Automotive Catalytic Converter," SAE Paper No. 2005-01-2160, Society of Automotive Engineers, Warrendale, PA.

53. Shamim, T. (2005b), "Dynamic Response of Automotive Catalytic Converters to Modulations in Engine Exhaust Compositions," *International Journal of Engine Research*, Vol. 6, pp. 557–567.

54. Shamim, T. and Medisetty, V. (2003), "Dynamic Response of Catalytic Converters to Changes in Air-Fuel Ratio," *ASME J. of Engineering for Gas Turbines and Power*, Vol. 125, No. 2, pp. 547–554.

55. Shamim, T., Shen, H., and Sengupta, S. (2000), "Comparison of Chemical Kinetic Mechanisms in Simulating the Emission Characteristics of Catalytic Converter," *SAE Transactions—Journal of Fuels and Lubricants*, Vol. 109, SAE Paper No. 2000-01-1953, Society of Automotive Engineers, Warrendale, PA.

56. Shamim, T., Shen, H., Sengupta, S., Son, S., and Adamczyk, A.A. (2002), "A Comprehensive Model to Predict a Three-Way Catalytic Converter Performance," *ASME Journal of Engineering for Gas Turbines and Power*, Vol. 124, pp. 421–428.

57. Siemund, S., Leclerc, P., Schweich, D.J., Prigent, M., and Castagna, F. (1996), "Three-Way Monolithic Converter: Simulations Versus Experiments," *Chemical Engineering Science*, Vol. 51, pp. 3709–3720.

58. Silveston, P.L. (1995), "Automotive Exhaust Catalysis Under Periodic Operation," *Catalysis Today*, Vol. 25, pp. 175–195.

59. Silveston, P.L. (1996), "Automotive Exhaust Catalysis: Is Periodic Operation Beneficial?," *Chemical Engineering Science*, Vol. 51, pp. 2419–2426.

60. Subramanian, B. and Varma, A. (1984), "Reactions of CO, NO, O_2, and H_2O on Three-Way and Pt/Al_2O_3 Catalyst," *Frontiers in Chemical Engineering*, Proceedings of the International Chemical Engineering Conference, Vol. 1, pp. 231–240.

61. Subramanian, B. and Varma, A. (1985), "Reaction Kinetics on a Commercial Three-Way Catalyst: the $CO-NO-O_2-H_2O$ System," *Ind. Eng. Chem. Prod. Res.*, Vol. 24, pp. 512–516.

62. Taylor, K.C. and Sinkevitch, R.M. (1983), "Behavior of Automobile Exhaust Catalysts with Cycled Feed Streams," *Industrial and Engineering Chemistry Product Research and Development*, Vol. 22, pp. 45–50.

63. Ullah, U., Waldram, S.P., Bennett, C.J., and Truex, T. (1992), "Monolithic Reactors: Mass Transfer Measurements Under Reacting Conditions," *Chemical Engineering Science,* Vol. 47, pp. 2413–2418.

64. Voltz, S.E., Morgan, C.R., Liederman, D., and Jacob, S.M. (1973), "Kinetic Study of Carbon Monoxide and Propylene Oxidation on Platinum Catalysts," *Ind. Eng. Chem. Prod. Res. Dev.*, Vol. 12, pp. 294–301.

65. Weltens, H., Bressler, H., Terres, F., Neumaier, H., and Rammoser, D. (1993), "Optimization of Catalytic Converter Gas Flow Distribution by CFD Prediction," SAE Paper No. 930780, Society of Automotive Engineers, Warrendale, PA.

66. Wendland, D.W. (1993), "Automobile Exhaust-System Steady-State Heat Transfer," SAE Paper No. 931085, Society of Automotive Engineers, Warrendale, PA.

67. Wendland, D.W. and Matthes, W.R. (1986), "Visualization of Automotive Catalytic Converters Internal Flows," SAE Paper No. 861554, Society of Automotive Engineers, Warrendale, PA.

68. White, F.M. (1999), *Fluid Mechanics, 4th Edition*, McGraw-Hill, New York.

69. Windmann, J., Braun, J., Zacke, P., Tischer, S., Deutschmann, O., and Warnatz, J. (2003), "Impact of the Inlet Flow Distribution on the Light-Off Behavior of a 3-Way Catalytic Converter," SAE Paper No. 2003-01-0937, Society of Automotive Engineers, Warrendale, PA.

70. Young, L.C. and Finlayson, B.A. (1976a), "Mathematical Models of the Monolithic Catalytic Converter, Part 1: Development of Model and Application of Orthogonal Collocation," *AIChE Journal*, Vol. 22, pp. 331–343.

71. Young, L.C. and Finlayson, B.A. (1976b), "Mathematical Models of the Monolithic Catalytic Converter, Part 2: Application to Automobile Exhaust," *AIChE Journal*, Vol. 22, pp. 343–353.

10.11 Appendix

The chemical reaction rate expressions for the reaction scheme are as follows. The values of different coefficients used in these expressions are listed elsewhere [Montreuil et al. (1992)].

$$\text{RATE (1)} = \frac{C_1 e^{-EA(1)/RT_S} X_{CO} X_{O_2}^{ex(1)}}{[(1 + C_4 X_{CO})^2 (1 + C_5 X_{C_3H_6})]} \qquad (10.101)$$

$$\text{RATE (2)} = \left[C_2 e^{-EA(6)/RT_S} + (1 - \beta_1) C_3 e^{-EA(3)/RT_S} \right] \cdot$$
$$\frac{3 X_{C_3H_6} X_{O_2}^{ex(2)} (1 + C_{31} X_{C_3H_6} X_{CO})}{(1 + C_{35} X_{C_3H_6})^2 (1 + C_{32} e^{-EA(7)/RT_S} X_{CO})} \qquad (10.102)$$

$$\text{RATE (3)} = \beta_1 \cdot \left[C_3 e^{-EA(3)/RT_S} \right] \cdot$$
$$\frac{3 X_{C_3H_6} X_{O_2}^{ex(2)} (1 + C_{31} X_{C_3H_6} X_{CO})}{(1 + C_{35} X_{C_3H_6})^2 (1 + C_{32} e^{-EA(7)/RT_S} X_{CO})} \qquad (10.103)$$

$$\text{RATE (4)} = \frac{C_6 e^{-EA(4)/RT_S} X_{H_2O} X_{O_2}}{\left[(1 + C_{47} X_{H_2})^2 (1 + C_4 X_{CO})^2 (1 + C_5 X_{C_3H_6}) \right]} \qquad (10.104)$$

$$\text{RATE (5)} = \frac{C_{46} X_{NH_3}^{ex(4)}}{\left(1 + C_{47} X_{H_2} + C_{48} X_{CO} + C_{49} X_{C_3H_6} \right)^2} \qquad (10.105)$$

$$\text{RATE (6L)} = \frac{(1 - \beta_1) C_{15} X_{CO} e^{-EA(2)/RT_S} X_{NO}^{ex(10)}}{\left[(1 + C_{16} X_{O_2})^{ex(11)} (1 + C_{50} X_{CO})^2 \right]} \qquad (10.106)$$

$$\text{RATE (6R)} = \frac{\beta_1 \cdot C_{11} e^{-EA(5)/RT_S} X_{CO} X_{NO}^{ex(5)} \left(1 + C_{12} X_{O_2}\right)^{ex(6)}}{[F1 + F2 + F3 - 2]^2}$$

$$F1 = \left(1 + \text{CMILL} \cdot X_{CO}\right)^{ex(7)}$$

$$F2 = \left(1 + C_{14} X_{H_2}\right)^{ex(8)} \tag{10.107}$$

$$F3 = \left(1 + C_{34} e^{EA(10)/RT_S} X_{C_3H_6}\right)^{ex(23)}$$

$$\text{RATE (7)} = \frac{\beta_1 \cdot C_{17} X_{CO} X_{NO}^{ex(13)} \left(1 + \text{CMILL} \cdot X_{O_2}\right)^{ex(14)}}{F4 \cdot F5 \cdot F6}$$

$$F4 = \left[1 + C_{19} |T_S - C_{20}|^{ex(12)}\right]$$

$$F5 = \left[1 + C_{13} X_{CO} + C_{33} X_{H_2}\right] \tag{10.108}$$

$$F6 = \left[1 + C_{30} X_{C_3H_6}\right]$$

$$\text{RATE (8L)} = \frac{(1 - \beta_2) C_{22} X_{H_2}^{ex(16)} X_{NO}^{ex(15)}}{\left(1 + C_{44} X_{O_2}\right)^2} \tag{10.109}$$

$$\text{RATE (8R)} = \frac{C_{21} X_{H_2} X_{NO}^{ex(9)} \left(1 + C_{23} X_{O_2}\right)^2 \cdot \beta_2}{[F7 + F8 + F9 - 2]^2}$$

$$F7 = \left(1 + C_8 X_{CO}\right)^{ex(7)}$$

$$F8 = \left(1 + C_9 X_{H_2}\right)^{ex(8)} \tag{10.110}$$

$$F9 = \left(1 + C_{34} e^{EA(10)/RT_S} X_{C_3H_6}\right)^{ex(23)}$$

$$\text{RATE (9L)} = \left(1 + \beta_2 \cdot \text{CMILL} \cdot X_{O_2}\right)^{\text{ex}(18)} \cdot \text{RATE (9R)} \qquad (10.111)$$

$$\text{RATE (9R)} = \frac{C_{25} X_{H_2} X_{NO}^{\text{ex}(17)}}{F10 \cdot F11 \cdot F12}$$

$$F10 = \left[1 + C_{27} |T_S - C_{28}|^{\text{ex}(26)}\right]$$

$$F11 = \left[1 + C_{24} X_{CO} + C_{18} X_{H_2}\right] \qquad (10.112)$$

$$F12 = \left[1 + C_{26} e^{EA(11)/RT_S} X_{C_3H_6}\right]$$

$$\text{RATE (10)} = \frac{3 C_{10} e^{-EA(8)/RT_S} X_{O_2} X_{C_3H_8}^{\text{ex}(26)}}{\left(1 + C_{51} X_{O_2}\right)^{\text{ex}(27)}} \qquad (10.113)$$

$$\text{RATE (11)} = \frac{\beta_1 \cdot 6 C_{38} X_{C_3H_6} X_{NO}^{\text{ex}(24)} \left(1 + C_{37} X_{O_2}\right)^{\text{ex}(3)}}{\left(1 + C_{39} X_{CO}\right)\left[F1 + F2 + F3 - 2\right]^2} \qquad (10.114)$$

$$\text{RATE (12)} = (1 - \beta_1) 9 C_{42} e^{-EA(9)/RT_S} X_{NO}^{\text{ex}(22)} X_{C_3H}^{\text{ex}(25)} \qquad (10.115)$$

$$\text{RATE (13)} = \frac{\beta_1 \cdot 2.4 C_{38} X_{NO}^{\text{ex}(20)} X_{C_3H_6} \left(1 + \text{CMILL} \cdot X_{O_2}\right)^{\text{ex}(21)}}{\left[F13 \cdot F14 \cdot F15\right]}$$

$$F13 = \left[1 + C_{40} |T_S - C_{41}|^{\text{ex}(19)}\right]$$

$$F14 = \left[1 + C_{29} X_{C_3H_6}\right]^2 \qquad (10.116)$$

$$F15 = \left[1 + C_{43} X_{CO} + C_{45} X_{H_2}\right]$$

where ß$_1$ and ß$_2$ are rich-to-lean blending functions;

$$0 \leq \beta_1 \leq 1 \quad \text{and} \quad 0 \leq \beta_2 \leq 1$$

$$\beta_1 = 2500\, R_1 + 0.5$$

$$\beta_2 = 2500\, R_2 + 0.5$$

if $\quad \beta < 0, \beta = 0$

if $\quad \beta > 0, \beta = 1$

Here,

$$R_1 = X_{CO} + 9X_{C_3H_6} + 1.5X_{NH_3} - X_{NO} - 2X_{O_2} - R_{flip} \qquad (10.117)$$

$$R_1 = X_{CO} + 9X_{C_3H_6} + 1.5X_{NH_3} - X_{NO} - 2X_{O_2} - H_{flip} + X_{H_2} \qquad (10.118)$$

R indicates rich-side (redox > 1) rate, and

L indicates lean-side (redox < 1) rate

CHAPTER 11

Evaporative Emissions Reduction

Jenny Spravsow and Christopher Hadre
DaimlerChrysler Corporation

11.1 Introduction

11.1.1 Overview of Evaporative Emissions Standards

The California Air Resources Board (CARB) was formed in 1967 to help combat the growing air pollution problem in southern California. Smog was prevalent in the Los Angeles area. The goal of the CARB is to maintain air quality, research the causes of pollution, and "attack the serious problem of motor vehicles" (see the CARB website). The CARB decided to regulate emissions from automobiles, both from the tailpipe and those due to evaporation.

Previously, the evaporative emissions standards conformed to the U.S. Environmental Protection Agency (EPA) standards of stricter evaporative emissions of 2.0 g of hydrocarbon (HC) for passenger cars and 3.0 g HC for trucks for a three-day test procedure. Beginning in the 2004 Model Year, the California Low Emissions Vehicle II (LEV II) standards began to be phased in. The passenger car standard became 0.50 g HC for cars and 0.90 g HC for light-duty trucks for a three-day test procedure. At the same time, the new evaporative standards also brought tighter controls on tailpipe emissions, called the Ultra-Low Emissions Vehicle II (ULEV II) standards. The vehicle must meet both criteria to be certified as LEV II and to be included in the phase-in.

An even more stringent standard, 0.054 g of fuel-based HC emissions, is a result of the Zero Emissions Vehicle (ZEV) mandate. The CARB has determined that by 2005, 10% of its fleet will be vehicles, preferably electric or fuel cell, that have inherently zero fuel-based HC emissions. Credits toward this can be earned by selling Partial Zero Emissions Vehicles (PZEVs). By using industry best practice methods, supporting the PZEV program with traditional internal combustion engine vehicles is possible.

The EPA also lowered the evaporative emissions standards beginning in the 2004 Model Year. The latest standards, Tier II, lower the allowable evaporative emissions from 2.0 g for a passenger car down to 0.95 g, for the three-day test procedure; however, these standards are noticeably less stringent than those required by the CARB.

11.1.2 Types of Evaporative Emissions

Evaporative emissions are defined as those given off from a vehicle when it is parked (at rest). The main sources of evaporative emissions can be classified into two categories: (1) fuel based, and (2) non-fuel based. Fuel-based evaporative emissions typically come from fuel vapor permeation through fuel system components and/or the migration of fuel vapor from the canister or air induction system. Non-fuel-based emissions are those from components such as tires, paints, sealants, and glues [Stewart and Hui (2000)]. Emissions from R134a air-conditioning refrigerant and from the alcohol in windshield washer solvent also are non-fuel based, and they may be subtracted from the PZEV emissions test results. The ZEV mandate has forced a close look at emissions in the engine compartment.

Evaporative emissions can be classified into three categories: (1) diurnal emissions, which are those emissions that are due to the evaporation of fuel during a daily thermal cycle, called a "diurnal"; (2) running losses, which are those emissions that occur during vehicle operation; and (3) refueling emissions, which are those emissions that occur as fuel vapor in the fuel tank is displaced with liquid during a refueling event.

11.1.3 Evaporative Emissions Test Procedures

The two-day test procedure for evaporative emissions consists of a Federal Test Procedure (FTP) drive cycle with a loaded canister, a one-hour soak at elevated temperature, followed by two "diurnal" days. The overall test result is the sum of the one-hour hot soak and the highest valued diurnal day. The canister is loaded to capacity with butane/nitrogen mix. This represents the condition of the canister if the vehicle is not driven for many days or on a vehicle recently refueled. The FTP driving cycle simulates starting the car after an extended (overnight) soak and driving through a specified speed trace to simulate a typical commute (as defined in the U.S. Code of Federal Regulations [CFR]). The hot soak immediately following the FTP is to simulate parking the vehicle at 72°F after a drive. A diurnal day is a simulation of a theoretical worst-case temperature delta that can be seen in California. First, the vehicle is soaked at 65°F. Then, the temperature is increased to 105°F in 12 hours and drops down to 65°F again in the remaining 12 hours. As a vehicle is heated, the fuel in the fuel tank can generate vapor. Any vapor must be contained and not allowed to escape to the atmosphere.

The three-day test procedure requires an additional driving cycle called the running loss test, which consists of a drive cycle performed at 105°F ambient temperature. Emissions are measured at the fill cap and canister to ensure emissions are contained while the vehicle is in operation. This also ensures the purge system of the vehicle is functioning adequately. Immediately after the running loss test, the vehicle is placed in a one-hour hot soak at 105°F, followed by three diurnal days, using the same temperatures as described here previously. The overall test result is the sum of the hot soak and the highest-valued diurnal day.

The apparatus used to determine the diurnal and hot soak emissions is called a sealed housing for evaporative determination (SHED). As shown in Figure 11.1, it is a sealed box with an instrument called a flame ionization detector (FID) to detect and measure HCs at incremental intervals. The SHED environment is sampled for HC concentration at the beginning and end of each test; thus, the emissions from the vehicle can be calculated for both the hot soak and each of the diurnal days. Because the FID detects all HCs, including those that are non-fuel based, another instrument called an Innova infrared (IR) detector may be used. An Innova IR detector can distinguish between fuel-based HCs and those coming from other sources. Among available filters for the Innova detector are R134a, methanol, and ethanol. This tool can aid in diagnosing sources and the root of problems in vehicles. In a PZEV test protocol, the Innova detector results may be used to isolate and understand the non-fuel-based HCs and exclude them from the final result.

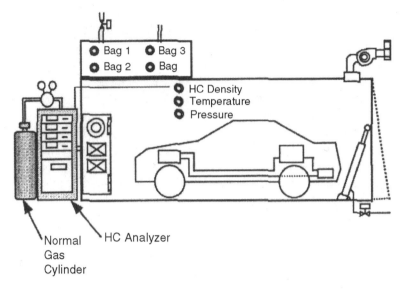

Figure 11.1 SHED equipment [Matsushima et al. (2000)].

The California certification test fuel, known as California Phase II, is 7 Reid vapor pressure (RVP), a measure of the volatility of fuel. It is less volatile than the EPA certification fuel, which is set at 9 RVP. The EPA diurnal cycle, however; is less severe in its temperature profile. The "day" begins at 72°F and peaks at 95°F before returning to 72°F. The next proposed California fuel, California Phase III, will have the same RVP as the current fuel but may contain some percentage of alcohol. A fuel containing alcohol typically has different permeation characteristics.

11.2 Types of Evaporative Emissions Control Systems

There are several types of fuel storage and delivery systems. One type involves maintaining positive pressure in the fuel tank during operation. Another allows little vapor generation because of a vacuum, and another is "free breathing," meaning open to the atmosphere and able to vent at all times. Some advantages to a pressure system are the ability to create vent paths to different areas of the tank and fill tube. A "remote vapor dome" can be placed in a reservoir anywhere convenient to packaging. Many packages choose the top of the fill tube for the reservoir to contain vapor [Matsushima *et al.* (2000)].

One advantage of low vapor generation is the reduced need for vapor management. To stop the creation of fuel vapor in a gas tank, the air above the fuel must be removed. One method to accomplish this is to create a variable volume tank using a bladder methodology. As fuel is consumed, the bladder collapses, removing the vapor dome. If there is little air to mix with the fuel, correspondingly less vapor is created. The materials to make the bladder can vary. Some possibilities are multilayer high-density polyethylene (HDPE), with a permeation barrier, or a monolayer material without a permeation barrier. The bladder may be enclosed in another permeation-resistant container [Arase *et al.* (2001)]. A fuel system that does not generate many vapors might be used on a hybrid electric vehicle (HEV), where space and weight are at a premium and the engine is small. See Section 11.3.4 of this chapter for further discussion about canister and engine control technology.

A free-breathing system involves fewer components such as valves (Figure 11.2). The fuel level in the tank during refueling is controlled by a valve welded on the tank (10). When the fuel reaches a defined height, a float is shut, and the fuel backs up the fill tube and shuts off the nozzle. Another valve (5) is welded to the tank and, with the inlet check valve (ICV) (9), prevents fuel from spilling during a rollover accident. The rollover valve (ROV) (5) allows venting when the tank or vehicle is tipped on an angle.

To reduce vapor generation during refueling, some of the fuel vapor from the tank is recirculated (6) to the top of the fill tube to be re-entrained with the dispensing fuel. Because the vapor is already saturated, the liquid fuel cannot generate too much more vapor. The recirculation orifice is sized using several vehicle-specific parameters. This orifice typically is located in the control valve (10), with flow management and rollover valve features. The canister is vented to a fresh-air port that is open to the atmosphere at all times. During diurnal events, the tank can breathe during warm-up and cool-down. Fuel vapor can migrate to the canister freely. During drive cycles, if the fuel is generating pressure faster than the purge is drawing it off, the pressure in the tank can rise and vent through the canister. A free-breathing system is less complex, in that the vapor dome is integral to the tank, and there are fewer components and less plumbing. A control strategy for such a system is simpler to implement.

11.3 Reducing Evaporative Emissions

To reduce evaporative emissions, the large vapor sources must be identified. Examples of large sources of permeation or micro leaks can be the seals in

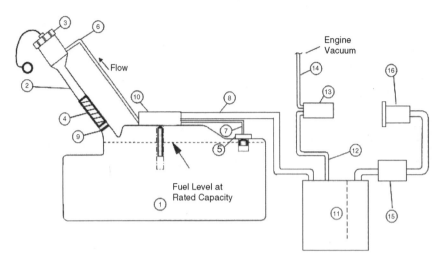

Figure 11.2 Free-breathing fuel storage and delivery system. (1) Fuel tank (steel or plastic). (2) Fuel filler tube. (3) Fuel cap (with pressure/vacuum relief). (4) Fill tube to fuel tank connector (elastomeric). (5) Tank grade vent/rollover valve (where applicable). (6) Vapor recirculation line. (7) Tank vapor line. (8) Vapor line to canister. (9) Check valve (ICV). (10) Multifunction control valve (fill limit, liquid discrimination, flow control orifice, rollover). (11) Canister. (12) Purge line. (13) Purge device. (14) Purge line to intake manifold. (15) Leak detection device. (16) Filter. [DaimlerChrysler Corporation]

the fuel system, specifically the pump module seal, as well as O-rings in the connectors. If overmolded valves are used for plastic tank applications, these likely will have a seal that protects a permeation path. The seal material must be chosen carefully, seal design and joint design must be done properly, and service concerns must be addressed to result in a robust seal design.

Reducing the number of joints and connectors in a fuel system is another key method for reducing the level of evaporative emissions. Choosing connectors that are permeation resistant and have long-term durability is necessary. Designing the connection to the proper dimensions also is critical. The secure connection of the fill tube to the tank body likewise must be addressed.

The fuel tank has the largest surface area of any fuel system component. For plastic tanks, that means it can be a large contributor to emissions through permeation. It is essential for the barrier layer of the fuel tank not to have voids or gaps in the main body of the tank or in the pinch region. The module flange and any welded-on valves also must be made of high-permeation-resistant plastics. Minimizing the number of openings in the tank, regardless if the tank is metal or plastic, is directionally correct for reducing evaporative emissions.

11.3.1 Seals

Due to serviceability and assembly concerns, several parts require seals. Pump modules and fill tubes are examples. Seals may be required on certain components to satisfy 15-year/150,000-mile requirements. Fluorocarbon elastomer (FKM) typically is the material of choice for LEV II and PZEV. Fluorocarbon elastomer is useful to 260°F; therefore, underbody and engine compartment temperatures must be studied and controlled. Proper clearances to hot components, such as exhaust pipes and catalysts, must be maintained. Future PZEV certification procedures may focus on engine emissions. Engine emissions typically come from fuel injectors and manifold seals and gaskets. Engine emissions may come through the permeable materials in an air intake system. Fluorocarbon elastomer may be necessary for the intake manifold seals and fuel injector seals for PZEV applications. The shape of the seal and the amount of compression on the seal are two critical design parameters. The seal must fill the gland area properly. Too much seal material can cause a failure by yielding other components due to the high pressures generated in the area. Too little material may not fill the area, leaving leakage paths. Overcompression of the seal can cause failure of the seal material by cracking or tearing. In FKM elastomers, the higher the fluorine content, the more impermeable the material is to fuel. Higher fluorine content also promotes better tensile properties [Aguilar and Kander (2000)].

Overmolded valves that are welded on tanks may have O-rings between the cover and the overmold. The seal is not removable and usually is composed of FKM. Quick connectors may have one, two, or three O-rings, depending on whether the joint carries vapor or liquid fuel. For seals in a fuel supply line, FKM material is ideal because it does not swell or permeate in liquid fuel. Nitrile material will swell considerably in liquid fuel, starting almost immediately on contact.

11.3.2 Connectors

To minimize connections, the closer the canister is to the tank and the closer the tank can be to the metal chassis fuel supply line, the better. This will minimize flexible jumpers that are good for assembly but are not directionally correct for permeation. For connections in which the hose is pushed over a hose barb, the hose and the barb should be designed for long life. For rubber hoses, the barb must not be designed in such a way as to cut the inner layers of the hose. A swage is an operation in which a nylon tube is pressed-fit over an end form of three barbs. The joint is not serviceable. An O-ring may be needed for PZEV applications to ensure sealing meets the long-term warranty and useful life goals. The O-ring will not improve the permeation characteristic, but it will prevent leaks as the vehicle ages.

A connection that is used for ease of assembly or may be disassembled in service might require a quick connect. The quick connect should have adequate O-ring material, and the connector body could be multilayer nylon. For connectors that carry liquid fuel, each O-ring typically is good for different temperature ranges. For cold temperatures, -40 to $+32°C$, use a material that is not brittle when cold. For high temperatures, use a material with a high melting point that retains its properties in high temperatures. Another connector type could involve a nonremovable band clamp over braided barrier hose. Still other types could have metal quick connector bodies or a spring and cage that lock together.

The largest connection in the fuel system is the fill tube to the fuel tank. For higher permeation resistance, the hose can be upgraded to low-permeation materials such as one having an FKM inner with an ozone-resistant cover.

11.3.3 Materials

Typical fuel tank materials are steel and a multilayer co-extruded HDPE. A steel tank can be deep drawn or stamped in two pieces: an upper shell and a lower shell. One advantage to steel is that components can be installed in the tank shell before it is welded, minimizing the amount of openings in the tank. Steel thicknesses typically are 0.7 mm, which has a slight volume advantage over a 6-mm total plastic thickness. However, steel cannot be deep drawn into the complex shapes that are possible in plastic molding. The steel tank shape must be determined early in the program to optimize the shell design. Radii and curves must be designed with the shell welding process in mind. Steel dies are long lead items and must be finalized earlier in the vehicle development process. The welding process must be controlled to ensure no microscopic gaps in the seam. Any discontinuity of the weld seam can result in permeation and, in extreme cases, fuel leakage. Tank strap materials must be chosen carefully to prevent galvanic corrosion.

Plastic tanks have been shown to permeate more than steel tanks. To reduce permeation in a plastic tank, the use of low-permeation plastic valves and modules is recommended. One concept for reducing permeation in HDPE plastic components is to sulfonate them. During the process of sulfonation, the parts are exposed to sulfur gas in high concentrations. The sulfur bonds with the plastic, creating a fuel vapor permeation barrier. The longer the parts are exposed to sulfur, the deeper the layer. This process has some drawbacks, namely, toxicity and durability of the sulfur layer. Cycle time to complete the sulfonation process can be lengthy. Next, sulfur gas is toxic, and this process must be performed with safety equipment. Because the sulfur does not penetrate the entire part, if the component is nicked or scratched, a permeation path could be created. This same concept can apply to any HDPE part, such as a fuel tank shell. It is not common practice, however, since the development of multilayer fuel tank shells.

Today's plastic fuel tanks are molded using a multilayer, co-extruded parasin consisting of an HDPE outer layer, a low-density polyethylene (LDPE) adhesive layer, a permeation barrier of ethyl vinyl alcohol (EVOH), another LDPE adhesive layer, a regrind layer, and an inner HDPE layer. The EVOH barrier does not allow fuel vapor molecules to pass through, but it is very brittle, which is why it is the middle layer. The upper and lower molds of a plastic tank are joined together in the "pinch seam." Because EVOH does not bond to itself, there is a small gap in the EVOH layers in this area. That gap is a primary permeation path.

Several new methods can improve or eliminate this source. An extra layer of plastic "tape" composed of the same multilayer tank material can be adhered to the tank in the pinch region. This will increase the length of the potential permeation path created by the EVOH gap in the pinch seam, thus reducing possible emissions. Another way to reduce permeation at the pinch seam is to design the pinch area with special consideration for bringing the EVOH layers as close as possible. To reduce the possibility of voids and to increase permeation resistance, increase the thickness of the EVOH layer. However, there will be a maximum thickness where any beneficial effect no longer results. Another technology involves putting valves inside the plastic tank and welding another multilayer nylon cover over the whole tank, including the pinch seam. This reduces the exposed welded components and has edge treatment in one concept.

A second material for reducing permeation through valves is the use of acetal copolymer (POM) plastic. Although acetal has high resistance to fuel vapor permeation, it does not weld to HDPE; therefore, special provisions must be made for attaching POM valves to an HDPE fuel tank. Pump module flanges also are made of acetal.

The vapor lines that connect the tank, canister, and purge harness should have the proper material specification. Fluorocarbon elastomer can be used as a barrier layer in the flexible vapor tubing. In systems where vacuum is maintained in the tank and the lines are FKM, the permeation does not increase with time [Nuiya et al. (1999)]. The pipeline of HC never fills up. When the fuel system is allowed to reach high pressures, the permeation value increases, as shown in Figure 11.3. The HC concentration present in the hose causes the HC pipeline to fill.

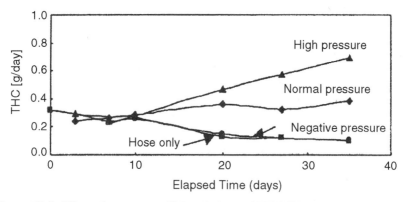

Figure 11.3 Effect of vacuum on HC emissions of FKM [Nuiya et al. (1999)].

11.3.4 Canister and Engine Control Technology

11.3.4.1 Canister Basics

On vented fuel systems that allow free breathing of the fuel tank to the atmosphere, a canister containing an adsorptive material (typically activated carbon) is used to capture fuel vapor and "clean" the air before it reaches the outside atmosphere. As fuel vapor is generated in the fuel tank under any condition (e.g., refueling, running loss, or diurnal), it is routed through the canister where the fuel vapor molecules condense into liquid on the porous surface of the activated carbon (adsorption), allowing only cleaned air to vent to the atmosphere. Hydrocarbon molecules are stored in the carbon pellets until purge air is drawn through the canister during engine operation, where they evaporate off the carbon pellets (desorption) into the airstream and are burned as part of the combustion process. The amount of air that flows through the canister determines how "clean" the canister will be for the next refueling or diurnal event.

The vapor canister is sized proportionally to the amount of vapor that is generated in the fuel tank during either a refueling (onboard refueling vapor recovery, or ORVR, system) or diurnal event (conventional system). Due to the conditions of loading rate and vapor generation to which a canister is subjected during a refueling event, ORVR becomes the worst case for canister sizing. A canister sized to meet ORVR requirements also will meet two- and three-day diurnal requirements, if enough purge air is provided to clean the canister prior to the diurnal event. The "working capacity" of a canister is defined as the mass of HC vapor that is loaded into the canister before 2 g breakthrough after a minimum of 10 and maximum of 100 purge-load cycles. See Section 11.2 of this chapter for more information on evaporative control systems and reducing vapor generation.

When the engine operates, it pulls air through the carbon bed, desorbing the HCs from the carbon bed and using them in the combustion process. Enough air must be passed through the canister to prepare it adequately for another refueling event or diurnal day (to prevent breakthrough). Large transients in vapor concentration of the purge airstream potentially can cause problems with the engine control system. A balance must be maintained to meet both exhaust and evaporative standards. The methodology for maintaining this balance is through proper metering of purge flow to the engine (purge valve) and restricting vapor flow from the fuel tank using orifi to prevent large transient vapor waves from entering the purge airstream. It is easier to calibrate the engine for a constant regulated flow than for dynamic transient concentrations. A device that has been developed for transient control is called a flow management valve. The valve places a small orifice restriction between the fuel tank and canister during engine operation but allows high flow during refueling. In Figure 11.2, the flow management valve is integral to valve 10.

11.3.4.2 Canister Emissions and Controlling Methodology

Canister bleed emissions (those diffusing from the canister during diurnal loading) are a main source of potential emissions that must be strictly controlled in a PZEV environment. Unfortunately, a canister designed for ORVR (low-flow restriction, high-working-capacity carbon) has characteristics opposite those that are necessary to control evaporative emissions. High-activity carbons that can adsorb higher quantities of fuel vapor per unit volume also have a tendency to hold onto them during purge, which can lead to higher emissions during the diurnal test. New developments in carbon that release stored HCs more easily with a small volume of purge air are beneficial in a PZEV environment due to the increased control over purge air with small engines and exhaust emissions. A smaller engine will have less purge air available and less opportunity to completely clean the canister. Unfortunately, these carbons that purge easier also have a lower specific working capacity, which makes them unsuitable for the entire canister; therefore, only a small volume strategically placed in the canister and combined with a larger quantity of conventional carbon is preferred. Another alternative for lowering canister emissions is through the use of a secondary HC trap optimized for adsorbing small quantities of emissions (therefore low working capacity) while not increasing system backpressure. One example of a secondary HC trap is an extruded open-cell carbon honeycomb structure.

Other ways to improve bleed emissions include increasing the canister length-to-diameter ratio (L/D), which effectively increases the localized airflow over the carbon pack during purge, thus cleaning it more efficiently [Itakura *et al.* (2000)].

11.3.4.3 Engine Emissions

The same principle may apply in the engine air intake system. The concentration at the cylinders and intake manifold may be higher than the concentration past the throttle body and into the air intake. Because the intake is open to the atmosphere, the particles will migrate to that area of low concentration outside the airbox. This also must be controlled in PZEV vehicles. The size and type of engine induction trap varies from a "brick" of carbon, or honeycomb structure, to carbon-coated paper, to thin panels of granulated carbon. The working capacity of these traps is extremely low because the trap is not intended to capture true HC breakthrough or to trap a large amount of HCs as if an injector is leaking into the manifold but "slow-moving bleed emissions" following the principle of diffusion.

11.4 Summary

Several technologies are available for reducing evaporative emissions. Some approaches involve reducing the HC-rich vapor in the tank. If only a small amount of vapor is generated, only a small amount must be controlled. This can be done by keeping a constant vacuum on the tank or by using a bladder technique that contracts as fuel is used, thereby reducing the vapor dome. Imparting a small pressure to the tank can reduce vapor generation by forcing the vapor to remain in a liquid form. These techniques will reduce the required working capacity of the canister.

Permeation through plastic tank walls and pinch and weld areas is a large contribution to evaporative emissions. Ensuring the EVOH barrier layers are in contact or having a treatment for them is critical. Valving that is welded-on should be of low-permeability plastic, such as acetal. The seals in the over-molded valves and pump module should be of low-permeability material and designed for the proper opening and compression. O-rings in swages may be included for sealing at various temperature ranges and for useful lifetime, and proper materials should be chosen for the application. Connections should be designed for serviceability where necessary, and they should be minimized for evaporative emissions reduction. Surface finish and mold parting lines should be specified to give the friendliest surface possible.

Due diligence to connections, seals, and engine interaction will reduce the overall evaporative emissions values. Viewing the overall vehicle as a system from fuel cap to engine air inlet will help minimize negative interactions of components while maintaining and optimizing system performance.

11.5 References

1. Aguilar, H. and Kander, R.G. (2000), "Fuel Permeation Study on Various Seal Materials," SAE Paper No, 2000-01-1099, Society of Automotive Engineers, Warrendale, PA.

2. Arase, T., Ishikawa, T., Kobayashi, M., Hyodo, Y., Kasai, M., and Nakano, N. (2001), "Development of Vapor Reducing Fuel Tank System," SAE Paper No. 2001-01-0729, Society of Automotive Engineer, Warrendale, PA.

3. Itakura, H., Kato, N., Kohama, T., Hyoudou, Y., and Murai, T. (2000), "Studies on Carbon Canisters to Satisfy LEVII EVAP Regulations,"

SAE Paper No. 2000-01-0895, Society of Automotive Engineers, Warrendale, PA.

4. Matsushima, H., Iwamoto, A., Ogawa, M., Satoh, T., and Ozaki, K. (2000), "Development of a Gasoline-Fueled Vehicle with Zero Evaporative Emissions," SAE Paper No. 2000-01-2926, Society of Automotive Engineers, Warrendale, PA.

5. Nuiya, Y., Hajime, U., and Suzuki, T. (1999), "Reduction Technologies for Evaporative Emissions in Zero-Level Emission Vehicle," SAE Paper No. 1999-01-0771, Society of Automotive Engineers, Warrendale, PA.

6. Stewart, S.J. and Hui, C. (2000), "Evaporative Leakage from Gas Caps," SAE Paper No. 2000-01-1171, Society of Automotive Engineers, Warrendale, PA.

CHAPTER 12

Onboard Diagnostics

Glenn Zimlich, Kathleen Grant, and Timothy Gernant
Ford Motor Company

12.1 Introduction

The state of California requires onboard diagnostics (OBD II) on all passenger cars, light-duty trucks, and medium-duty vehicles [California Code Regulations, Title 13, Section 1968.2]. OBD II is a set of diagnostic algorithms in the powertrain control strategy that detects when an emissions-critical component has failed or reduced efficiency. When an emissions-related component is detected to have failed, the malfunction indicator light (MIL) must be illuminated on the vehicle instrument panel. In addition, the OBD II system must record the status and test results of the various diagnostic tests so that the results can be reviewed at inspection/maintenance (I/M) stations. The diagnostics work toward improving air quality by detecting when a vehicle produces more emissions than allowed. The OBD II system also aids the service community in identifying and isolating vehicle faults [Ogawa (2002)]. A vehicle operator is expected to take the vehicle for service to repair the faulty component when the MIL is illuminated.

The diagnostics monitor any component that affects the emissions system performance or is an input or output to the electronic powertrain control module. It is not practical to cover the entire OBD II system in this chapter. Rather, this chapter will focus on the eleven main monitored subsystems, as outlined in Table 12.1.

TABLE 12.1
MONITORED SYSTEMS/COMPONENTS EXPLORED IN THIS CHAPTER

Section	Monitored Component/System	Acronym (If Applicable)
12.2	Catalyst System Monitor	–
12.3	Comprehensive Component Monitor	CCM
12.4	Cold-Start Emissions Reduction Control Strategy Monitor	–
12.5	Engine Misfire Monitor	–
12.6	Evaporative System (Leak Detection) Monitor	–
12.7	Exhaust Gas Recirculation System Monitor	EGR
12.8	Fuel System Monitor	–
12.9	Oxygen Sensor Monitor	–
12.10	Secondary Air System Monitor	SAIR
12.11	Variable Valve Timing/Control Monitor	VVT
12.12	In-Use Performance Tracking	–

These monitoring systems are affected by new low emissions standards (LEV II) or new regulatory requirements. Due to the complexity of the OBD II requirements, it is impossible to cover all aspects of the monitors. Many of the subtleties such as the availability of monitor/requirement phase-in and interim monitoring thresholds will not be addressed here. Additionally, the individual emissions standards will not be part of this assessment, but the interactions of the LEV II requirements as related to OBD II will be explored. This work focuses on fuel-injected gasoline engines rather than exploring hybrid electric, diesel, alternate fuel, or direct injection applications.

12.1.1 Emissions Failure Thresholds for Diagnostic Monitors

As identified in Section 12.1, diagnostic failures are, in most cases, related to an increase in vehicle emissions measured on the Federal Test Procedure (FTP) cycle. Generally, the monitors are expected to make a determination over this vehicle drive cycle. The new In-Use Monitor Performance Ratio Tracking requirement will give an indication of how often monitors execute the diagnostic tests in the field. This will be discussed further in Section 12.12 of this chapter. The emissions constituents of interest are carbon monoxide (CO), nonmethane organic gases (NMOGs), and oxides of nitrogen (NOx). All the monitors, except the comprehensive component monitor (CCM), catalyst, and evaporative system monitors, must illuminate the MIL before the vehicle exceeds 1.5 times the applicable tailpipe emissions standard (for CO, NMOG, or NOx). There is relief for the super ultra-low emissions vehicles/partial zero emissions vehicles (SULEV II/PZEV). The emissions failure thresholds increase to 2.5 times the SULEV II emissions standards. The evaporative system monitor is not calibrated to a tailpipe emissions standard. The monitor detects a leak of a prescribed orifice diameter in the fuel/evaporative system (e.g., fuel tank, fuel vapor lines, fittings, connectors). The catalyst monitor is a special case, in that the monitor must detect a 1.75 times (2.5 times SULEV II) emissions degradation of only NMOG or NOx constituents.

In some cases, the malfunction of an emissions control device does not lead to an OBD emissions failure. The manufacturer is required to monitor the component for function alone (i.e., functional monitor). If the deterioration of a component leads to an OBD emissions failure, the component monitor must be calibrated to illuminate the MIL at the specified OBD II emissions threshold. This type of diagnostic monitor will be referred to as a threshold monitor.

The diagnostic system is impacted by low emissions technology primarily in two ways. First, OBD emissions standards are linked (multiples of 1.5 times for most monitors) by the regulatory tailpipe standard to which the vehicle is certified. As the certified tailpipe standards are reduced, so is the delta from the "useful life'" to the OBD failure criteria. This delta is the detection window available to the OBD engineer to identify the fault. Figure 12.1 depicts the detection window for NMOG emissions. The reduced detection window as a function of decreasing emissions standards has the net effect of reducing the available distinction between good and failed components. This affects the ability of the OBD engineer to separate the two conditions and to eliminate both alpha (α) and beta (β) errors (as defined in Section 12.1.2). When discussing the OBD window of opportunity, of particular interest is the reduction in the NOx standards on LEV II vehicles. This significant reduction affects the overall detection level (relative to the LEV I standards) as well as the detection window.

A second direct impact of LEVs is the increased hardware required to achieve lower emissions standards. The components must be monitored either due to the direct impact to tailpipe emissions or because the component is an input or output of the powertrain control module. The complexity of the control system and the ability to monitor new components must be managed to provide a complete diagnostic system.

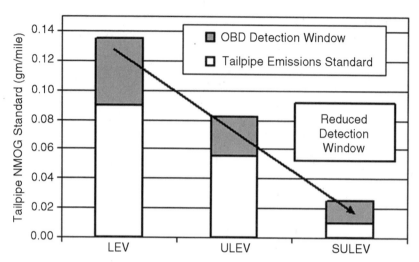

Figure 12.1 Reduction in the OBD detection window due to reduced emissions standards.

12.1.2 Proper Identification of Diagnostic Failures

The challenge to the OBD engineer is the reliable detection of the fault at the appropriate time. The ability to detect a fault is important but not the only condition that must be satisfied. Automobiles are mass produced and, as such, are subject to varying environmental conditions, drive cycles, and usage patterns. Also, the manufacturing and component tolerances stack up to ensure that, similar to humans, no two vehicles are exactly the same. The detection of a fault should be insensitive to these varied conditions. When a good part is identified as bad, an alpha (α) error (also known as a Type I error) is identified. When a failed part is not detected, a beta (β) error (also known as a Type II error) is introduced. With these factors in mind, the OBD engineers strive to identify a fault at the appropriate time. Figure 12.2 contains an example of the challenges. Case 1 can be calibrated to identify a faulty component exclusively. Case 2 contains signal noise that reduces the capability of the monitor to properly distinguish all good parts from failed or faulty parts. In this instance, the detection system is susceptible to both α and β errors. If a fault threshold (or detection level) was chosen exactly between the two distributions, some of the good parts would be identified as bad, and some of the failed parts

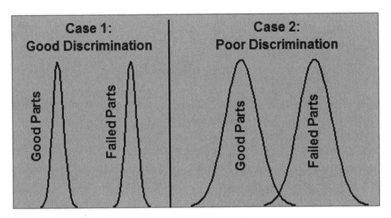

Figure 12.2 Two cases demonstrating the challenge of setting accurate fault thresholds.

would be identified as good. At this threshold, both α and β errors would be encountered in the field.

The motor vehicle manufacturer must balance unexpected warranty of monitored components, customer dissatisfaction from improper MIL illumination, and the possibility of regulatory noncompliance. Developing robust monitors is a great concern for proper fault detection and customer satisfaction.

12.2 Catalyst System Monitor

12.2.1 Theory, Application, and Regulatory Implications

The catalytic converter is one of the primary means of reducing tailpipe emissions. This does not imply that more stringent emissions standards can be met solely by utilizing a larger catalyst. Quite the opposite is true. To meet the lower emissions standards, more effective utilization of the catalyst is required, with a reduction of emissions prior to catalyst warm-up. Automotive engineers must reduce cold startup emissions, the engine combustion should support stable combustion while heating the catalyst, and the air/fuel ratio (A/F) deviations should be minimized during warmed-up operation.

In the past, the catalyst monitor had been required to detect only a failure of the catalyst before the emissions exceeded 1.75 times the NMOG standard. For LEV II applications, the monitor also must detect an increase in NOx emissions. The incremental requirement is being phased-in with the LEV II requirements. OBD catalyst monitoring involves correlating the depletion of catalyst oxygen storage content (OSC) to the degradation of tailpipe emissions. The catalyst OSC provides a mechanism to store excess oxygen during lean operation and release oxygen during rich conditions. This feature tends to improve three-way catalyst (TWC) performance and dampen the A/F fluctuations after the catalyst. As the catalyst OSC is depleted, there is an increase in the A/F fluctuations that leads to increased voltage activity on the oxygen sensor mounted after the monitored catalyst. The increased voltage activity then is correlated to the increase in tailpipe emissions. Figure 12.3 demonstrates the change in sensor voltage activity. The catalyst with greater OSC has smaller voltage swings than the catalyst with lower OSC.

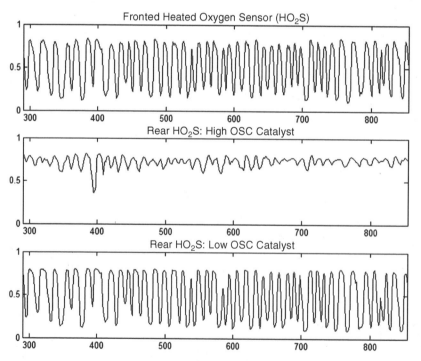

Figure 12.3 Demonstration of how oxygen storage affects oxygen sensor signal.

Although most catalyst monitors rely on the depletion of catalyst OSC, they can further be divided into two major categories: (1) passive catalyst monitoring, and (2) active (or intrusive) catalyst monitoring. Active catalyst monitoring involves perturbing the catalyst inlet fuel-air ratio to quantify the catalyst oxygen storage capacity. When performed under sufficiently controlled conditions, this method can be more accurate due to the direct measurement of the catalyst OSC. One drawback of active monitoring is the potential to increase emissions during the test. The fuel-air perturbation can cause degradation in emissions if the magnitudes of the swings are high or the duration is sufficiently long. The result is a loss of catalyst conversion efficiency. Passive catalyst monitoring relies on controlled signal analysis between the fuel control oxygen sensor (mounted before the catalyst) and the post-catalyst oxygen sensor. The signal analysis can be simple, such as comparing the pre- and post-sensor voltage swings (sensor voltage switches), or it can be complex, such as integrating the signal area.

12.2.2 Catalyst Monitor Operation

For a catalyst to operate effectively, it must be maintained at a sufficient temperature. Temperature variations, particularly low temperatures, will interact with both active and passive catalyst monitors. The lower catalyst temperatures can increase the oxygen breakthrough on the catalyst, simulating reduced catalyst efficiency. Catalyst monitor operation generally is limited until the catalyst is up to operating temperature to avoid these inaccurate diagnostic detections. The rate of engine exhaust gas recirculation (EGR) also can interact with the catalyst monitor. The oxygen sensor signal can change under varying amounts of EGR flow. With the increased focus on lower NOx emissions, the changes in EGR flow rate should be validated to ensure accurate monitor performance. Both catalyst temperature and EGR flow rate must be properly accounted for on LEVs.

Low emissions vehicles rely in part on faster catalyst light-off. The correlation of tailpipe hydrocarbon (HC) performance to catalyst OSC during warm-engine steady-state conditions is more challenging because the degradation of vehicle emissions during warm operation is small. Rieck, Collins, and Moore (1998) reported that the catalyst formulation (and OSC type and durability) could be adjusted to improve the OBD performance of the catalysts for LEVs. The presumption is that the catalyst has oxygen storage that can be monitored for efficiency. Catalysts with little or no OSC are difficult (if not impossible) to diagnose using dual oxygen sensor methodologies. The use of a temperature sensor to directly measure the catalytic activity is more appropriate.

12.3 Comprehensive Component Monitor

Comprehensive component monitoring (CCM) is an umbrella requirement that addresses the many other components that are part of the powertrain control system. These components may or may not significantly increase the emissions of the vehicle. Most of the tests are circuit continuity, out of expected range, in-range rationality diagnostics (inputs), or functionality diagnostics (outputs). A significant impact from new low emissions technologies is powertrain complexity. As the number of inputs and outputs increases, so does the number of components that must be monitored. Considering the umbrella monitoring requirements of the CCM tests, the additional components must be monitored if the input or output can affect any emissions constituent during any reasonable in-use driving cycle or if the input or output can affect any of the other monitored systems or components (including data on vehicle data links).

12.4 Cold-Start Emissions Reduction Control Strategy Monitor

A new requirement for the 2006 Model Year and beyond LEV II vehicles is the Cold-Start Emissions Reduction Control Strategy Monitor [California Code Regulations, Title 13, Section 1968.2]. Engine startup control strategies that reduce emissions are to be monitored. A cold-start strategy that elevates the engine speed and retards the ignition timing to generate more exhaust heat is an example of a start routine that potentially must be monitored. These strategies are designed to bring the catalyst system up to operating temperature quicker and thus reduce tailpipe emissions. Available published information is limited due to newness of the requirement.

The cold-start emissions reduction control strategy monitor is required to meet the applicable OBD emissions-based thresholds. The sensitivity of the vehicle to changes in spark timing, engine speed, or other emissions reduction control parameters determines if the monitor will require a functional or a threshold failure determination. Because engine operation during startup is exposed to many uncontrollable influences, designing a threshold monitor for this requirement will be challenging. Developing systematic approaches to startup strategies, aftertreatment design, and OBD monitoring algorithms will be essential to successful monitor implementation.

12.5 Engine Misfire Monitor

12.5.1 Theory, Application, and Regulatory Implications

A common method for determining engine misfire uses the signal from an encoder mounted to the crankshaft to calculate the acceleration of the engine due to combustion events. The encoder often is used to determine the proper delivery time of ignition spark and fuel injection pulse. When a cylinder misfires, the crankshaft slows down instantaneously. This deceleration is measured and compared to a reference value to make the misfire determination. Other methods to detect misfire also can be used, such as measurement of the combustion ionization or direct cylinder pressure measurement. The lack of engine misfire is not only crucial to meeting low emissions levels, but also to protecting the catalyst system from excess heat that might damage the system. The misfire criteria that is used to declare a fault using catalyst damage criteria is called Type A. The misfire level that is related to when emissions exceed the applicable OBD II threshold is called Type B. The regulation allows for a minimum misfire rate of 5% for Type A and 1% for Type B. The expediency required to make a Type A or Type B determination also is different. Table 12.2 summarizes the minimum misfire rates relevant to the two types of misfire conditions.

TABLE 12.2
MINIMUM MISFIRE RATE FOR FAULT THRESHOLD

Engine No. of Cylinders	Misfire Minimum Clip			
	Type A		Type B	
	Rate	Misfire Events/ 200 Rev.	Rate	Misfire Events/ 1000 Rev.
4	5%	20	1%	20
6	5%	30	1%	30
8	5%	40	1%	40

12.5.2 Misfire Monitor Operation

Emissions standards for LEVs are less tolerant to misfiring conditions. It is possible to exceed emissions thresholds under very low levels of engine misfire. Therefore, Type B thresholds may need to be set at the minimum clip of 1% on these applications. To enhance catalyst light-off, many types of exhaust heat management techniques also are being applied, with the ultimate goal of increased catalyst efficiency. Engineers often will adjust spark, air, EGR flow rates, or camshaft timing (VVT) to help achieve and maintain high catalyst efficiency. If high catalyst operating temperatures are the result, then the 5% Type A misfire clips may need to be employed.

12.6 Evaporative System Monitor

12.6.1 Theory, Application, and Regulatory Implications

Vehicular emissions traditionally are thought of as those relating to the products of engine combustion. Another possible source of HC emissions is the evaporation of gasoline in the fuel tank to the atmosphere. To prevent this, the fuel tank vapors pass through an activated charcoal canister. The charcoal adsorbs the unburned HCs while the vehicle is not in operation. Later, the vapors are purged from the canister and pulled into the intake manifold to undergo combustion in the engine.

Most of the emissions-based OBD II regulations dictate that the MIL must illuminate when the tailpipe emissions exceed a 1.5-times tailpipe standard in terms of grams per mile. Those dealing with the evaporative system monitor handle things differently by specifying a given leak size within the evaporative emissions system that must illuminate the MIL. The most stringent requirement dictated by the California Air Resources Board (CARB) requires the OBD II system detect a 0.020-inch-diameter leak.

12.6.2 Initial Vacuum Decay-Based Method for Leak Detection

One common method for leak detection uses the same system hardware as the vehicle uses to control evaporative emissions, with the addition of a valve to seal the vent at the carbon canister. By sealing the system while actively pulling vapors into the intake manifold, a vacuum is pulled on the tank. When the system reaches a target vacuum, the system also is sealed from the intake manifold, and the pressure is monitored over time. In a vehicle with good system integrity, the pressure will remain nearly constant over time. However, if the system has a leak, the influx of air from the atmosphere into the system will cause the pressure to rise back to atmospheric over time. The rise in system pressure is correlated to leak size to determine when to illuminate the MIL.

This method has several pitfalls, such as dynamic vehicle operation, vehicle grades, fuel volatility, and large tank volumes. In general terms, anything that causes a pressure buildup during monitor operation will reduce the distinction between no leak and leak conditions. Uncompensated vapor generation during monitor execution could increase the pressure buildup. The grades in a road confound the system because the pressure sensor used to measure the pressure in the system normally is referenced to atmosphere. As the vehicle is driven up a grade, the atmospheric pressure decreases. To the sensor, this scenario appears the same as if the atmospheric pressure remained constant while the system pressure increased, as is the case with a system that has a 0.020-inch-diameter leak. Additionally, leak detection can be confounded by volatile fuel in combination with fuel slosh within the tank. This condition promotes fuel evaporation, thus creating a pressure rise in a sealed system. Finally, this method of leak detection is much more difficult to accurately perform on a large-capacity tank. Given a set amount of time, the amount of air inducted into the evaporative system through a 0.020-inch-diameter leak is a smaller percentage of the monitored vapor space in a large tank. In absolute terms, this leads to a delta pressure during the span of monitor operation that is smaller in magnitude. This leads to a reduction in the ability to separate a system with a leak from one with good system integrity. The example data in Case 2 of Figure 12.2 demonstrates this system distribution overlap problem. Without additional action, this could lead to both α and β errors.

The preceding discussion details several of the primary system interaction problems of diagnostic systems that operate during either static or dynamic conditions of vehicle operation. All of these pitfalls have caused evaporative system leak detection to be performed under a select set of engine and vehicle operating conditions. These restrictive monitoring schemes led to a change in the regulations on the in-use performance criteria. The requirements evolved from "reasonably occurring in-use" into specific performance metrics described in Section 12.12 of this chapter. Combined with the inability to detect the mandated 0.020-inch-diameter orifice leak on some large tanks, this drives the need for new technology.

12.6.3 Positive Pressure Decay Leak Detection

One method employed to overcome some of these obstacles was to incorporate the use of additional system hardware, in the form of an electronically controlled pump. The pump is used to apply a positive pressure to the sealed system. Perry and Delaire (1998) chronicled the development of positive pressure decay diagnostics in their SAE paper. Initial systems would apply a positive pressure until the system pressure came into equilibrium with the internal spring load of the pump. A leak-free system would maintain the target pressure. The presence of a leak would cause the pump to duty cycle at a given rate to replace the pressure loss associated with the leak. Then the control system evaluates the duty cycle to determine if the leak exceeds a threshold. To isolate the noise factors due to grade changes and fuel slosh, the test is performed only while the vehicle is stationary. Based on this criterion, the initial test times were considered excessively long. Through the implementation of new pump cycling, preconditioning the vapor space, and an early detection algorithm, test times were greatly reduced. Figure 12.4 shows a pressure trace over time of a positive pressure decay system.

The same noise factor associated with fuel volatility that plagues the vacuum decay method is present on the pressure decay systems and, as expected, has the opposite effect. Previously, where volatile fuel would create a false detection, the pressure rise associated with fuel evaporation could now cause the diagnostic to yield a false pass. Additionally, the application of a positive pressure to the fuel system will cause unburned HCs to be forced out of the system through any leaks that are present in the system. Therefore, the diagnostic itself causes an increase of evaporative emissions through the leak it is designed to detect. This is one advantage of the vacuum-based method. During the vacuum-based diagnostic, fresh air is pulled into the system from the atmosphere through any potential leak in the system. Although the pump-based methods may not be as environmentally friendly in practice, they are sensitive enough to detect 0.020-inch leaks where traditional vacuum decay methods are not [Perry and Delaire (1998)].

12.6.4 Natural Vacuum-Based Leak Detection

Another method used to achieve the 0.020-inch standard has its basis in the ideal gas law. Instead of using either the intake manifold vacuum or a pump, this method uses the natural thermodynamics of the fuel system following vehicle operation. Assuming that the volume of the system is held constant, the

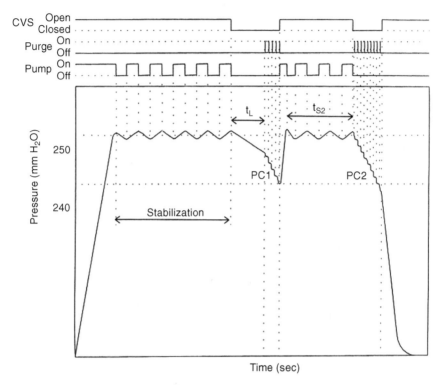

Figure 12.4 Pressure trace of the leak detection pump system [Perry and Delaire (1998)].

system pressure will change with the associated temperature changes brought about during the soak period after vehicle operation. Initially, after shutdown, heat will be transferred to the tank in the form of waste heat from other vehicle components. Assuming a constant volume, this heat addition will cause a pressure rise in the tank. Eventually, the vehicle will begin to cool, which results in a corresponding drop in fuel system pressure. Although the magnitude of pressure changes generated in the tank is not as large as methods that use a pump or the intake manifold as a vacuum source, the change in pressure can be viewed over a much longer time scale by allowing the test to run while the engine is off. This method potentially could have design implications to the charging system because it requires a power draw to cycle the canister vent solenoid while the vehicle is not running. Through the use of a volatility, pressure, and vacuum phase, test times have been successfully limited to 40 minutes after vehicle shutdown (DeRonne *et al.* (2003). Because the vehicle is tested at rest, some of the common system interaction problems are no longer an issue. These systems should execute more often and improve in-use performance tracking metrics.

12.7 Exhaust Gas Recirculation System Monitor

An EGR system is used on many vehicles to reduce engine-out NOx emissions or as an aid to improve fuel economy. Some vehicle operating conditions are adjusted based on the level of EGR being deployed, which also can affect emissions levels. Regulations require that an EGR malfunction must be recognized prior to exceeding the applicable OBD II emissions standards. If a nonfunctional EGR system does not exceed the OBD II malfunction standard, then a functional monitor is sufficient to meet the regulations.

Numerous techniques have been used to monitor the EGR system. Sensors such as pressure sensors (manifold or EGR pressure) or temperature sensors can be used to either directly or indirectly measure EGR flow. When implementing an EGR system on LEVs, there is a greater risk that a malfunction to this system could result in an exceedance of the standard, which would require an emissions-based monitor.

12.8 Fuel System Monitor

12.8.1 *Theory, Application, and Regulatory Implications*

Most vehicle manufacturers use a set of algorithms that allow the control system to adapt to a particular vehicle. The control strategies adjust the vehicle fueling to maintain optimal catalyst efficiency, compensate for manufacturing tolerances, and adapt to normal vehicle aging. The control systems typically contain both a long-term fuel trim and a short-term fuel trim. The design must be robust enough to allow these fuel trims to adjust for normal sources of variability, as identified in Table 12.3. When these adjustments (or adaptation) have reached their useful limit or result in an increase of emissions to the OBD II threshold, an OBD II fault must be triggered.

The fuel system monitor has been in use for decades and has been used in applications ranging from closed-loop carburetion control to high-pressure fuel injection including alternate fuels (e.g., liquefied petroleum gas [LPG], natural

**TABLE 12.3
POSSIBLE SOURCES OF VARIABILITY
ON FUEL SYSTEM ADAPTATION**

Sources of Potential Variability	
Carbon buildup on injectors	Component tolerances
Evaporative system purge delivery	Positive crankcase ventilation (PCV)
Fuel type	Inlet airflow variation or resonance
Fuel flow variation or resonance	Component aging

gas vehicles [NGV]). During the early applications of this diagnostic, prior to OBD II, the thresholds were based on driveability issues and as an aid to technicians. As time progressed and vehicle emissions requirements became more stringent, it has matured into an emissions-based monitor.

12.8.2 Fuel System Monitor Operation

For LEVs, the fuel system adaptation must be capable of "learning" quickly. The reduction of catalyst efficiency due to non-adaptation can be crucial to meeting low emissions levels. The demonstration cycle for the fuel monitor entails applying a fuel bias and allowing the adaptive learning to compensate for the shift. Figure 12.5 demonstrates an FTP and the quick fuel system adaptation required to maintain high catalyst efficiencies.

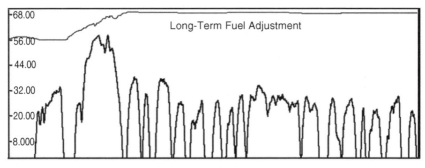

Figure 12.5 Adaptive learning events over the FTP.

Disregarding the cold-start emissions, maintaining the A/F ratio in the optimum catalyst operating window is not optional; it is required to meet the regulated emissions levels. Typical LEVs must be able to achieve an A/F ratio with a fuel bias applied that nearly replicates the A/F ratio variation without any fuel bias applied. It has been observed that the high NOx efficiencies required of LEV II vehicles require thorough and robust fuel system adaptation. Figure 12.6 shows an example of the emissions sensitivity of fuel adaptation requirements. In this case, the vehicle had not fully adapted to the fuel bias induced. At a time of approximately 190 seconds, the vehicle begins to accelerate. To perform the acceleration, a higher load is placed on the engine. This action drives the vehicle into a region where the adaptive fuel algorithm has not had a chance to compensate for the fueling error in the long-term fuel trim. As a result, the short-term fuel trim must compensate for the error. When the vehicle completes the acceleration and enters an area where the long-term fuel trim has adapted sufficiently lean, the short-term fuel trim is now overcompensating for the induced error. The result is increased tailpipe NOx emissions as displayed in the NOx concentration measurements.

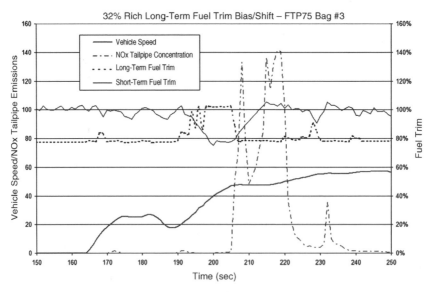

Figure 12.6 *Effect on tailpipe emissions of incomplete adaptive fuel trim learning.*

12.9 Oxygen Sensor Monitor

12.9.1 Theory, Application, and Regulatory Implications

Oxygen sensors are used in the emissions control system to maintain a stoichiometric A/F ratio in the exhaust stream. The sensors measure the oxygen content of vehicle exhaust by referencing it to the concentration of oxygen in the atmosphere. If the concentration of oxygen in the exhaust differs greatly from that of the atmosphere, the result will be an output voltage. A rich mixture will produce a higher voltage (~0.8v) than the voltage (~0.1v) produced by a lean mixture. A limitation of traditional oxygen sensors is that they are capable of reporting only a rich or a lean mixture. They cannot discern the magnitude to which the mixture is rich or lean. This leads to the sensor voltage switching back and forth, as the A/F ratio fluctuates around stoichiometry. As an oxygen sensor ages, it may become contaminated, and its response to a change in the oxygen content of the exhaust will become slower. This reduction in response time of the oxygen sensor will lead to less precise A/F control of the engine. This will lead to degradation in vehicle emissions.

12.9.2 Oxygen Sensor Monitor Operation

The monitoring of the oxygen sensor response can be intrusive or non-intrusive, similar to the catalyst monitor. An intrusive monitor will artificially ramp the short-term fuel trim lean-to-rich and rich-to-lean at a given frequency, while measuring the response of the oxygen sensor. On a V-engine, each bank can be cycled 180° out of phase with each other to minimize the impact on emissions performance. A non-intrusive monitor takes advantage of the binary nature of the sensor to develop a response frequency or other metric to correlate the deterioration of the sensor response rate to the increase in vehicle emissions. As emissions standards decrease, the degree to which the response of the sensor can slow before the OBD II standard is exceeded can be greatly diminished. The magnitude of the difference in frequency response decreases between a sensor at the end of the useful life of a vehicle, and one that causes the emissions of the vehicle to exceed the OBD II standard. This causes an increase in the probability of α and β errors. Additionally, the demands of lower emissions standards may limit the duration of intrusive monitors. The shorter diagnostic test times may lead to an increase in variability.

New wide-range or universal exhaust gas oxygen (UEGO) sensors are beginning to become commonplace at reduced emissions levels. A UEGO sensor does not have the "switching" response of the traditional oxygen sensor. It has A/F magnitude information so the actual exhaust A/F ratio can be determined. This leads to more precise A/F control, which in turn can reduce emissions. Monitoring algorithms for these sensors are still evolving. Maloney (2001) reported that dynamic response rate (intrusive) diagnostics are capable of detecting UEGO sensor degradation on direct injection gasoline engine vehicles. The diagnostic was conducted close to stoichiometric operation, so use on a traditional port fuel injected (PFI) gasoline engine appears promising.

12.10 Secondary Air System Monitor

A secondary air (SAIR) system allows for air to be pumped into the exhaust stream to allow for an increased exothermic reaction. The heat generated lights off the catalyst quicker. Currently, a functional monitor is required, irrespective of the emissions standard. The functional diagnostic test is executed on the FTP cycle, not necessarily during the intended operation. One commonly used method to test the SAIR system for functionality is by using the exhaust gas oxygen sensor or universal exhaust gas oxygen (UEGO) sensor. After the system has warmed up, the air pump can intrusively be commanded on, and the oxygen sensor can be measured to ensure that excess oxygen is seen in the exhaust stream. The monitor operates during fully warmed-up conditions to avoid any emissions breakthrough. In 2006, this requirement will undergo a significant modification. The secondary air system must then be monitored while it is being used (i.e., cold start), and it must be emissions based. Due to these considerable changes, a more direct measurement of the system may be required. Use of a mass airflow (MAF) sensor, pressure switches, or a UEGO sensor has allowed for the secondary air test while cold. Measuring the airflow or the pressure changes from the pump operation will give a signal that can more easily be correlated to emissions. Use of an oxygen sensor to diagnose the system requires attention to sensor warm-up. Whichever diagnostic methodology is chosen, interactions with the environment will be more of a concern because monitoring must occur during the startup operation.

12.11 Variable Valve Timing/Control System Monitor

Active research continues into true "camless" engines. The most commonly accepted means of varying the intake and exhaust valve timing is through the use of the camshaft. Many different types of VVT control exist. The various types of architecture can include controlling only the intake cam, only the exhaust cam, or both the intake and exhaust cams, either together or independently. The dynamic control can be on the valve timing or lift. Obviously, the relationship to the OBD system is concerned with the level of an emissions impact to the failed hardware. When the control manipulates the residual fraction of burned gases and this results in an impact on engine NOx emissions, the OBD concerns must be addressed. Current monitoring requirements for the VVT systems are in accordance with the comprehensive component requirements described earlier in this chapter. Specifically, the system must be monitored for a misalignment between the camshaft and crankshaft of one or more cam/crank sprocket cogs [California Code Regulations, Title 13, Section 1968.2]. Beginning in the 2006 Model Year on LEV II applications, the regulations expand the degree to which the VVT systems must be monitored. The impact on emissions of both the target error and the response rate of the system will need to be monitored. Target error refers to the steady-state difference between the requested valve timing and the actual valve timing that is observed during operation. Diagnosing the response rate of the system may prove to be especially troublesome due to not only the transient nature of the control but also environmental conditions. Both engine oil temperature and viscosity are two system noise factors that can cause the response rate of the system to vary for a given engine operating condition. Both of these variables must be considered when designing the diagnostic test.

12.12 In-Use Performance Tracking

OBD II monitors can be separated into two major categories: (1) continuous, and (2) noncontinuous. Continuous monitors, such as circuit checks (e.g., open, short, out of range), execute constantly during engine operation. Noncontinuous monitors execute a specific test within a given set of entry conditions once per trip. Previously, regulations were quite specific that these noncontinuous monitors had to complete on the first engine start portion of the FTP. The expectation was that these monitors would run reasonably often during real-world operation. Not all vehicle operators drive in the same manner as an FTP cycle. When this is the case, the monitor(s) may not complete often during real-world driving patterns. Similarly, execution of the traditional method of monitoring the evaporative emissions system can be lower than desired due to the many interactions with dynamic vehicle operation. The pitfalls of the monitor have been detailed in Section 12.6.2. The latest regulations [California Code Regulations, Title 13, Section 1968.2] stipulate performance metrics on the frequency of execution of some of the noncontinuous monitors.

The OBD II system will now calculate and track in-use performance data. Table 12.4 summarizes the target tracking ratios for the required noncontinuous monitors. The OBD system of each vehicle will calculate the numerator and denominator of the ratio. The performance ratios then can be calculated offboard the vehicle. The numerator of this ratio is defined as the number of times the correct window of opportunity has been encountered, such that each monitor would have detected a malfunction. The denominator is defined by the number of times the vehicle has been operated in a CARB-defined manner that monitors should have completed. The regulations are specific with regard to cumulative vehicle operation, ambient temperature, and ambient pressure.

TABLE 12.4
MONITORS ASSESSED BY THE IN-USE
PERFORMANCE TRACKING ALGORITHM

In-Use Performance Tracking Requirement	Tracking Completion Ratio
Catalyst Monitor Bank #1	0.336
Catalyst Monitor Bank #2	0.336
Oxygen Sensor Monitor Bank #1	0.336
Oxygen Sensor Monitor Bank #2	0.336
0.020-Inch Evaporative Leak Detection	0.26
0.040-Inch Evaporative Leak Detection	0.52
EGR Monitor	0.336
VVT Monitor	0.336
SAIR Monitor	0.26

12.13 Summary

Onboard vehicle diagnostics are an ever-expanding regulatory effort. The California regulations continue to evolve, as do the U.S. federal diagnostic requirements. The expansion is not limited to the United States; Europe and Japan have implemented diagnostic standards (EOBD and JOBD, respectively). The trend will most likely continue into other countries. Just as there are multiple OBD regulations that must be satisfied, many variations of the diagnostic algorithms are covered here. We have attempted to highlight the principle methods employed and to identify emerging trends as appropriate.

Do not consider this work as the final chapter for OBD; it is merely an attempt to capture the current state as related to LEVs. Clearly, the regulations will continue to evolve, in both the quest for lower vehicle emissions and the refinement of diagnostic requirements.

The complexity and challenges of OBDs for LEVs has been addressed here. It should be apparent that the requirements of the diagnostics are beginning to rival the complexity of the new technology required to meet the lower emissions standards. This necessitates having OBD II considerations as part of the development process of new low emissions technologies and components. Depending on the situation, an emissions-reducing component without an available diagnostic algorithm may not meet regulatory requirements.

12.14 References

1. California Code Regulations, "Malfunction and Diagnostic System Requirements for 2004 and Subsequent Model-Year Passenger Cars, Light-Duty Trucks, and Medium-Duty Vehicles and Engines (OBD II)," Final Regulation Order, Title 13, Section 1968.2.

2. Rieck, J.S., Collins, N.R., and Moore, J.S. (1998), "OBD-II Performance of Three-Way Catalysts," SAE Paper No. 980665, Society of Automotive Engineers, Warrendale, PA.

3. DeRonne, M., Labus, G., Lehner, C., Gonsiorowski, M., Western, B., and Wong, K. (2003), "The Development and Implementation of an Engine-Off Natural Vacuum Test for Diagnosing Small Leaks in Evaporative Emissions Systems," SAE Paper No. 2003-01-0719, Society of Automotive Engineers, Warrendale, PA.

4. Maloney, P.J. (2001), "A Production Wide-Response Diagnostic Algorithm for Direct-Injection Gasoline Application," SAE Paper No. 2001-01-0558, Society of Automotive Engineers, Warrendale, PA.

5. Ogawa, T. and Morozumi, H. (2002), "Diagnostics Trends for Automotive Electronic Systems," SAE Paper No. 2002-21-0021, Society of Automotive Engineers, Warrendale, PA.

6. Perry, P.D. and Delaire, G. (1998), "Development and Benchmarking of Leak Detection Methods for Automobile Evaporation Control Systems to Meet OBDII Emission Requirements," SAE Paper No. 980043, Society of Automotive Engineers, Warrendale, PA.

CHAPTER 13

Emissions Measurements

Michael Akard
Horiba Instruments, Inc. USA

13.1 Introduction

There are two types of regulated emissions for gasoline-powered vehicles: (1) exhaust, and (2) evaporative. This chapter will focus on exhaust emissions. Exhaust emissions are measured from a vehicle driving a regulated test procedure on a dynamometer. Each of the major automotive markets (i.e., Europe, Japan, and the United States) has specified driving cycles. In addition to exhaust emissions, evaporative emissions monitor the amount of fuel (hydrocarbons [HCs]) released by the vehicle that is not included in the tailpipe emissions.

The earliest regulation for emissions from mobile sources was in the United States. The capabilities of the analyzers and the sample collection systems were more than sufficient to measure the early regulated levels of pollutants. Although numerous regulatory updates have occurred since the initial regulations, the biggest change in the regulations has been lowering the allowed pollutant emissions. As the allowed emissions levels dropped, analyzer manufacturers were making improvements in the analysis. The U.S. regulations describe a bag sample collection method using a constant volume sampler (CVS). Sample collection systems also were undergoing refinement but until recently were not fundamentally different than the earliest CVSs. Without some modification, the standard CVS technology is not sufficient to measure exhaust levels that are, in some cases, lower than the ambient air.

To accurately and repeatably measure the near zero emissions levels currently required, many factors must be addressed. This chapter is intended to give an overview of the changes in the utilities, analyzers, sampling systems, verification procedures, and sample transfer techniques intended to improve the analysis for modern vehicles. The different approaches used for the sample handling systems have generated the most debate. Errors associated with the dynamometer are not covered in this chapter.

13.2 Exhaust Emissions

The most stringent emissions requirements are currently in California, the largest automotive market within the United States. The following discussion is specifically for the California low emissions vehicle (LEV II) regulations but can be applied with minor modifications to any of the major automotive markets. Vehicles that are compliant with these regulations are more commonly known as super ultra-low emissions vehicles (SULEVs). A SULEV that also demonstrates zero evaporative emissions is classified as a partial zero emissions vehicle (PZEV). The following description is a brief overview of the steps used to determine vehicle emissions.

Tailpipe emissions for certification testing first are diluted and then are collected into a sample bag. Many research facilities also measure the raw tailpipe emissions and even the gas exiting the engine before or from the middle of the catalytic converter. Sampling from these points removes some of the exhaust emissions and requires corrections to correlate with a test without these sampling points. These raw measurements and the corrections to the bag measurements are beyond the scope of this chapter. The concentration for the regulated components is measured from the bag. The dilution factor is calculated in the U.S. Code of Federal Regulations (CFR) for the CVS, assuming stoichiometric conditions for a typical gasoline (H:C = 1.85), as

$$C_8H_{15} + 11.75 O_2 + 44.19 N_2 \rightarrow 8 CO_2 + 7.5 H_2O + 44.19 N_2 \quad (13.1)$$

Using this fuel, the carbon dioxide (CO_2) concentration at stoichiometry would be given as

$$[CO_2]_{stoich} = \frac{8}{8 + 7.5 + 44.19} = 13.4\% \quad (13.2)$$

The dilution factor (DF) is then calculated as shown in Eq. 13.3, taking into account the fuel that does not undergo complete combustion as carbon monoxide (CO) and total hydrocarbons (THCs):

$$DF = \frac{[CO_2]_{stoich}}{[CO_2]_{samp} + [CO]_{samp} + [THC]_{samp}} \qquad (13.3)$$

$$= \frac{0.134}{[CO_2]_{samp} + [CO]_{samp} + [THC]_{samp}}$$

The concentration of each component in the sample bag, C_{samp}, is then converted to the mass exiting the tailpipe by using a total flow measurement, V_{mix}, and the density for each component, ρ,

$$M = \rho \times \left(C_{samp} - C_{amb} \times \left(1 - \frac{1}{DF}\right)\right) \times V_{mix} \qquad (13.4)$$

This mass, M, is determined for each bag (phase of the drive cycle). The weighted mass per mile, Y, is then calculated using the distance driven in the test cycle, D, and a weighting factor, as

$$Y = 0.43 \times \left(\frac{M_{phase1} + M_{phase2}}{D_{phase1} + D_{phase2}}\right) + 0.57 \times \left(\frac{M_{phase2} + M_{phase3}}{D_{phase2} + D_{phase3}}\right) \qquad (13.5)$$

The tailpipe emissions limits are 20 mg/mile for the nitrogen oxides (NOx), nitric oxide (NO), and nitric dioxide (NO_2). Carbon monoxide emissions limits are 1000 mg/mile. The tailpipe emissions limit for THC is adjusted to nonmethane hydrocarbons (NMHCs). In addition, the California Air Resources Board (CARB) has defined nonmethane organic carbon (NMOG) for oxygenated fuels. The NMOG is equal to the NMHC with the oxygenated compounds added to the measurement. The tailpipe emissions limits are 10 mg/mile for NMOG. The oxygenated compounds are not measured from the sample bag. These compounds either can be measured directly or can be collected in impingers or cartridges and analyzed separately. California also regulates formaldehyde to be less than 4 mg/mile.

The Federal Test Procedure (FTP-75) consists of three drive phases. The first phase, the cold start, contributes most of the emissions for the vehicle. The second phase produces exhaust concentrations very close to the ambient air concentration. This test cycle will be particularly difficult to measure accurately. The hot start, or third phase, will contribute slightly more emissions than the second phase but still much less than the first phase.

At these low emissions limits, each source of error must be examined and reduced. To accurately and reproducibly measure these emissions, the complete system must be evaluated for possible errors. An error in the NOx measurement for a TLEV test of 0.01 g/mi is only 2.5% of the emissions limit. *This same error would be almost 50% of the SULEV emissions limit.*

13.3 Constant Volume Sampler

The original CVS used for certification is described in the CFR. This device diluted the exhaust gas with ambient air and pumped the diluted gas into a sample bag. The total diluted flow is controlled by a critical flow venturi (CFV) and the blower. With the pressure and temperature measured at this device at the CFV, an overall flow can be determined. The sample bags are filled with this diluted exhaust (before the CFV). The ambient air used for the dilution is collected into an ambient bag. Figure 13.1 shows the basic steps of the emissions measurements with a CVS [Sherman *et al.* (2004b)]. The process has been broken down into six steps, and each step can contribute measurement error.

13.3.1 Dilution Air

The earliest SULEV testing was done in Japan. To improve the accuracy of the CVS, the dilution air was first refined before use. This technique will use a high-volume sample with the HCs, CO, and NOx significantly reduced to consistently low concentrations. The dilution air refinement (DAR) system used by Tayama *et al.* (1998) produced a dilution gas with a concentration of no greater than 0.1 ppm C and 0.1 ppm NO. This is a significant reduction from the typical ambient air values for THC of 2–5 ppm C. The CO_2 was not removed. Figure 13.2 is a representation of the system used by Tayama *et al.* (1998). To refine the large volume of air required, the dilution refinement system is expensive and requires a large amount of power for operation.

There are three significant advantages to using refined dilution air. The first is a lower ambient bag concentration. This lower concentration will reduce the

Emissions Measurements

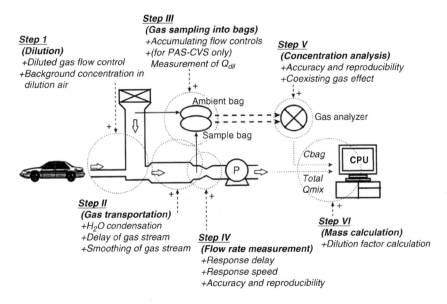

Figure 13.1 *Configuration and error factors of a CVS system [Inoue et al. (1999)].*

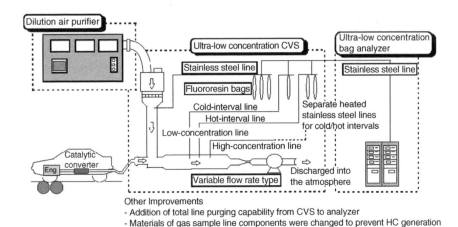

Figure 13.2 *System configuration for a CVS with dilution air purification [Tayama et al. (1998)].*

error associated with the dilution factor measurement shown in Eq. 13.4. The second advantage is that by using a dilution gas significantly cleaner than the exhaust, the analysis is simplified. Figure 13.3 shows the relative decrease in measurement error determined by Tayama *et al.* (1998). The third advantage is that by using a stable dilution gas, there is little variability in the contribution of the dilution gas to the sample bag concentrations. This error is described in more detail in Section 13.3.4 on bag sampling.

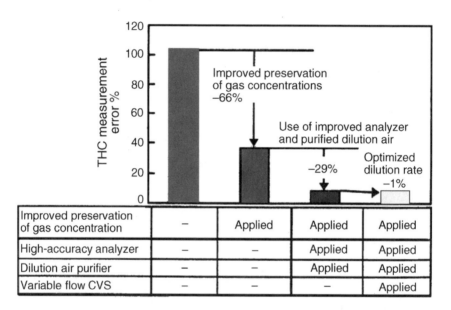

Figure 13.3 Increased measurement accuracy for improvements at 1/10 ULEV emissions limits [Tayama et al. (1998)].

Some more common and less expensive dilution air treatments in use generate less dramatic accuracy improvements. Maintaining a constant humidity is important for the dilution air. Constant humidity is important to determine the necessary dilution ratio to avoid condensation. A charcoal filter also can be used to help stabilize the HC values. A charcoal filter will have no effect on methane concentrations and has been shown to cause error in some cases [Silvis *et al.* (1999)]. Most systems should have a sampling port for background measurement. This is useful to compare dilution gas readings to ambient bag measurements. This will help the user differentiate between dilution gas contamination and bag or system contamination. Heating the dilution air can allow the use of lower dilution rates.

13.3.2 Exhaust Dilution

Modern systems use a remote mixing tee located near the tailpipe of the vehicle. There are many advantages to diluting close to the exhaust source, such as reduced chance for HC hang-up, reduced heating requirements, and reduced chance for water condensation. Also, it is easier to maintain the required pressure within 5 inches of water [40 CFR 86.109-94(c)(1)]. Silvis and Lewis (1999) have demonstrated with a fluid dynamics model that a design can take advantage of the "elbow meter" effect to maintain the tailpipe pressure to well below this limit, even under high flow conditions. They also observed during the testing of this low-loss remote mix tee that the flow measurements became less noisy.

Austin and Caretto (1998) have pointed out three basic errors to the dilution factor calculation in Eq. 13.3, as follows:

1. The equation does not properly account for pollutants in the dilution air.

2. Excursions from stoichiometric operation in the vehicle will cause an error in the calculation.

3. Any measurements made on a dry basis will be more concentrated and can introduce error into the DF calculation.

The New York Department of Environmental Conservation has adopted a modification to Eq. 13.3 for its inspection/maintenance (I/M) program that corrects the first error as

$$DF_{NY} = \frac{13.4}{\left([CO_2]_{samp} + [CO]_{samp} + [HC]_{samp}\right) - \left([CO_2]_{dil} + [CO]_{dil} + [HC]_{dil}\right) \times \left(1 - \frac{1}{DF}\right)}$$

(13.6)

The terms in Eq. 13.6 are the same as those shown in Eq. 13.3 with the addition of the dilution gas terms denoted with a subscripted "dil." The new dilution factor DF_{NY} still uses the assumption of stoichiometric vehicle operation. The average dilution factor is dictated by the need to avoid condensation during the test.

13.3.3 Dilution Ratio Optimization

The analysis of the sample is improved at lower dilution ratios. This is especially true for very low emissions. It is always analytically easier and more accurate to measure a large difference between two numbers. Inherently, the CVS will produce a variable instantaneous dilution ratio. The overall flow is controlled so that as the vehicle accelerates during the test phase and the tailpipe flow increases, the dilution flow will decrease. As the vehicle decelerates or idles, the tailpipe flow decreases, and the dilution flow increases. The traditional CVS determined the averaged dilution ratio from Eq. 13.3. The addition of a smooth approach orifice (SAO) to measure the instantaneous dilution flow is an improvement in the dilution ratio determination (as previously discussed in the proportional ambient sampling [PAS] CVS).

Figure 13.4 from Behrendt *et al.* (2002) shows the changing DF over the first phase of the FTP-75. Figure 13.5 from the same paper shows the humidity measurements and the calculated humidity from the CO_2 concentration during the same phase. The portions of the test where large DFs are achieved show corresponding lower levels of humidity. The idle states or decelerations will not cause any condensation. The lower the DFs that dip near the value of 1 (undiluted exhaust) in Figure 13.4, the greater the chance for condensation. In Figure 13.6, Behrendt *et al.* (2002) indicate the dilution flow required to avoid condensation at various temperatures. By maintaining an elevated temperature, the dilution ratio can be reduced significantly. Use of this approach requires that the system avoid any colder regions that could cause condensation. Minimizing the DF for the enhanced CVS also will introduce oxygen interference concerns. Oxygen interference will be discussed later in this chapter.

13.3.4 Bag Sampling

During each test phase, two bags are filled: (1) the ambient bag, and (2) the sample bag. The difference in the readings is crucial to the measurement. The ambient bags are filled with the dilution gas upstream of the mixing

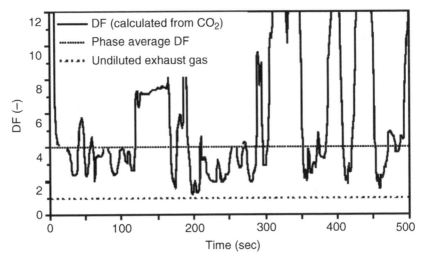

Figure 13.4 *Time-resolved DF for an average DF = 4 (calculated from CO_2 for Phase 1 of FTP-75) [Behrendt et al. (2002)].*

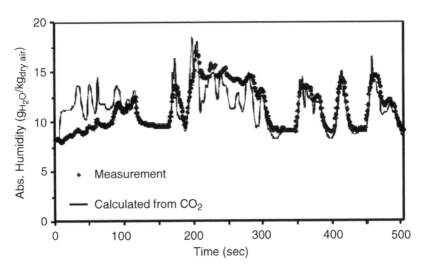

Figure 13.5 *Absolute humidity in diluted exhaust at a sample venturi in Phase 1 of FTP-75 (calculated from CO_2 and a capacitive humidity sensor) [Behrendt et al. (2002)].*

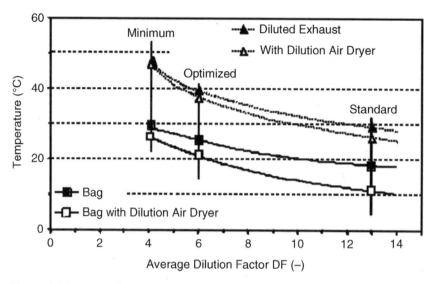

Figure 13.6 Required temperature to avoid water condensation in diluted exhaust [Behrendt et al. (2002)].

tee. The sample bags are filled through venturis after the mix tee. One of the serious drawbacks to this system is the error of the DF calculation increases dramatically during nonstoichiometric operation. Lean-burn direct injection or fuel shutoff strategies will cause the dilution factor to be underestimated.

A strategy to reduce the dilution factor error and to reduce the effect of concentration variations in the dilution gas was developed by Silvis and Chase (1999) and is called the PAS strategy. Instead of simply filling the ambient bag at a constant rate through a venturi, the dilution gas is pumped through a mass flow controller (MFC). The inlet pressure of the MFC is controlled by a pressure regulator. The MFC is controlled by a feedback loop to fill the ambient bag proportionately to the dilution gas flow. The dilution gas flow is measured by an SAO housed in the remote mix tee (RMT), as shown in Figure 13.7. The advantage of this method is to eliminate the effect of concentration variations in the dilution gas by integrating the dilution gas in the ambient bag in the same ratio as the dilution gas in the sample bag. This approach also allows the user to substitute the calculation for the DF (Eq. 13.3 or Eq. 13.6) with a measurement of the dilution air (measured by the SAO) and the total flow (measured

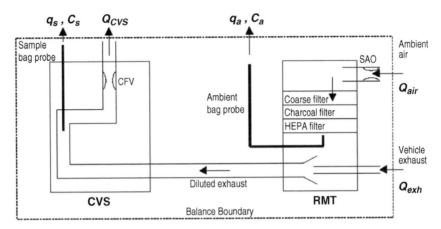

Figure 13.7 Constant volume sampler system with an RMT containing an SAO for inlet airflow measurement and a CFV for total flow metering [Silvas and Chase (1999)].

by the CFV before the blower). This flow measurement can eliminate the error associated with nonstoichiometric vehicle operation and can produce an accurate DF measurement.

13.3.5 Bag Materials

Two basic sources of error are associated with the sample bag. The first error occurs when the diluted exhaust in the sample bag is not mixed thoroughly. The sample bag will use a loop within the bag to fill evenly and draw sample evenly. This loop helps to avoid sample stratification. The concentrations in the diluted exhaust will change over the time of the sampling. This is especially true for the Phase I bag (cold start). For an accurate measurement, it is assumed that the bag will have evenly mixed the sample from the beginning of the test cycle with the sample from the end of the test cycle. Erratic driving of the test cycle still can cause short-term stratification within the bag.

The second source of error involves sample interaction with the bag materials. An ideal sample bag would not change the concentrations in the diluted exhaust from the time it was pumped into the bag until the sample was measured. In practice, the bag will show some permeability to CO_2 and the other constituents. Carbon dioxide is the component with the largest concentration difference with the ambient air surrounding the bag. Therefore, CO_2 would show the greatest loss of sample if the bag were equally permeable to all components. As CO_2 permeates through the bag, the reading will decrease over time. This will improve the calculated mileage of the vehicle. The CO_2 concentration also is used to determine the DF. Any error in this DF will affect the ambient bag correction, as shown in Eq. 13.4.

Ohtsuki *et al.* (2002) investigated the loss of CO_2 for three different bag materials in less than two hours, as shown in Figure 13.8. Note the ambient temperature and the CO_2 concentration (2%) used for this study. Using a higher concentration gas will increase the relative loss of CO_2 as the concentration gradient between the bag and the ambient conditions is increased. Comparison tests use the same size bags filled with the same amount of test gas.

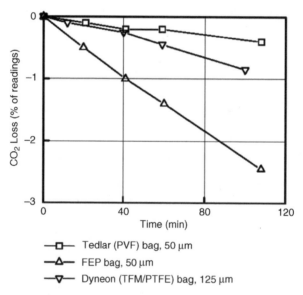

Figure 13.8 *Carbon dioxide permeation (room temperature, CO_2 2% by volume) [Ohtsuki et al. (2002)].*

Edward Sun and Wayne McMahon at the CARB investigated eight different bag materials for CO_2 loss [Sun and McMahon (2001)]. Figure 13.9 shows the percentage loss of CO_2 after one hour. The bags were filled with dry span gas in both studies. In practice, bags are read in less than one hour.

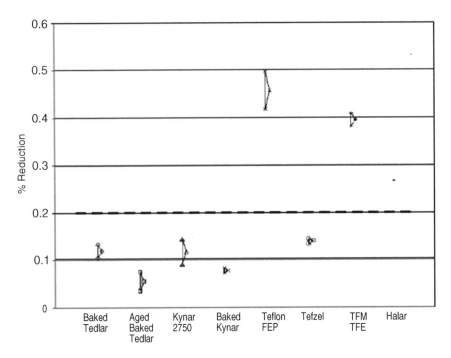

Figure 13.9 Carbon dioxide (1.8% by volume) reduction after 60 minutes [Sun and McMahon (2001)].

Another interaction with the bag material is HC contamination from the bag, or outgassing. The level of contamination has been investigated for various bag materials independently in all three major automotive markets [Behrendt *et al.* (2002); Hill *et al.* (2002); Ohtsuki *et al.* (2002); Sherman *et al.* (2004b); Sun and McMahon (2001)]. The findings differ somewhat, but the overall contribution from a baked Tedlar® bag filled with dry synthetic air is no more than approximately 60 ppb in approximately 30 minutes. The effect of water and CO_2 on the HC background also has shown some different results, depending on the study. A comprehensive study of bag materials undertaken at Southwest Research Institute [Sherman *et al.* (2004b)] found three bag materials to be adequate performance for low emissions testing.

In practice, it is always recommended to replace both the sample bag and the ambient bag at the same time. It also is important to purge and evacuate all new bags until the bag readings filled with zero air come down to typical values seen in the test cell. It has been shown by Behrendt *et al.* (2002) that the level of outgassing increases when the bag temperature increases (Figure 13.10). The CVS for this testing used a heated cabinet to house the sample bags. These results show that operating with bag temperatures above 30°C can add a significant amount of bag outgassing. The heated bag rack also will allow the user to more readily clean new bags by operating the purge and evacuation cycles under an elevated temperature. In addition, heating the collection system, bags, and sampling system to 30°C allows the user to operate at reduced dilution rates.

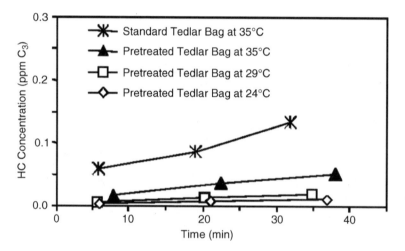

Figure 13.10 Outgassing from Tedlar bags at different temperatures [Behrendt et al. *(2002)*].

13.3.6 Flow Rate Measurement and Control

The conversion of the concentration in the bags to mass requires an accurate flow measurement (Step IV in Figure 13.1). Care taken in the design and the mix tee can reduce the signal noise. If the DF is measured directly with an SAO in the mix tee measuring the dilution gas, the CVS design should ensure that the dilution gas transport time to the total flow measurement at the venturi is short enough to avoid the need to time align the flow measurements. Otherwise, DF errors will be introduced with temperature differences at the CFV during the test. A heat exchanger in the CVS can be used to maintain a relatively constant temperature at the CFV and sample venturis. This will mean the overall dilution flow will be kept constant over the complete test phase. A modern CVS uses a range of venturis and bag sizes to ensure sufficiently filled bags collected at a constant rate over the different lengths of the test phases. The sample venturis and the CFV are located near one another to help prevent different sampling conditions between the two. Alternatively, an MFC can be used to collect the diluted exhaust into sample bags. Use of an MFC allows the user to select the fill rate for each phase. If alcohols and carbonyls (aldehydes and ketones) also are being tested, they would have a separate sampling line.

13.3.7 Sample Transfer from Bags to Analyzers

In addition to the bags changing the sample, the pumps, valves, sample transfer lines, and flow control devices can contribute contamination. The pollutant most easily affected by the sample delivery system is the THCs. Although the other pollutants have boiling points well below the ambient conditions of a test cell, the average boiling point for HCs in gasoline is above room temperature.

Gasoline consists primarily of HCs with 5 to more than 12 carbons per molecule. Vehicle manufacturers are required to speciate the HCs remaining in the exhaust with each engine family and verify that the ratios of the targeted toxic compounds within the exhaust do not change significantly. Discussion of the method of speciation is beyond the scope of this chapter, but it is important to realize the broad boiling-point range of the exhaust THCs. Any of the heavier HCs can adsorb on the bag material or any of the transfer components. This could mean that HCs from previous bags or even previous tests could contribute to the reading. This is termed hang-up. The amount of contamination or hang-up is larger when a higher THC concentration sample is collected.

To reduce the contribution of this error to the measurements, bags can be split into two groups: (1) dirty, and (2) clean [Behrendt *et al.* (2002); Krenn *et al.* (2000); Tayama *et al.* (1998)]. Each set of bags has its own sample delivery system from the bags to the analyzers. In some cases, the THC analyzer will have two separate sampling lines. The advantage of splitting these gas delivery systems is to avoid carryover from the dirty bag, Phase I, into either the ambient bags or the cleaner phases. Manufacturers have investigated with various materials to ensure minimal interaction with the surfaces [Behrendt *et al.* (2002); Schiefer *et al.* (2000)]. Electropolished 316 stainless steel [Guenther *et al.* (2002)] or brightly annealed tubing [Behrendt *et al.* (2002)] have been used effectively. Schiefer *et al.* (2000) summarized the physical interactions possible at each component in the sampling system.

An easily implemented contamination test is described by Ohtsuki *et al.* (2002) and is summarized in Figure 13.11. The exhaust connection is sealed after a phase has been collected. Both sample and ambient bags are then evacuated and filled with dilution air. This dilution air is measured in the normal configuration. The sample and ambient bags are then switched and measured again. Figure 13.12 shows the results of five of these tests. In a properly designed and uncontaminated system, the results should indicate that the sample bag is the primary source of carryover between tests. As the sample bag is moved to the ambient read line, the readings of the five tests remain approximately the same. The ambient bag reads lower concentrations, even when sampled through the sample read line. Therefore, most of the contamination for this system is in the sample bag.

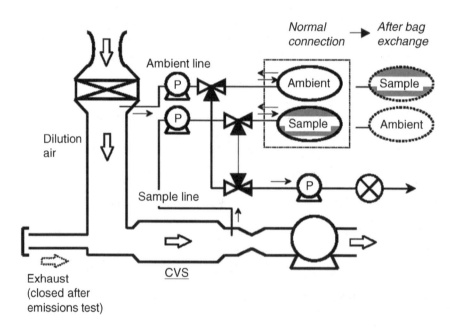

Figure 13.11 Experimental setup for confirming HC contamination [Ohtsuki et al. (2002)].

13.4 Bag Mini-Diluter

A new sampling system technique was developed that is fundamentally different from the CVS approach. The bag mini-diluter (BMD) was developed by

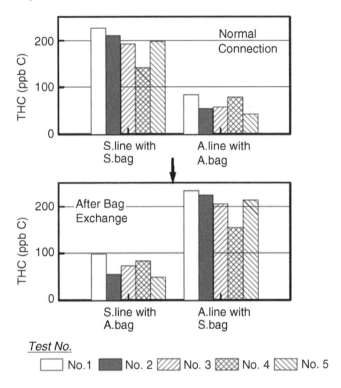

Figure 13.12 Results of five tests for confirming HC contamination [Ohtsuki et al. (2002)].

the American Industry/Government Emissions Research (AIGER) [Guenther *et al.* (2000) and (2002); Luzenski *et al.* (2002); Nagy *et al.* (2000); Sherman *et al.* (2001)]. Both the CARB [Silvis *et al.* (1999)] and the U.S. Environmental Protection Agency (EPA) [Environmental Research Consortium Technical Report (2001)] have allowed the use of the BMD for certification testing. The primary difference between the BMD and the CVS is that the BMD is a partial flow sampling system. The CVS dilutes the entire exhaust stream and then collects a constant amount of diluted exhaust into the bag. The CVS dilution rate changes over the course of the test cycle, so an average dilution ratio is used. The BMD samples a portion of the exhaust and dilutes this sampled exhaust at a constant dilution ratio. The amount of sampled exhaust is proportional to the total exhaust flow rate from the vehicle. Figure 13.13 shows a basic design schematic of the BMD. The direct vehicle exhaust (DVE) flow measurement is taken after the sample has been collected. Although this configuration requires a correction to the total flow for the amount sampled, it will avoid collection errors due to pressure drops or other effects from the meter on the exhaust stream. Figure 13.13 indicates that both the sample collection point and the DVE are in a heated area, but the temperature is reduced at the temperature measurement device [Luzenski *et al.* (2002)].

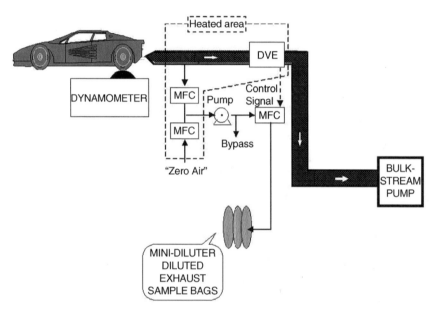

Figure 13.13 *Basic design elements of a BMD [Nagy et al. (2000)].*

This constant dilution rate is a fundamental advantage for the BMD because a minimal dilution rate is used that does not vary during the driving cycle. While previous studies have looked at minimizing dilution rates for CVS, this is always an averaged dilution rate. As the exhaust flow decreases, the dilution gas increases to provide a higher dilution rate in the CVS. As the exhaust increases, the dilution gas flow will decrease, and an excursion will form where condensation could occur. To avoid these points, the overall dilution must be increased (Figure 13.4).

The BMD also uses less dilution gas, because only a portion of the exhaust is diluted. This allows the use of cylinder dilution gas of zero air or nitrogen as well as a zero air generator [Rabellino (2002); Rabellino and Sherman (2003)]. By using a zero grade dilution gas, the advantages previously discussed with the dilution air refinement are gained [Behrendt *et al.* (2002)]. With zero grade dilution gas and low dilution rates, the concentrations in the bags will be very low, requiring better analyzers. One way to look at the effective methods for improving sampling methods is to compare the magnitude of the numbers measured. The use of refined dilution air results in a small difference (typical dilution ratio) in small numbers (refined dilution gas). The enhanced CVS results in a large difference (low dilution ratio) between large numbers (ambient air dilution gas). The BMD results in a large difference (low dilution ratio) between small numbers (refined dilution gas).

New requirements for a BMD compared to a CVS include a direct exhaust flow measurement with a fast response and heated exhaust sampling [Guenther *et al.* (2003); Thiel *et al.* (2003); Yassine *et al. (2003)*]. The signal from the flow meter is used to control the proportional sampling of the exhaust. Guenther *et al.* (2003) generated a list of the requirements for a direct exhaust meter. These requirements are summarized in Table 13.1. The E-Flow device developed by Flow Technologies was used for the initial BMD work as the most accurate and robust system. The device operates with two diametrically opposed ultrasonic sensors and is described in detail by Yassine *et al.* (2003). One concern about any flow measurement device is the operating temperature. Figure 13.14 shows the exhaust temperature profiles for three different vehicles during the US06 cycle. Exhaust temperatures reached a maximum of 462°C.

Fast response thermocouples can be used to correct meters for any temperature effects, but the system also must be robust at these temperatures. In the case of the E-Flow and some other devices, this requires heat exchangers or other techniques to reduce and stabilize the temperature of the exhaust before reaching the meter. Another concern with flow meters is the accuracy of the measurement at idle conditions.

A large set of data supports the correlation between the BMD and the CVS for low emissions vehicle (LEV) and even ULEV emissions [Guenther *et al.* (2000 and 2002); Luzenski *et al.* (2002); Nagy *et al.* (2000); Sherman *et al.* (2001)]. Figure 13.15 summarizes the correlation for four components (CO, CO_2, THC, and NOx). This study also used both nitrogen and air as a diluent. The only obvious consistent deviation from good correlation in this data can be seen in the THC values for the S phase.

TABLE 13.1
PERFORMANCE REQUIREMENTS FOR EXHAUST FLOW MEASUREMENT

Meter Performance Criteria	Requirement
a. System accuracy	±1.0% of point throughout measurement range
b. Drift	Less than ±0.5% of point drift from month to month
c. Signal processing	Measurement frequency and data acquisition rate ≥10 Hz
d. Response time	T_{90} response ≤0.5 seconds for step change
e. Dynamic performance at idle	Better than ±5% accuracy during idle operating conditions
f. Insensitive to gas properties	No accuracy impact from changes in temperature, pressure, or gas composition
g. Tailpipe pressure	≤ ±1.0 inch of water for FTP and ±5.0 inches of water for US06
h. Robustness	Durability proven during actual vehicle testing
i. Size and cost	Smaller is better; system cost will be the primary driver

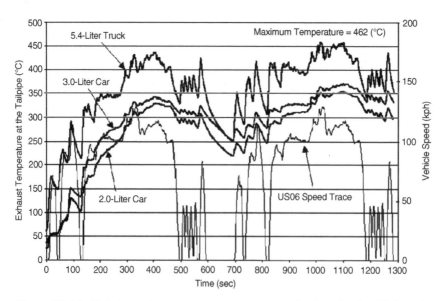

Figure 13.14 Tailpipe exhaust temperatures on two back-to-back US06 cycles [Guenther et al. (2003)].

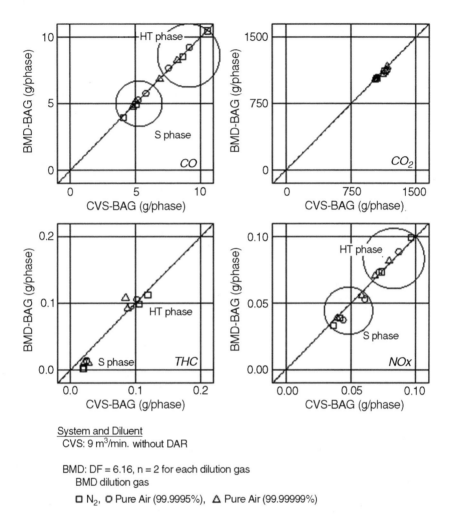

Figure 13.15 Comparison between BMD and CVS for LEV with FTP-75.

13.4.1 Bag Mini-Diluter Dilution Gas

The reduced amount of dilution gas required by the BMD makes the selection between zero grade air and zero grade nitrogen possible. Most work to date has used zero grade air as the BMD diluent. Nitrogen as a diluent will increase the collected sample differences from a sample collected with a CVS. Oxygen interference with the nondispersive infrared (NDIR) and the flame ionization detector (FID) potentially could be increased with the use of nitrogen for dilution gas. When air is used for the BMD dilution gas, the same gas source also is used as the FID and methane analyzer's zero gas for zero/span calibration and complete calibration (sometimes termed linearization). This eliminates the calculation error associated with DF error in the last term in Eq. 13.4. This dilution gas must meet the CFR zero gas impurity requirement of 1 ppm C [40 CFR 86.114-94(a)(5)]. This prohibits the use of unrefined air. The dilution gas must not have impurities of a level that will cause the linearization to fail a mid-span check. This level is dependent on the analyzer range used (Table 13.2).

TABLE 13.2
ZERO GAS IMPACT ON MID-SPAN READING

Analyzer Range (ppm C)	Highest Mid-Span (ppm C)	Lowest Mid-Span (ppm C)	Zero Gas Impurity	
			Highest (ppm C)	Lowest (ppm C)
10	5	1.5	0.20	0.035
5	2.5	0.75	0.10	0.018
3	1.5	0.45	0.06	0.011
1	0.5	0.15	0.02	0.004

The dilution gas must have a stable level of impurities. This is critical for performance. In the CVS system, zero gas was previously supplied independently of the dilution gas. The zero gas was supplied by cylinders that would have significant supply variation only at or just before a cylinder change. Now, relatively small variations in zero gas impurities over the 20- to 30-minute time frame will cause the analyzer linearization to fail. Larger variation in zero gas impurity could cause the zero/span drift check to fail, invalidating the test. This criteria is $\pm 2\%$ of full scale (FS). Variation in the oxygen concentration of the dilution gas can affect the FIA reading (see Section 13.4.2 on oxygen interference). There is no significant effect on the methane measurement. This analyzer is based on a separation process (gas chromatography [GC]) that will separate the sample oxygen from the methane before detection.

Impurity level drift can be increasing or decreasing relative to the zero calibration. The actual zero check includes both zero air impurity drift and analyzer drift. Figure 13.16 illustrates the worst case where the impurity level drift is increasing, adding to the measured exhaust pollutants. All the drift shown is being estimated as zero air generator drift and not analyzer drift. The average reading from dilution gas shown in the figure assumes constant dilution over the bag fill. This is *not* true because only the dilution factor is maintained over the bag fill, not the dilution flow. Figure 13.17 shows the actual dilution flow for a test. The impurity contribution to the measured pollutants will be the product of the density, dilution factor, dilution gas impurity concentration, and dilution flow. To estimate the effect of the dilution gas impurity changing during a test, the impurity concentration is estimated as a straight line between the zero calibration and the zero check. The actual impurity changes could have a significantly different pattern. Investigation of the zero air generator for typical trends can be done to refine this model.

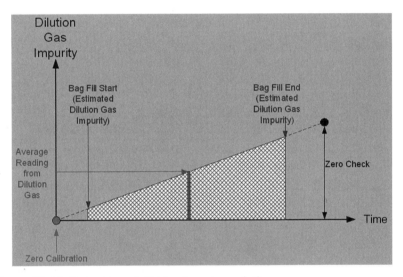

Figure 13.16 Estimation of dilution (zero) gas drift.

Using these estimates, a change in the average concentration of 0.1 ppm C could result in a higher reading of approximately 0.2 mg. For these calculations, the bag fill start begins immediately after the zero calibration. Table 13.3 lists a summary of a simulation. The mass emissions were calculated by multiplying the flow measurement values collected at 0.1-second intervals with the estimated zero gas impurity, a DF of 6, a time step of 0.001667 minutes, and the CFR listed density of 0.5768 kg/m^3. Figure 13.18 shows a plot of this calculated

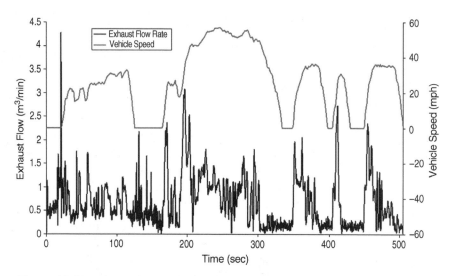

Figure 13.17 Exhaust flow and vehicle speed.

TABLE 13.3
DILUTION GAS IMPURITY EFFECT ON MASS FOR THC

Ending Dilution Gas Impurity (ppm C)	Mass Emissions Error THC (mg)
0.01	0.016
0.05	0.078
0.10	0.156
0.20	0.312
0.60	0.936
1.00	1.560
1.30	2.028

mass contribution from zero gas impurity drift in the lower line. The upper line is the integrated mass over the test phase from the zero gas impurity.

Analyzers that do not use air for a zero gas include the NDIR (CO and CO_2 analyzers) and the chemiluminescence analyzer (CLA) (NOx). Any NOx or

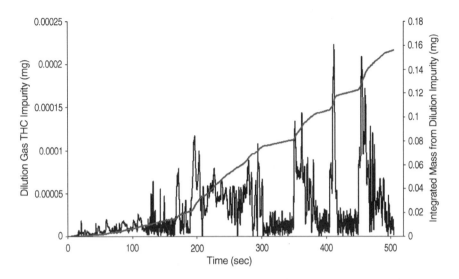

Figure 13.18 Mass emissions contribution from a constant dilution gas impurity drift (up to 100 ppb C).

CO in the dilution gas will contribute to the reported pollutants. Table 13.4 lists the calculated milligrams per phase for different dilution impurities using typical values for Bag 2. These calculations do not include a drift but an absolute impurity level. By not using the dilution gas for a zero gas, any impurity will

TABLE 13.4
DILUTION GAS IMPURITY EFFECT ON MASS FOR CO AND NOx

Diluent Impurity (ppm)	Bag 2 CO (mg/phase)	Bag 2 NOx (mg/phase)
1.000	38.60	59.89
0.500	19.30	29.94
0.200	7.72	11.98
0.100	3.86	5.99
0.050	1.930	2.99
0.010	0.386	0.599
0.005	0.193	0.299

be added to the BMD error. The amount of CO_2 in the dilution gas will have a similar effect on the measured CO_2 and will decrease the fuel efficiency.

13.4.2 Oxygen Interference

The oxygen concentration in the diluted exhaust will have an effect on the FIA, called oxygen interference. There also is an effect on the CO_2 measurement. These effects are important because the oxygen concentration in a BMD sample is different from the oxygen concentration from a CVS collected sample. The primary reason for this difference is simply a much lower dilution factor. Enhanced CVS techniques also reduce the dilution factor compared to traditional CVS samples. Typical oxygen concentrations are approximately 19.5% for a CVS bag but are as low as 18% for a BMD. These values are dependent on the dilution rates and the dilution gas conditions. Currently, there is no regulatory requirement for minimizing or correcting for oxygen interference differences among sampling methods in the light-duty vehicle industry. The heavy-duty on-road vehicle industry has been required to reduce the oxygen interference to less than 3% since 1979.

Oxygen interference corrections increase the concentration (mass emissions). The diluted sample oxygen concentration can be directly measured with an oxygen analyzer and used to correct the FID measurement. This is not a straightforward correction. Oxygen interference for an FID can be broken down into two components: (1) zero quench, and (2) span quench. The zero quench is simply an offset due to flowing a different amount of oxygen into the analyzer. Note that while the sample flow makes up only a small amount of the total oxygen supplied to the flame, the sample flow is introduced into the flame in the richest region of the flame. This fact could account for the disproportionate amount of influence of the sample gas oxygen concentration on the measurements. The zero quench is significant only at relatively small concentrations. The zero quench can be corrected simply by determining the response of the analyzer to synthetic air mixed through a gas divider with nitrogen [Sun *et al.* (2004)] and correcting the response with a third- or fourth-order curve.

The difficulty comes in dealing with the span quench portion of the oxygen interference. The approach used by Sun *et al.* (2004) is to optimize the FID not at the point of highest sensitivity but at the point where span quench is minimized. This point is determined by using two gas blends of propane at low concentrations: one balanced in nitrogen, and the other balanced in air. While maintaining a constant air and sample flow rate, vary the fuel flow rate and measure both blends at incremental pressure settings. The nitrogen balanced curve shown in Figure 13.19 shows a higher response than the air balanced from approximately 80- to 125-mL/min fuel flow. From 90- to 110-mL/min fuel flow rates, an appreciable difference in sensitivity can be observed. This is the span quench effect. To minimize this effect, the FID was optimized at 125 mL/min where the two curves cross but still exhibit high sensitivity. The span quench is not totally eliminated because this curve was generated at a single propane concentration. Sherman *et al.* (2004a) generated a set of data varying both the oxygen and the propane concentrations, as shown in Figure 13.20. Note that the two FIDs (from Figures 13.19 and 13.20) are optimized differently. The increasing span quench effect can be seen to dominate the error as the HC concentration increases.

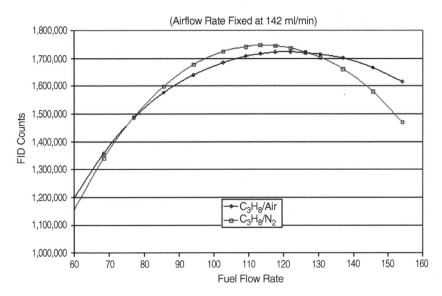

Figure 13.19 *Flame ionization detector sensitivity at 100 ppm C [Sun et al. (2004)].*

In addition to the oxygen quench error, FIDs optimized differently can have different response factors. Although the methane response factor is used to correct the subtraction of methane for NMHC values, all HCs other than propane have a relative response different than 1.00. Sun *et al.* (2004) demonstrated that sample flow adjustments can be used to adjust the methane relative response factors without changing the oxygen quench (air/fuel [A/F] ratio for the flame).

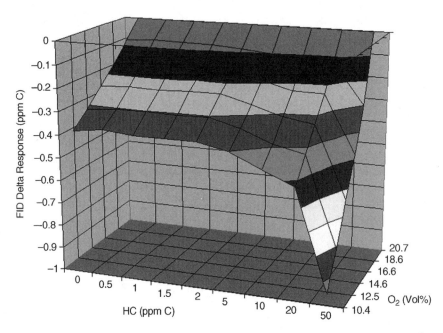

Figure 13.20 *Flame ionization detector response to a range of oxygen and propane concentrations [Sherman et al. (2004a)].*

13.4.3 Modeled Performance for Sampling Systems

The primary goal of using a BMD is to reduce the uncertainty of measurement during the measurement of ultra-clean vehicles. Sherman *et al.* (2001) have modeled the error associated with a variety of sampling systems. Figure 13.21 shows the percentage error contribution differences for seven different configurations for NMHC measurements at 8 mg/mile. Each type of error is estimated and indicated by different shading on the bar graph. Although the standard CVS shows an unacceptable level of error, the BMD, dilution air refinement, and heated enhanced CVS all show dramatic reductions in the modeled error.

13.4.4 Differences Among Bag Mini-Diluters

Two methods are used for collecting the exhaust for the proportional sampling. The first approach utilizes MFCs to control the flow of the raw exhaust and the dilution gas [Luzenski *et al.* (2002); Nagy *et al.* (2000)]. The second approach

Emissions Measurements

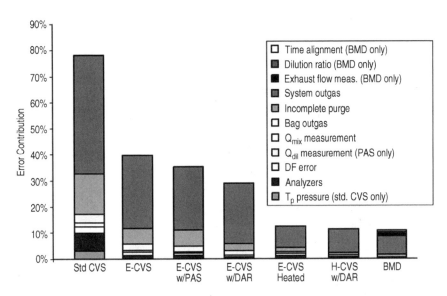

Figure 13.21 *Assessment of NMHC sensitivity at 8 mg/mile [Sherman et al. (2001)].*

uses CFVs [Guenther *et al.* (2000 and 2002); Silvis *et al.* (1999)] to maintain the ratio of exhaust and dilution gas. The BMD can be verified by leak check tests, response time checks, and DF checks. The diluted gas is metered into the sample bag with an MFC in both approaches. Figure 13.22 shows the basic design element schematic for the CFV-controlled BMD.

There are some advantages with the second design over the design shown in Figure 13.13. An MFC has an error associated with sampling gases with variable composition. The three-MFC approach must use average CO_2 and water (H_2O) concentration corrections. The dual-CFV approach has the advantage of being simpler with less chance for maintenance problems with failed MFCs trying to control raw exhaust. By maintaining the pressure into both CFVs at the same value, a consistent dilution ratio will be generated.

13.4.5 System Verification

The CFR requires only a static system verification—a critical flow orifice (CFO) check with propane. Although this test verifies the system integrity with respect to a leak, it does not provide a dynamic test. The BMD is a device

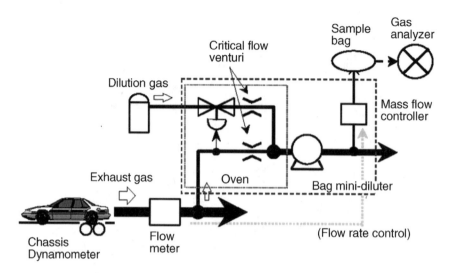

Figure 13.22 Alternative style BMD basic design elements [Guenther et al. (2002)].

with a rapidly changing sampling flow. To improve testing for the BMD (and the CVS) and to help with cell-to-cell correlation, a vehicle exhaust emissions simulator was developed. This device has been used to help verify the recovery for multiple components [Landry *et al.* (2001); Usmen *et al.* (2003)]. In addition, the vehicle exhaust emissions simulator can introduce components to the sampling system at a constant concentration, a ramped concentration, or a programmed sequence that changes the concentration and total flow to more closely simulate what would be exiting a vehicle.

13.5 Analyzer Accuracy

In the automotive emissions industry, analyzer accuracy typically has been defined by stating available ranges. The analog analyzers that were the basis for the written regulations were expected to produce accurate ($\pm 2\%$ of the measured value) down to 15% of the range. Current high-end automotive analyzers can accurately measure well below 15% of the full-scale range. A more appropriate measurement of accuracy could be described as the limit of linearization [Akard *et al.* (2002)]. Using this method, some analyzers can read accurately well below the typical expectations for a range. In addition, analyzers can be successfully calibrated to cover a wider range of measurement. This approach has been termed dynamic single range and is demonstrated by Sherman *et al.* (1999a and 1999b).

Although the theoretical advantage of this single range has been discussed [Kampelmuhler *et al.* (2000); Sherman *et al.* (1999b)], the major advantage for the user is the capability to analyze bag readings without having to "sniff" the bag. The sniff procedure is required for multiple-range analyzers to determine the appropriate range to be used for the measurement. By skipping this step and proceeding to the measurement directly after a calibration, the overall time the sample is in the bag is reduced, and the analyzer can spend more time in measurement mode to obtain a stable reading. In addition, the user will save calibration time and gases by using a single span concentration.

13.5.1 Calibration Gas Requirements

Gases of interest in the sample bag at low concentrations include THCs, CH_4 (methane), NO, NO_2, and CO. Table 13.5 shows the lowest commercially available ranges. The lowest ranges of all analyzers are limited by the same calibration gas requirements considerations. An investigation into the commercial availability of National Institute of Standards and Technology (NIST) traceable standards has been done for the U.S. market [Akard *et al.* (2003)]. In the Japanese market, lower traceable gas concentrations are commercially available for most gases.

**TABLE 13.5
HORIBA ANALYZER RANGES**

Component	Analyzer Model	Specified Lowest Range
CO(LE)/(SLE)	AIA-721LE	10 ppm
CO_2	AIA-722	5000 ppm
THC(LE)/(SLE)	FIA-726LE	1 ppm C
NOx(LE)/(SLE)	CLA-750LE	1 ppm
CH_4(LE)/(SLE)	GFA-720LE	1 ppm

Carbon dioxide and oxygen are both at fairly large concentrations in a bag. Therefore, the calibration gases required are easily accessible. The oxygen typically is not needed in the bag measurement unless the user is attempting to correct for oxygen quench. The CO_2 in the ambient bag typically is approximately 400–450 ppm. This measurement can be important for the dilution factor calculation (see Eq. 13.3 and Eq. 13.5) and for the fuel economy measurement.

The CFR specifies four gases to be used in a calibration with a gas divider: (1) zero, (2) span, (3) calibration, and (4) mid-span. The calibration gases must be NIST traceable to $\pm 1\%$ accuracy. The span gas must be only $\pm 2\%$ NIST traceable. The following discussion will deal with the calibration and the mid-span gases that are more difficult for manufacturers to produce.

The CFR requires a mid-span verification for any calibration curve generated with a gas divider [40 CFR 86.114-94(a)(7)]. The mid-span concentration must be between 15 and 50% of the full-scale range. The mid-span gas for the lowest range used is the lowest concentration blended gas cylinder used. Therefore, the lowest range that can be set up is determined by the analyzer specifications and the lowest concentration of gas a supplier can produce to meet the EPA protocol $\pm 1\%$ NIST traceable. To determine the lowest range that can be set up on an analyzer, the cylinder concentration should be doubled. This cylinder then would be used for a 50% mid-span cylinder.

To produce the greatest accuracy for ambient measurements of purified dilution gases, the lowest possible range should be used. After purchasing the mid-span cylinder, the range is set to twice the concentration of this gas. The calibration gas then should be purchased at approximately 90% of this range value.

13.5.2 Utilities and System Components

Utilities can be critical to performance. A clean power supply is important for the analyzers. Most modern benches have a significant level of local filtering. Vibration also can generate significant noise, especially for the low-range NDIR analyzers. Calibration and span gases should be located as near to the bench as possible. The NO span gas and FID fuel lines should not be made of Teflon® for safety reasons. The NO span gas line often will need to be flushed out after long periods of inactivity, such as weekends or overnights.

It was previously shown that the sample bag is a major contributor to THC contamination (Figure 13.12). This is true only if all other system components used have been selected for inertness and rigorously cleaned. Tayama *et al.* (1998) demonstrated the need to select proper materials for components in the sample line (Figure 13.23). Three iterations of materials were tested for this regulator location. The background contamination was tested for each configuration,

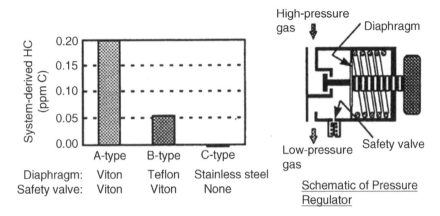

Figure 13.23 Regulator contamination [Tayama et al. (1998)].

and the stainless steel diaphragm with no safety valve was chosen. Emissions measurement manufacturers must evaluate the design and each individual component to achieve the requisite cleanliness and inertness for testing.

Equipment manufacturers also have experimented with different flush/back-flush techniques to speed cleaning of the system after measurement of more polluted exhaust. In addition, heated purges have been shown to decrease the time needed to remove contamination from the system.

13.6 Summary

Accurate emissions measurements have become more difficult as regulation levels have been lowered. Current emissions levels often can be lower than the ambient air used for sample dilution. Different sampling techniques have been used to eliminate error and improve reproducibility. Care must be taken at each step of the process to eliminate errors. The sampling system and the analyzers must be evaluated as a complete system.

13.7 References

1. Akard, M., McDonnough, D., Porter, S., Oestergaard, K., and Tsurumi, K. (2003), "System and Utility Considerations for Analyzer Calibration and Measurement of Low Concentrations in Automotive Exhaust," SAE Paper No. 2003-01-3154, Society of Automotive Engineers, Warrendale, PA.

2. Akard, M.L., Tsurumi, K., Oestergaard, K.E., and Inoue, K. (2002), "Why the Limit of Detection (LOD) Value Is Not an Appropriate Specification for Automotive Emissions Analyzers," SAE Paper No. 2002-01-2711, Society of Automotive Engineers, Warrendale, PA.

3. Austin, T.C. and Caretto, L. (1998), "Improving the Calculation of Exhaust Gas Dilution During Constant Volume Sampling," SAE Paper No. 980678, Society of Automotive Engineers, Warrendale, PA.

4. Behrendt, H., Moersch, O., Seiferth, C.T., Seifert, G.E., and Wiebrecht, J. (2002), "Studies on Enhanced CVS Technology to Achieve SULEV Certification," SAE Paper No. 2002-01-0048, Society of Automotive Engineers, Warrendale, PA.

5. Environmental Research Consortium Technical Report (2001), "The Bag Mini-Diluter Alternative to CVS Sampling," May 1, 2001, USCAR, Southfield, MI; http://www.uscar.org.

6. Guenther, M., Henry, T., Silvis, W.M., Wu, D., and Nakatani, S. (2000), "Improved Bag Mini-Diluter Sampling System for Ultra-Low Level Vehicle Exhaust Emissions," SAE Paper No. 2000-01-0792, Society of Automotive Engineers, Warrendale, PA.

7. Guenther, M., Vaillancourt, M., and Polster, M. (2003), "Advancements in Exhaust Flow Measurement Technology," SAE Paper No. 2003-01-0780, Society of Automotive Engineers, Warrendale, PA.

8. Guenther, M., Vaillancourt, M., Sherman, M.T., Carpenter, D., Rooney, R., and Porter, S. (2002), "Advanced Emissions Test Site for Confident PZEV Measurements," SAE Paper No. 2002-01-0046, Society of Automotive Engineers, Warrendale, PA.

9. Hill, J., Loo, J.F., and Swarin, S.J. (2002), "Evaluation of New Bag Sampling Materials for Low Level Emissions Measurements," SAE Paper No. 2002-01-0051, Society of Automotive Engineers, Warrendale, PA.

10. Inoue, K., Ishihara, M., Akashi, K., Adachi, M., and Ishida, K. (1999), "Numerical Analysis of Mass Emission Measurement Systems for Low Emission Vehicles," SAE Paper No. 1999-01-0150, Society of Automotive Engineers, Warrendale, PA.

11. Kampelmuhler, F., Schimpl, H., Weidinger, C.R., and Kreft, N. (2000), "High-Performance Linearization Procedure for Emission Analyzers," SAE Paper No. 2000-01-0798, Society of Automotive Engineers, Warrendale, PA.

12. Krenn, M., Kampelmuhler, F., Weidinger, C.R., and Poscharnig, G. (2000), "Evaluation of a New Design for CVS-Systems Meeting the Requirements of SULEV and Euro IV," SAE Paper No. 2000-01-0800, Society of Automotive Engineers, Warrendale, PA.

13. Landry, M.P., Guenther, M., Isbrecht, K.S., and Stevens, G.D. (2001), "Simulation of Low-Level Vehicle Exhaust Emissions for Evaluation of Sampling and Analytical Systems," SAE Paper No. 2001-01-0211, Society of Automotive Engineers, Warrendale, PA.

14. Luzenski, D.J., Bedsole, K.T., Hill, J., Nagy, D.B., and DeCarteret, S.S. (2002), "Evaluation of Improved Bag Mini-Diluter System for Low Level Emissions Measurements," SAE Paper No. 2002-01-0047, Society of Automotive Engineers, Warrendale, PA.

15. Nagy, D.B., Loo, J., Tupla, J., Schroeder, P., Middleton, R., and Morgan, C. (2000), "Evaluation of the Bag Mini-Diluter and Direct Vehicle Exhaust Volume System for Low Level Emissions Measurement," SAE Paper No. 2000-01-0793, Society of Automotive Engineers, Warrendale, PA.

16. Ohtsuki, S., Inoue, K., Yamagishi, Y., and Namiyama, K. (2002), "Studies on Emission Measurement Techniques for Super-Ultra-Low-Emission Vehicles," SAE Paper No. 2002-01-2709, Society of Automotive Engineers, Warrendale, PA.

17. Rabellino, L. and Sherman, M. (2003), "Oxygen Content Variation in 'Zero' Grade Air Due to Production, Treatment and Distribution," SAE Paper No. 2003-01-0393, Society of Automotive Engineers, Warrendale, PA.

18. Rabellino, L.A. (2002), "Zero Gases for Emission Monitoring—Production, Storage, Treatment and Usage," SAE Paper No. 2002-01-2712, Society of Automotive Engineers, Warrendale, PA.

19. Schiefer, E., Schindler, W., and Schimpl, H. (2000), "Study on Interferences for ULEV-CVS Measurement, Related to the Complete Measuring System, Discussion of Error Sources, Cross-Sensitivity and Adsorption," SAE Paper No. 2000-01-0796, Society of Automotive Engineers, Warrendale, PA.

20. Sherman, M., Mauti, A., and Rauker, Z. (1999a), "Evaluation of Mass Flow Controller Gas Divider for Linearizing Emission Analytical Equipment," SAE Paper No. 1999-01-0148, Society of Automotive Engineers, Warrendale, PA.

21. Sherman, M.T., Akard, M.L., and Nakamura, H. (2004a), "Flame Ionization Detector Oxygen Quench Effects on Hydrocarbon Emission Results," SAE Paper No. 2004-01-1960, Society of Automotive Engineers, Warrendale, PA.

22. Sherman, M.T., Chase, R.E., and Lennon, K.M. (2001), "Error Analysis of Various Sampling Systems," SAE Paper No. 2001-01-0209, Society of Automotive Engineers, Warrendale, PA.

23. Sherman, M.T., Chase, R.E., Mauti, A., Rauker, Z., and Silvis, W.M. (1999b), "Evaluation of Horiba MEXA 7000 Bag Bench Analyzers for Single Range Operation," SAE Paper No. 1999-01-0147, Society of Automotive Engineers, Warrendale, PA.

24. Sherman, M.T., Huron, G.D., Whitney, K., and Hill, J. (2004b), "An Investigation of Sample Bag Hydrocarbon Emissions and Carbon Dioxide Permeation Properties," SAE Paper No. 2004-01-0593, Society of Automotive Engineers, Warrendale, PA.

25. Silvis, W.M. and Chase, R.E. (1999), "Proportional Ambient Sampling: A CVS Improvement for ULEV and Lean Engine Operation," SAE Paper No. 1999-01-0154, Society of Automotive Engineers, Warrendale, PA.

26. Silvis, W.M., Harvey, N., and Dageforde, A.F. (1999), "A CFV Type Mini-Dilution Sampling System for Vehicle Exhaust Emissions Measurement," SAE Paper No. 1999-01-0151, Society of Automotive Engineers, Warrendale, PA.

27. Silvis, W.M. and Lewis, G. (1999), "The Control of TP Pressure in Emissions Sampling Systems," SAE Paper No. 1999-01-0152, Society of Automotive Engineers, Warrendale, PA.

28. Sun, E.I. and McMahon, W.N. (2001), "Evaluation of Fluorocarbon Polymer Bag Material for Near-Zero Exhaust Emission Measurement," SAE Paper No. 2001-01-3535, Society of Automotive Engineers, Warrendale, PA.

29. Sun, E.I., McMahon, W.N., Peterson, D., Wong, J., and Tsurumi, K. (2004), "Oxygen Quench Effect on Flame Ionization Detector for Hydrocarbon Emission Measurements," SAE Paper No. 2004-01-1431, Society of Automotive Engineers, Warrendale, PA.

30. Tayama, A., Kanetoshi, K., Tsuchida, H., and Morita, H. (1998), "A Study of a Gasoline-Fueled, Near-Zero-Emission Vehicle Using An Improved Emission Measurement System," SAE Paper No. 982555, Society of Automotive Engineers, Warrendale, PA.

31. Thiel, W., Hornreich, C., Mörsch, O., and Seifert, G. (2003), "Problems of Partial Sample Systems for Modal Raw Exhaust Mass Emission Measurement," SAE Paper No. 2003-01-0779, Society of Automotive Engineers, Warrendale, PA.

32. Usmen, R., Yassine, M.K., and Hartrick, M.D. (2003), "Vehicle Exhaust Emissions Simulator—A Quality Control Tool to Evaluate the Performance of Low Level Emission Sampling and Analytical Systems," SAE Paper No. 2003-01-0391, Society of Automotive Engineers, Warrendale, PA.

33. Yassine, M.K., Marji, M.S., Berndt, R.W., and Laymac, T.D. (2003), "Parameters Affecting Direct Vehicle Exhaust Flow Measurement," SAE Paper No. 2003-01-0781, Society of Automotive Engineers, Warrendale, PA.

34. 40 Code of Federal Regulations 86.109-94(c)(1).

35. 40 Code of Federal Regulations 86.114-94(a)(5).

36. 40 Code of Federal Regulations 86.114-94(a)(7).

CHAPTER 14

Near-Zero-Emission Gasoline-Powered Vehicle Systems

Fuquan (Frank) Zhao
Brilliance Jinbei Automobile Corporation

14.1 Introduction

The ultra-clean emissions standards such as PZEV (partial zero emissions vehicle) require a reduction of more than 90% in tailpipe emissions of hydrogen (HC), nitrogen oxide (NOx), and carbon monoxide (CO) from the current vehicle level. Lower tailpipe emissions are only part of the emissions reduction requirements for those near-zero-emissions gasoline-powered vehicles. Due to the fact that tailpipe emissions are getting lower day by day, evaporative emissions are becoming an increased fraction of the total vehicle system emissions. The PZEV vehicles also must meet the zero evaporative emissions requirement. As a result, zero fuel-based evaporative emissions are required for a PZEV vehicle where 0.054 g per test of fuel-based evaporative emissions or less rounds to zero. The remaining evaporative emissions are non-fuel-based, which are usually derived from paint, tires, seats, washer fluid, sealants, and adhesives. The near-zero-emissions vehicles also must meet the onboard diagnostics (OBD) requirements for detecting deterioration in emissions control systems. The challenge for the PZEV OBD system is to develop the capability of detecting the difference between an adequately functioning emissions control system meeting the emissions standard and one with a deterioration such that the emissions are about to exceed the required threshold, defined as a multiple of the emissions standard. This multiple is 2.5 for PZEV vehicles. In addition, PZEV vehicles are required to meet the durability requirement of either 15 years or 150,000 miles for all emissions-related components. Above all these, PZEV vehicles must retain all customer attributes, such as rapid starting in all ambient conditions, stable idle, good driveability, high fuel efficiency, and good performance with an affordable price. Clearly, all these requirements put a tremendous challenge on developing such a vehicle for mass production.

14.2 System Requirements for a Near-Zero-Emissions Gasoline-Powered Vehicle

Because many vehicle design and operating parameters affect emissions, a system approach must be taken to meet the increasingly stringent emissions standard, which includes the optimization of the engine system, control system, and catalyst system to reduce engine-out emissions and improve catalyst conversion efficiency. For a cold engine, the emphasis should be placed on reducing cold-start emissions and improving catalyst light-off performance.

All engine design and operating parameters that determine mixture preparation, combustion, emissions formation, and catalyst conversion efficiency must be carefully examined from the system perspective to minimize engine-out emissions while enhancing catalyst light-off performance. Those parameters include, but are not limited to, fuel preparation, spray characteristics (e.g., droplet size distribution, cone angle, targeting), fuel injection timing, valve timing, airflow, cranking speed, spark timing, burn rate, idle speed, engine warm-up rate, coolant distribution around the intake valves, crevice volume, and exhaust manifold design. The benefits of various types of injectors such as single-fluid, multi-stream, air-assisted, and heated injectors must be assessed to take advantage of each technology. The piston ring top-land crevice must be minimized to reduce HC emissions. The excess coolant on the intake side must be minimized to enhance the mixture preparation process during cold start and the warm-up period. The exhaust manifold design for enhancing the mixing process between the cylinder events and reducing the heat loss requires careful consideration, particularly when implementing the secondary air injection strategy. Engine startup and shutdown strategies also must be carefully investigated to maximize the gain of each strategy.

Many factors affect catalyst performance, including heat flux to the catalytic converter; heat loss from the engine to the exhaust system; catalyst formulation, loading, volume, and aging (thermal degradation and chemical poisoning); substrate configuration for reduced thermal mass; catalyst configuration and utilization; operating temperature; and air/fuel (A/F) ratio control, including with an aged catalyst. The catalyst type and the amount of catalyst loading have a significant impact on catalyst light-off temperature. The thermal mass of the catalyst substrate configuration also significantly affects the catalyst light-off time. An ultra-thin-wall substrate with a high cell density can significantly reduce the time constant to light-off the catalyst. As the thermal resistance of catalysts is improved, chemical poisoning primarily from the phosphorus in the engine oil is increasingly important. Reducing engine oil consumption is becoming particularly important in developing near-zero-emissions gasoline-powered vehicles. The reduced phosphorus loading on the catalyst is effective in reducing catalyst light-off time.

The most important factor that determines the catalyst light-off time is the engine-out heat flux to the converter during cold start. Strategies used to achieve a rapid catalyst heating can be classified into two approaches. The first is the lean starting approach, which typically requires a slight lean-out of the A/F mixture in combination with significantly retarded spark timing. How to maintain stable combustion with a slightly lean mixture and an extremely retarded spark timing is a daunting task, particularly with a high-driveability-index (DI) fuel. The other is the rich starting approach, which is achieved by operating the engine with an extremely rich mixture and injecting secondary air into the exhaust port via an air pump during cold start. Rich combustion produces a large amount of engine-out emittants of H_2, CO, and HC, which will react with the secondary air inside the exhaust manifold to reduce the converter-in emissions and significantly boost catalyst light-off performance due to the exothermic reaction inside the exhaust system. Because there is no need to over-retard the spark timing, this rich starting approach is more robust and tends to be less sensitive to fuel properties. Both approaches have been applied to mass-production vehicles.

Fuel-based evaporative emissions must be reduced using a system approach that should be focused on both the powertrain and the fuel storage and delivery sources. The powertrain source includes those from both the base engine and the air induction system. Evaporative emissions of fuel storage and delivery systems are from the fuel tank, fuel lines, vapor lines, charcoal canister, fuel filler cap, and control valves. To eliminate nearly all fuel-based evaporative emissions, a high rate of purging of vapor emissions is necessary during nearly all operating conditions. In addition to providing high purging rates, new materials are a mandate for reduced permeability and leakage under the required lifetime of durability.

Several OEMs have succeeded in launching mass-produced vehicles that are certified to meet the near-zero-emissions standards such as the California PZEV standard. Several representative systems are outlined in subsequent sections of this chapter to explain the basic elements and general layouts of these near-zero-emissions gasoline-powered vehicle systems.

14.3 BMW Partial Zero Emissions Vehicle System

The BMW PZEV system was described in detail by Landerl *et al.* (2003). The key elements in this near-zero-emissions gasoline-powered vehicle system are summarized as follows:

- **Engine**: Six-cylinder inline with continuous adjustment of the intake and exhaust camshafts to realize internal exhaust gas recirculation (EGR).

- **Engine compression ratio**: 10.1 (compared to the standard practice of 10.5 for the baseline engine).

- **Piston top-land height**: 3 mm.

- Two half-shell manifolds with a minimum gas travel and compact surface area are used to reduce the wall heat losses.

- A rich starting approach via a secondary air system is used during the warm-up phase of the catalyst.

- **Fuel injection pressure**: 5 bar (compared to the standard practice of 3.5 bar for the baseline engine).

- **Fuel spray pattern**: Four-hole injector (compared to the standard practice of a two-hole injector for the baseline engine).

- Optimized piston rings to reduce oil consumption.

- **Fuel system**: The tank with filler pipes is being changed to stainless steel, as well as the tank ventilation system with the fuel tank vent valve and vent valve line. Flexible sections of corrugated stainless steel tubing are welded into the stainless steel lines at the necessary places.

Near-Zero-Emission Gasoline-Powered Vehicle Systems

- To reduce evaporative emissions, modifications were made to the engine positive crankcase ventilation (PCV) system, including the integration of an oil separation facility and PCV pressure control system into the cylinder head cover and welding the uniform distribution strip for blowby gas discharge into the intake manifold.

- **Converter system**: Two close-coupled catalysts (cell density of 1200 cpsi) and two underfloor catalysts (UFCs).

- Two wide-range oxygen sensors.

- A radiator with a catalytic coating was used.

- An earlier intake closing strategy was chosen. This implemented intake camshaft lock-up position leads to an increase in the cylinder charge during the starting operation. In addition, this evacuates the intake manifold faster, which is desirable with respect to improved mixture control as a consequence of faster fuel evaporation and rapid opening of the secondary air valve.

- An optimized start synchronization is used to shorten the synchronizing period during the starter phase and to achieve emissions-optimized injection timing, particularly preventing fuel from being injected into the intake valve open period.

- High-resolution engine speed detection is used in the engine management system to enable better resolution of the engine speed gradient during engine startup, which can selectively optimize the fuel injection timing, ignition timing, and mixture preparation for every individual power stroke.

14.4 Ford Partial Zero Emissions Vehicle System

The Ford PZEV system is based on a rich starting approach via secondary air injection to achieve the catalyst light-off performance and has been described in detail by several publications [Carney (2003); Kunde *et al.* (2003); Wade (2003)]. The fundamental elements of the Ford PZEV system are summarized as follows:

- **Engine system**: 2.3 L, four-valve, four-cylinder engine.

- **Compression ratio**: 9.7.

- **Piston ring top-land height**: 4 mm.
- Reduced piston ring tension.
- New horning specification for improving bore cylindricity and wall surface finish.
- Dual-wall exhaust manifold.
- **Fuel injector**: Twelve-hole, low-leakage (1.8-mm^3/min) injector with fine droplets (compared to that of the four-hole baseline injector).
- **Spray targeting**: 90% of the fuel targeted on the intake valve (goal).
- **Airflow control**: Two-position charge motion control valve (CMCV).
- Electric air pump for delivering secondary air.
- An HC trap was incorporated into the air intake system between the mass airflow sensor and throttle body.
- A coil-on-plug ignition system with iridium-tipped, thin-wire spark plugs.
- An electric, stepper-motor, water-cooled EGR system.
- To minimize oil incursion into the combustion chamber by way of the valves, upgraded seals (sharp-edge lip design) were incorporated. To optimize the benefit associated with the new seal and to ensure high-mileage robustness with minimum wear, the stems were polished for both the intake and exhaust valves.
- Zero-permeability gaskets/joints.
- A steel fuel tank.
- The fuel rail-mounted pulse damper was modified to include a nonpermeable diaphragm as well as a sealing cap.
- An increased volume of carbon canister.
- For the fuel system, the length of flexible tubes was reduced. The material of the filler pipe was changed to stainless steel. A quick-on cap with a fluoroelastomer seal was added. The elastomeric hose was replaced with a convoluted steel pipe.

- **Catalyst system**: Close-coupled catalyst with two bricks (first brick 900/2.5 and 0.69 L; second brick 400/4.0 and 0.69 L); UFC with two bricks (first brick 400/4.0 and 0.84 L; second brick 400/4.0 and 0.84 L).

- Three heated exhaust gas oxygen (HEGO) sensors for pre- and post-catalyst oxygen control. The first sensor is positioned in front of the first catalyst for lambda control. The second sensor is placed between the first and second catalysts and is used by the OBD II system to infer catalyst efficiency and to report any problems. Finally, downstream of the Number 2 catalyst is a third oxygen sensor that is used to ensure that the catalysts are operating at the optimized A/F ratio.

- Mass airflow sensor.

- **Control system**: x-Tau transient fuel compensation is used for A/F ratio control; $\pm 0.5\%$ A/F ratios for cylinder-to-cylinder airflow distribution (goal); $\pm 0.5\%$ A/F ratio (warmed up) for A/F ratio control; open-loop A/F ratio control adaptively updated based on the most representative speed/load cell in a closed loop. Oxygen sensors prior to and/or after the catalyst are used for high-mileage A/F ratio control. The EGR valve is controlled by a stepper motor. Purge control is via a vapor management valve and purge compensation for A/F ratio control during purging.

14.5 Honda Ultra-Clean Gasoline-Powered Vehicle System

Honda ultra-clean gasoline-powered vehicle systems are described in several recent publications [Kitagawa (2000); Yamazaki et al. (2004)]. Figure 14.1 shows an overview of the Honda super ultra-low emissions vehicle (SULEV) system [Kitagawa (2000); Kitagawa et al. (2001)], which was based on a lean A/F ratio control at cold start. The key technologies used in the Honda clean vehicle systems are summarized briefly as follows:

- Variable valve timing and lift electronic control (VTEC) is used to generate a high-swirl intake flow by opening only one intake valve, which will enable a lean A/F mixture.

- Air-assisted injector.

- Low-heat-capacity exhaust manifold and pipe.

- Electronically controlled EGR valve.

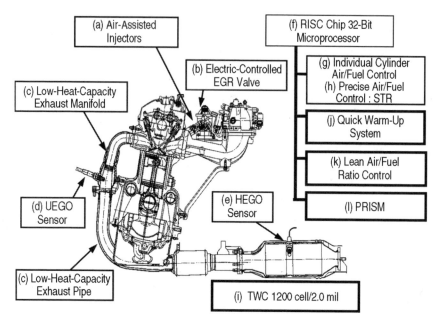

Figure 14.1 Overview of the Honda SULEV system [Kitagawa (2000)].

- Universal exhaust gas oxygen (UEGO) sensor in the exhaust manifold.
- HEGO sensor in the middle of the UFC.
- Underfloor high-cell-density (1200/2 mil; 1.7 L; 330 g/ft^3; Pt:Pd: Rh = 1:15:1) three-way catalyst (TWC).
- Lean A/F ratio control approach.
- A secondary O_2 feedback control system was developed to realize individual cylinder A/F ratio control and precise A/F ratio control. Conversion performance is optimized by the constant identification of changes to the three-way catalyst conditions that are affected by engine operating conditions, catalyst temperature, and deterioration. This system calculates the future value of the HEGO sensor and determines the optimum fuel feedback target level, resulting in higher fuel feedback system precision.

- A quick warm-up system (QWS) was developed to quickly warm the engine. Figure 14.2 illustrates the basic operating principle for this system. The horizontal axes represent time, and the vertical axes represent the rotary air control valve (RACV) opening, engine speed, and ignition timing. While the engine is at idle after startup, the QWS uses the RACV by means of feed-forward control to supply large quantities of intake air. As the engine warms up, the amount of intake air gradually is reduced. When the engine speed exceeds a preset level, ignition timing feedback control will kick in. The ignition timing is controlled to maintain the target engine speed. For example, when the intake air is increased, ignition timing is controlled to retard the timing to achieve a timing that can maintain the target engine speed.

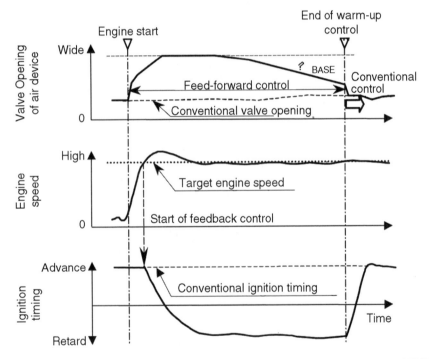

Figure 14.2 Operating principle of the Honda quick warm-up system (QWS) [Kitagawa (2000)].

- **HC adsorption system**: Figure 14.3 illustrates the system layout that can meet SULEV standards via an HC adsorber. An active-type HC adsorption system has been adopted, in which the functions of the adsorbent and the catalyst are separated. The HC adsorber is positioned downstream of the catalyst, and the desorbed HCs are returned to the intake manifold through the EGR system by operating the exhaust changeover valve. The adsorber consists of metallic honeycomb substrate, which is coated with zeolite as HC adsorbent and is surrounded by the main exhaust passage. Figure 14.4 gives a conceptual diagram of the exhaust changeover valve operation, exhaust gas flow, and operation of HC adsorption and desorption. The exhaust changeover valve is located in front of the HC adsorber (downstream of the catalyst), and the negative pressure in the intake manifold generated during engine operation activates the exhaust valve. When the exhaust valve opens, the exhaust gas flows into the HC adsorber. This kind of condition continues until the temperature of the HC adsorber reaches the temperature at which HC starts to desorb. Under this condition, the negative pressure acting on the exhaust changeover valve is released to the atmosphere just before the adsorbed HC begins to desorb. Then the exhaust valve shuts off, and the flow passage of the exhaust gas to the HC adsorber is closed. As a result, the exhaust gas flows only through the main exhaust passage. As part of the diagnostic requirement to detect the performance deterioration of the HC adsorber that could lead to emissions system failure, a new sensor capable of measuring the humidity in the exhaust gas has been developed based on the correlation between the HC adsorption performance and moisture adsorption ability.

14.6 Nissan Partial Zero Emissions Vehicle System

The Nissan PZEV is based on a lean starting approach to achieve the catalyst light-off performance in combination with an HC trap to adsorb HC emissions before the catalyst is lit-off [Nishizawa *et al.* (2000 and 2001); Oguma (2003)]. Figure 14.5 shows the system layout of the Nissan PZEV system. The basic elements of the Nissan PZEV system are summarized as follows:

- **Engine**: Four-valve, four-cylinder.

- Fine-spray injector with an average droplet size of 71 μm.

Near-Zero-Emission Gasoline-Powered Vehicle Systems

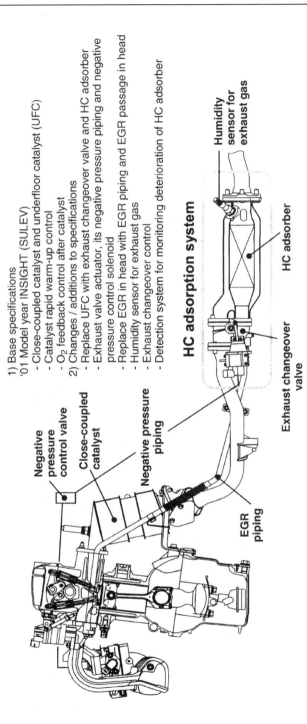

1) Base specifications
'01 Model year INSIGHT (SULEV)
- Close-coupled catalyst and underfloor catalyst (UFC)
- Catalyst rapid warm-up control
- O_2 feedback control after catalyst

2) Changes / additions to specifications
- Replace UFC with exhaust changeover valve and HC adsorber
- Exhaust valve actuator, its negative pressure piping and negative pressure control solenoid
- Replace EGR in head with EGR piping and EGR passage in head
- Humidity sensor for exhaust gas
- Exhaust changeover control
- Detection system for monitoring deterioration of HC adsorber

Figure 14.3 The Honda SULEV system layout based on an HC adsorber [Yamazaki et al. (2004)].

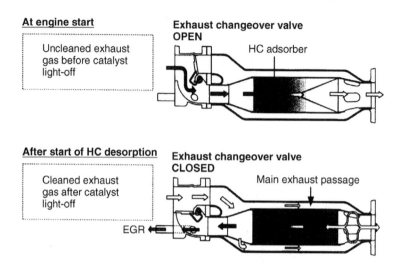

Figure 14.4 Operating principle of the Honda HC adsorber [Yamazaki et al. (2004)].

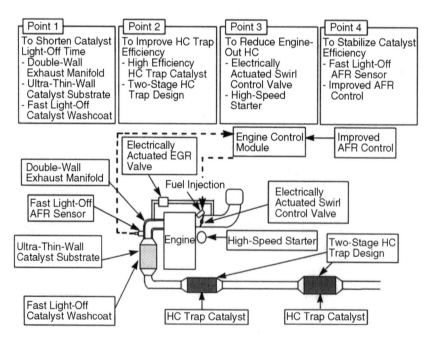

Figure 14.5 Overview of the Nissan PZEV system [Nishizawa et al. (2000)].

- For the second-generation PZEV system, the exhaust cam opening timing has been retarded, thereby holding combustion gases at a higher temperature for a longer duration to enhance in-cylinder HC oxidation to reduce engine-out emissions.

- An electrically actuated swirl control valve was used to increase the charge motion level in the intake port to improve mixture preparation and enhance combustion stability, particularly with a leaner mixture. Figure 14.6 shows the effect of intake port airflow velocity on the ignition retardation capability, catalyst temperature, and engine-out HC emissions. The results indicate that the air velocity must be raised above a certain level to realize the full benefit. Strong intake airflow provides sufficient kinetic energy to enhance fuel atomization and vaporization. In addition, throttling the intake also is effective in reducing the intake air volume, which is another reason to raise the exhaust gas temperature. The application of an electrically activated swirl control valve provides a variable capability that allows the airflow to be throttled more in a desired direction.

- Stainless steel dual-wall exhaust manifold for the first and second generations of PZEV (the inner pipe has a thinner wall thickness of 0.8 mm); cast iron exhaust manifold for the third generation of PZEV. To balance the power and catalyst light-off characteristics, the method of joining the branches of the exhaust manifold from each cylinder was improved for the third-generation PZEV. Figure 14.7 illustrates the exhaust manifold designs for the first-generation (Sentra CA) and third-generation PZEV systems. Instead of joining the Number 1 and Number 2 branches and the Number 3 and Number 4 branches as was done in the first-generation PZEV, the Number 1 and Number 4 branches and the Number 2 and Number 3 branches are now joined for the third-generation PZEV system. This arrangement makes it possible to minimize the extra length added to the exhaust manifold branches.

- A high-speed starter was used in the system. Figure 14.8 shows the effect of engine starting speed on cold-start peak HC emissions. Clearly, the peak HC emissions were reduced by 10% when the engine speed was increased from 200 to 250 rpm. Increasing the engine speed can evacuate the intake port faster, which is effective in creating a combustible mixture with a fixed amount of fuel injected. In addition, the higher kinetic energy via a higher engine speed also is beneficial in enhancing fuel atomization and thus vaporization.

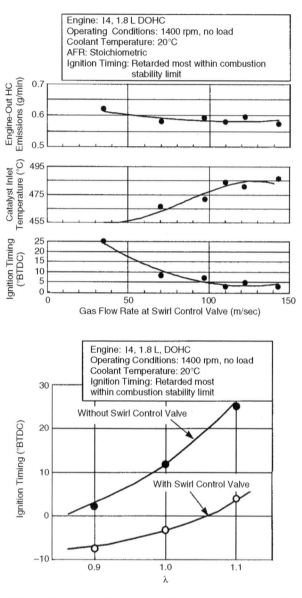

Figure 14.6 Effect of intake airflow on ignition retardation capability, catalyst temperature, and HC emissions [Nishizawa et al. (2000)].

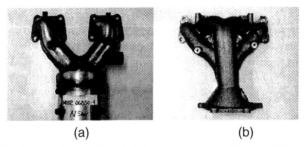

Figure 14.7 *Comparison of the (a) first-generation (Sentra CA) and (b) third-generation Nissan PZEV exhaust manifold designs [Nishizawa et al. (2001)].*

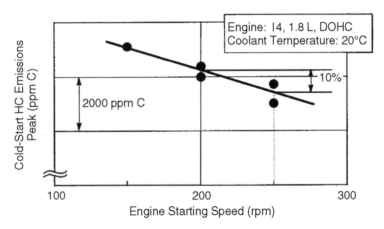

Figure 14.8 *Effect of engine starting speed on cold-start peak HC emissions [Nishizawa et al. (2000)].*

- Electrically actuated EGR valve for only the first and second generations of PZEV; continuously variable valve timing (VVT) for the third generation of PZEV to replace the EGR system.

- **Control system**: Fast light-off A/F ratio sensor; improved A/F ratio control; increased purging air for the canister. Figure 14.9 shows the effect of improved A/F ratio control on NOx and HC conversion efficiency. Clearly, the NOx conversion efficiency was improved significantly without any degradation of HC and CO conversion performance. When comparing the

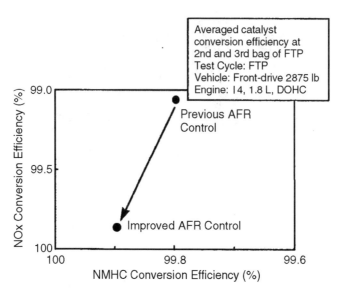

Figure 14.9 *Effect of improved A/F ratio control on NOx and HC emissions [Nishizawa et al. (2000)].*

efficiency of the catalyst between the previous A/F ratio control and the improved A/F ratio control, the improvement is more than four times that of the baseline, which means the tailpipe NOx emissions were reduced by more than 80% simply with this improved A/F ratio control. In addition, this improved A/F ratio control made it possible to increase the amount of purge air for the canister, which is required to realize the zero evaporative emissions without sacrificing exhaust emissions.

- **Catalyst system**: Fast light-off catalyst substrate (900-cell-density substrate); ultra-thin-wall (2-mm) catalyst substrate for the first and second generations of PZEV. Figure 14.10 shows the effect of ultra-thin-wall catalyst substrate on HC emissions reduction during cold start. The wall thickness of the third-generation PZEV is 1.8 mm; two-stage HC trap; three three-way catalysts (close-coupled catalyst, two two-stage HC traps) for the first-generation PZEV; two three-way catalysts (first HC trap integrated inside the close-coupled catalyst; second trap integrated inside the UFC) for the second and third generations of PZEV. The total catalyst capacity was significantly reduced from 3.9 L for the first-generation Nissan Sentra CA system to 1.8 L for the third-generation PZEV system. Figure 14.11 shows

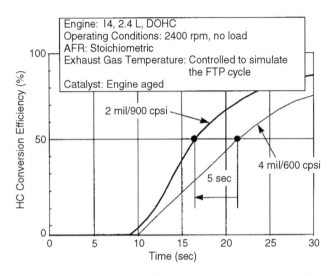

Figure 14.10 Effect of ultra-thin wall on catalyst light-off and HC conversion efficiency [Nishizawa et al. (2000)].

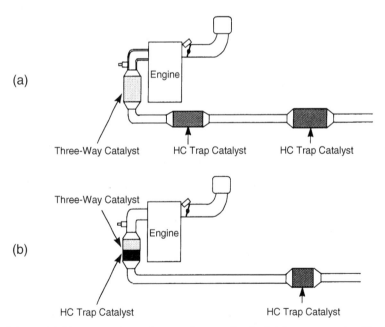

Figure 14.11 Comparison of system layouts for the (a) first-generation (Sentra CA) and (b) third-generation Nissan PZEV systems [Nishizawa et al. (2001)].

a comparison of the PZEV system layouts between the first-generation (Sentra CA) and third-generation systems. To improve the thermal resistance, the wall thickness of only the three outermost cells was increased to enhance the strength of the outer surface, and the geometry of the exhaust manifold diffuser angle was optimized. (See Figure 14.12 for the effect of exhaust manifold diffuser angle on the internal temperature distribution of the catalyst.)

- Catalyst-coated radiator.

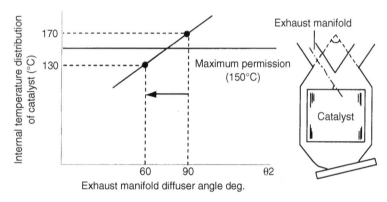

Figure 14.12 Effect of exhaust manifold diffuser angle on the internal distribution of catalyst temperature [Nishizawa et al. (2001)].

14.7 Toyota Partial Zero Emissions Vehicle System

The Toyota PZEV system was based on a lean starting approach via an underfloor HC-adsorbing three-way catalyst and has been described by Kidokoro et al. (2003). Figure 14.13 illustrates the system layout. The major features of the Toyota PZEV system are summarized briefly as follows:

- **Engine**: 2.4 L, four-cylinder engine.

- **Compression ratio**: 9.6.

- **Fuel injector**: Twelve-hole injector with a Sauter mean diameter (SMD) of 70 μm. For the injector hole contour, a new taper-punched hole replaced

	'03MY (PZEV)	'02MY (ULEV)
Displacement	2362 cc	→
Bore × Stroke	88.5 × 96.0 mm	→
Number of Cylinders	4	→
Number of Valves	Intake (IN) 2, Exhaust (EX) 2	→
Valve Timing (IN Open/Closed, EX Closed/Open)	VVT-i $\frac{3\mid3}{60\mid45} \leftrightarrow \frac{46\mid3}{17\mid45}$ (deg.)	VVT-i $\frac{-4\mid3}{60\mid45} \leftrightarrow \frac{46\mid3}{10\mid45}$
Compression Ratio	9.6	→
Fuel Injector	295 cc/min (Improved 12 holes)	265 cc/min (12 holes)
Intake Manifold	With IACV	Without IACV
Exhaust Manifold	Stainless Steel, Compact Double Wall	Stainless Steel, Single Wall
Close-Coupled Catalyst	1.1 L, Ceramic 2 mil-900 cpsi	1.1 L, Ceramic 3 mil-600 cpsi
Underfloor Catalyst (UFC)	1.3 L, Ceramic 3 mil-600 cpsi (HC Adsorbing 3-Way Catalyst)	0.9 L, Ceramic 4 mil-400 cpsi
Air-Fuel (A/F) Ratio Sensor	Planar Type (Fast Light-Off)	Cup Type
Sub-Oxygen (O_2) Sensor	Super Stability Type	Normal Type

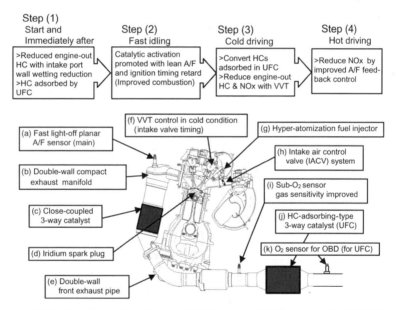

Figure 14.13 Overview of the Toyota PZEV system [Kidokoro et al. (2003)].

the conventional straight-punched hole. By making a thinner fuel film inside the hole, fuel atomization was promoted immediately below the hole.

- Iridium spark plug.

- Variable valve timing control in the cold condition (intake valve timing). At the most retarded VVT, the valve overlap timing is maximized to enhance the hot backflow of residual into the intake port, which is effective in promoting fuel atomization. Due to the difficulty in operating the VVT immediately after cold start because of the high engine oil viscosity, a new control method of setting the VVT neutral position at the overlap angle of approximately 6° was adopted. When the oil pressure is high enough to control, the VVT is positively operated.

- The intake air control valve (IACV) system was adopted to intensify the charge motion to improve combustion stability and to reduce wall wetting on the intake port and cylinder wall surface. An approximately 15% cutout relative to the entire intake port cross section was provided at the top of the intake port to create a strong intake tumble motion. Figure 14.14 shows the effect of IACV on combustion stability as a function of A/F ratio for two types of fuel injectors. It is evident that with the closed IACV system, the lean limit A/F ratio at the same torque fluctuation was greatly increased when compared with that of the opened IACV system. Further expansion

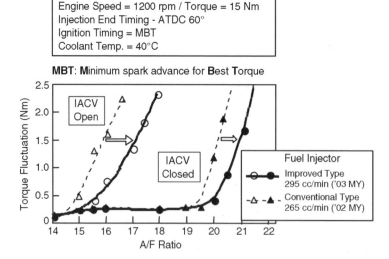

Figure 14.14 Effect of intake air control valve on lean limit for different types of fuel injectors [Kidokoro et al. (2003)].

of the lean A/F limit was obtained with the improved fuel injector that produces a finer droplet size.

- A high-efficiency close-coupled catalyst integrated with a stainless steel, double-wall compact exhaust manifold. Figure 14.15 shows a comparison of the exhaust manifold for applications to ultra-low emissions vehicle (ULEV) and PZEV systems. Major improvements from the ULEV include shorter branches, more compactness, and a double-wall structure to prevent heat loss from the exhaust gas. Because the new PZEV exhaust manifold is designed to minimize heat loss, the close-coupled catalyst is easily exposed to high-temperature exhaust gases during high-power operation. To retain the structural reliability of the substrate and efficient catalyst conversion performance, a space was provided in front of the catalytic substrate for uniform exhaust gas flow into the catalyst. As a result, the temperature distribution in the center and circumference of the catalyst front area was improved significantly due to the spacer in front of the substrate.

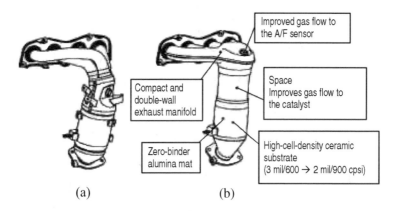

Figure 14.15 Comparison of exhaust manifold designs for applications to (a) ULEV (2002 MY) and (b) PZEV (2003 MY) [Kidokoro et al. (2003)].

- **Converter system**: Close-coupled catalyst (1.1 L, ceramic, 900 cpsi/2 mil); UFC (1.3 L, 600 cpsi/3 mil, adsorbing three-way catalyst).

- Located under the floor, the HC adsorber can adsorb HCs during cold start.

- Fast light-off planar A/F sensor (main).

- Sub-O_2 sensor (prior to the UFC) gas sensitivity improved.

- O_2 sensor for OBD downstream of the UFC.

- To reduce exhaust emissions variability under the cold condition, intake air amount feedback control was added for stable engine speed and early lean A/F feedback control to stabilize the exhaust gas amount during fast idle. In addition, to prevent a decrease in the catalyst conversion performance from variance in the A/F ratio among the cylinders after warm-up, the A/F sensor installation position in the exhaust manifold was optimized to make a drastic improvement in gas exposure from each cylinder for the sensor. Considering the mixed exhaust gas exposure, the sub-O_2 sensor position was moved downward from just downstream of the close-coupled catalyst to the front face of the UFC. As a result, the sub-O_2 sensor has an equal exposure to the exhaust gas from each cylinder.

- If hydrogen is present in the exhaust gas, the controlled A/F ratio by the O_2 sensor shifts leaner from the stoichiometry due to the fact that hydrogen diffuses faster, which leads to emissions variability, particularly for NOx. To minimize this impact, the Toyota PZEV system adopted an improved O_2 sensor that applies a layer of catalyst to the top of the sensor coating layer to convert the hydrogen in the exhaust stream. As a result, the lean shift was eliminated, and an accurately controlled stoichiometric A/F ratio was realized. Figure 14.16 shows the schematic of the improved O_2 sensor and its hydrogen sensitivity compared to the conventional sensor.

14.8 Toyota Ultra-Clean Hybrid Vehicle System

The Toyota ultra-clean hybrid vehicle system has been discussed in several publications [Inoue *et al.* (2000); Kuze *et al.* (2004); Muta *et al.* (2004)]. The basic elements of the Toyota hybrid ultra-clean emissions system are summarized as follows:

- **Converter system**: A high-performance starter catalyst with an ultra-thin wall and high cell density (900 cpsi/mil) that is close-coupled to the exhaust manifold. Farther downstream, a UFC was used that incorporated an active HC adsorber. Figure 14.17 shows the operating principle of the Toyota active HC adsorber for its hybrid application to meet the SULEV emissions standard. When the engine is started cold, the intake manifold vacuum is applied to the diaphragm to close the two-way valve.

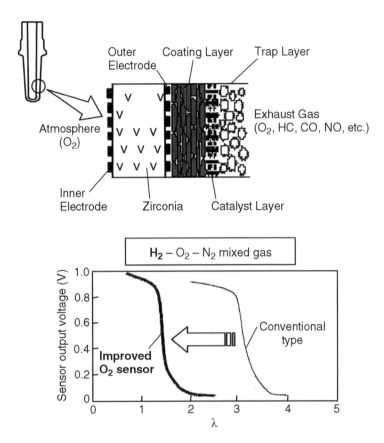

Figure 14.16 Schematic of the Toyota improved O_2 sensor and its hydrogen sensitivity, compared to a conventional sensor [Kidokoro et al. (2003)].

As a result, the exhaust gases are introduced to the HC adsorber that is installed around the periphery to physically adsorb the HCs (see the top part of Figure 14.17). After the starter catalyst has been fully warmed up and the proper HC conversion efficiency has been ensured, the vacuum that is applied to the actuator is released to open the two-way valve, thus introducing the exhaust gases directly into the UFC. Due to the heat of the UFC that is received by the HC adsorber, and the small amount of exhaust gases that flow into the HC adsorber, the temperature of the HC adsorber rises gradually, allowing the HCs to be desorbed. The UFC then cleans the desorbed HCs.

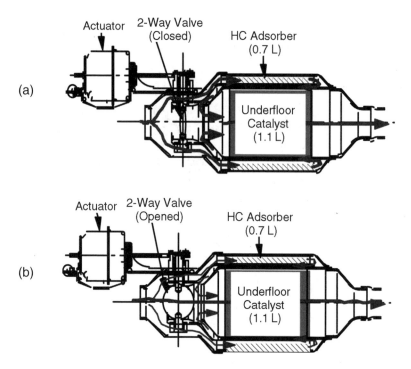

Figure 14.17 Operating principle of the active HC adsorber used in the Toyota hybrid SULEV system: (a) HC adsorbing process, and (b) HC desorbing process [Inoue et al. (2000)].

- A dual O_2 sensor system was used, with one before and the other after the starter catalyst.

- By taking advantage of motor assist, the engine runs mainly to warm the catalyst shortly after starting. By maintaining a constant volume of intake air by the engine, the A/F ratio fluctuation has been restrained to minimize the exhaust emissions.

- The fuel cutoff immediately before the engine is stopped is delayed for hybrid operation to its limit in order to restrain the lean condition in the catalyst, thus preventing the three-way catalyst oxygen storage from saturation. Furthermore, the A/F mixture is enriched for a short time following a restart in order to maintain the O_2 storage at its proper level, thus considerably reducing the NOx emissions (Figure 14.18).

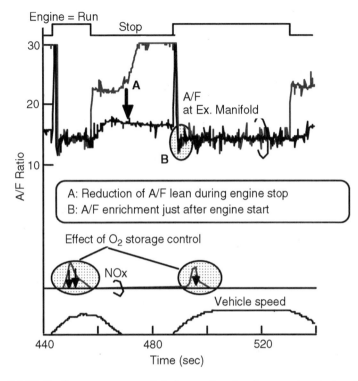

Figure 14.18 Fueling strategy to minimize engine shutdown and restart emissions [Inoue et al. (2000)].

- Figure 14.19 illustrates the vapor-reducing fuel tank used in the Toyota hybrid system. The system consists of a bladder membrane that expands and contracts in accordance with the volume of the fuel that remains in the tank. As a result, the amount of fuel vapor that is generated in the fuel tank while the vehicle is either running or stopped has been dramatically reduced. This system made it possible for the evaporative emissions to be purged in the hybrid system where the manifold vacuum is very low and has a minimal chance to realize the purge control.

- An approach called coolant heat storage system (CHSS) was developed for application to the Toyota hybrid ATPZEV (Advanced Technology PZEV) system without relying on an HC adsorber. This system recovers hot engine coolant in a heat-insulated reservoir and uses it for the next cold

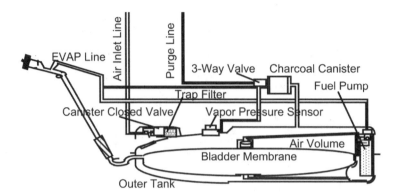

Figure 14.19 Overview of the vapor-reducing fuel tank for the Toyota hybrid ATPZEV system *[Inoue et al. (2000)]*.

start. This approach contributed greatly to the reduction of engine-out HC emissions during engine cold start and warm-up phases through rapid heating of the intake port wall with stored hot coolant that was recovered during hot driving conditions.

14.9 Summary

Great success has been achieved in developing ultra-clean gasoline-powered vehicles to meet the world's most stringent regulation of PZEVs. This is largely attributed to the system approach that has been realized to optimize all engine and vehicle design and operating parameters that can affect tailpipe emissions. Note that all applications today are limited to vehicles powered by small-displacement, four-cylinder engines. Applying these technologies to vehicles powered by large-displacement (e.g., vee) engines presents significant challenges. Continued efforts must be directed toward improving the existing PZEV technologies and exploring new enablers for simplifying the system and enhancing its capability for a broader range of applications.

14.10 References

1. Carney, D. (2003), "Global I4 Goes PZEV," *Automotive Engineering*, No. 7, pp. 74–78.

2. Inoue, T., Kusada, M., Kanai, H., Hino, S., and Hyodo, Y. (2000), "Improvement of a Highly Efficient Hybrid Vehicle and Integrating Super Low Emissions," SAE Paper No. 2000-01-2930, Society of Automotive Engineers, Warrendale, PA.

3. Kidokoro, T., Hoshi, K., Hiraku, K., Satoya, K., Watanabe, T., Fujiwara, T., and Suzuki, H. (2003), "Development of PZEV Exhaust Emission Control System," SAE Paper No. 2003-01-0817, Society of Automotive Engineers, Warrendale, PA.

4. Kitagawa, H. (2000), "L4-Engine Development for a Super-Ultra-Low Emissions Vehicle," SAE Paper No. 2000-01-0887, Society of Automotive Engineers, Warrendale, PA.

5. Kitagawa, H., Urata, Y., Yasui, Y., and Mibe, T. (2001), "Development of Super Ultra Low Emission System," *J. of Japan Automotive Technology* (in Japanese), Vol. 55, No. 9, pp. 69–74.

6. Kunde, O., Zanna, T.D., Brinkmann, B., and Zabkiewicz, G. (2003), "2.3L PZEV Ford Focus," *Advanced Engine Design and Performance*, Proceedings of the Global Powertrain Conference (GPC) '03, pp. 179–186.

7. Kuze, Y., Kobayashi, H., Ichinose, H., and Otsuka, T. (2004), "Development of New Generation Hybrid System (THS)—Development of Toyota Coolant Heat Storage System," SAE Paper No. 2004-01-0643, Society of Automotive Engineers, Warrendale, PA.

8. Landerl, C., Liebl, J., Hofmann, R., and Melcher, T. (2003), "The BMW SULEV (PZEV) Concept—Uncompromised Emissions Reduction," *Advanced Engine Design and Performance*, Proceedings of the Global Powertrain Conference (GPC) '03, pp. 172–178.

9. Muta, K., Yamazaki, M., and Tokieda, J. (2004), "Development of New-Generation Hybrid System THS II—Drastic Improvement of Power, Performance, and Fuel Economy," SAE Paper No. 2004-01-0064, Society of Automotive Engineers, Warrendale, PA.

10. Nishizawa, K., Momoshima, S., Koga, M., Tsuchida, H., and Yamamoto, S. (2000), "Development of New Technologies Targeting Zero Emissions for Gasoline Engines," SAE Paper No. 2000-01-0890, Society of Automotive Engineers, Warrendale, PA.

11. Nishizawa, K., Mori, K., Mitsuishi, S., and Yamamoto, S. (2001), "Development of Second Generation of Gasoline P-ZEV Technology," SAE Paper No. 2001-01-1310, Society of Automotive Engineers, Warrendale, PA.

12. Oguma, H., Koga, M., Nishizawa, K., Momoshima, S., and Yamamoto, S. (2003), "Development of Third Generation of Gasoline P-ZEV Technology," SAE Paper No. 2003-01-0816, Society of Automotive Engineers, Warrendale, PA.

13. Wade, W.R. (2003), "Near Zero Emission Internal Combustion Engines," 2003 Soichiro Honda Lecture, Proceedings of ICEF03, 2003 Fall Technical Conference of the Amercian Society of Mechanical Engineers (ASME) Internal Combustion Engine Division, September 7–10, 2003, Erie, PA.

14. Yamazaki, H., Ueno, M., Endo, T., and Sugaya, S. (2004), "Research on HC Adsorption Emission System," SAE Paper No. 2004-01-1273, Society of Automotive Engineers, Warrendale, PA.

Acronyms

AAI	Air-assisted injector (or injection)
AAM	Alliance of Automobile Manufacturers
AAPFI	Air-assisted port fuel injection
A/F	Air/fuel (ratio)
AFI	Air-forced injector (or injection)
AFPFI	Air-forced port fuel injection
AIGER	American Industry/Government Emissions Research
ALO_4	Aluminum tetraoxide
AQIRP	Auto/Oil Air Quality Improvement Research Program
ATDC	After or above top dead center (of piston travel)
ATPZEV	Advanced technology partial zero emissions vehicle (Toyota)
BDC	Bottom dead center (of piston travel)
BMD	Bag mini-diluter
BMEP	Brake mean effective pressure
BTDC	Before or below top dead center (of piston travel)
CAD	Crank angle degree
CARB	California Air Resources Board
CAT	Catalyst
CC	Close-coupled (catalyst)
CCD	Charge-coupled device
CCM	Comprehensive component monitor
CE	Conversion efficiency (of hydrocarbon)
CEC	Coordinating European Council
CFI	Central fuel injection
CFO	Critical flow orifice
CFR	U.S. Code of Federal Regulations; *also* Corporate Fuels Research
CFV	Critical flow venturi
CHSS	Coolant heat storage system

CICT	Color imaging capturing technique
CLA	Chemiluminescence analyzer (or detector)
CMCV	Charge motion control valve
COP	Crossover point (NOx)
COV	Coefficient of variation
CP2-RFG	California Phase II reformulated gasoline
CRC	Coordinating Research Council
CVI	Closed-valve injection
CVS	Constant volume sampler
CV-SCV	Continuously variable swirl control valve
D	Diffusion coefficient (cm^2/sec)
DAR	Dilution air refinement (system)
DF	Dilution factor
DI	Driveability index; $DI = 1.5 \times T_{10} + 3 \times T_{50} + T_{90}$; temperatures in degrees Fahrenheit (°F); *also* Direct injection
DISI	Direct injection spark ignition
DVE	Direct vehicle exhaust
ECM	Engine control module
ECU	Engine control unit
EDI	Evaporation driveability index
EGI	Exhaust gas ignition
EGR	Exhaust gas recirculation
EHC	Electrically heated catalyst
EOBD	European onboard diagnostics (standards)
EPA	U.S. Environmental Protection Agency
ETBE	Ethyl-tertiary butyl ether
EVC	Exhaust valve closing (timing)
EVO	Exhaust valve opening (timing)
E_{XX}	Percent evaporated at temperature XX
FFID	Fast-response flame ionization detector; *also* FRFID
FIA	Flame ionization analyzer
FID	Flame ionization detection (or detector)
FKM	Fluorocarbon elastomer
FTP	U.S. Federal Test Procedure; *also* FTP-75

GC	Gas chromatography
GDI	Gasoline direct injection (engine)
GIMEP	Gross indicated mean effective pressure
GRNN	Generalized regression neutral network
GSA	Geometric surface area
H	Henry's constant (kPa)
H/C	Hydrogen-to-oxygen ratio (of fuel)
HDPE	High-density polyethylene
HEGO	Heated exhaust gas oxygen (sensor)
HEV	Hybrid electric vehicle
HPIV	Holographic particle image velocimetry
I4	Inline four-cylinder (engine)
IACV	Intake air control valve
ICV	Inlet check valve
ILIDS	Interferometric laser imaging for drop sizing
I_m	Mixing index
I/M	Inspection/maintenance (station)
IMEP	Indicated mean effective pressure (kPa)
IR	Infrared
ISHC	Indicated specific hydrocarbon (emissions)
IVC	Intake valve closing (or closed)
IVO	Intake valve opening (or open)
JOBD	Japanese onboard diagnostics (standards)
L/D	Length-to-diameter (ratio)
LDPE	Low-density polyethylene
LDV	Laser Doppler velocity (or velocimetry)
LEA	Laser extinction and absorption
LEV	Low emissions vehicle; *also* LEV II
LIEF	Laser-induced exciplex fluorescence
LIF	Laser-induced fluorescence
LPG	Liquefied petroleum gas

MAF	Mass airflow (sensor)
MAP	(Intake) manifold absolute pressure (kPa)
MBT	Minimum spark advance for best torque
MFB	Mass fraction burned
MFC	Mass flow controller
MIL	Malfunction indicator light
MMT	Methylcyclopentadienyl manganese tricarbonyl
MON	Motor Octane Number
MPI	Multipoint injection
MTBE	Methyl-tertiary butyl ether
NDIR	Nondispersive infrared
NGV	Natural gas vehicle
NIST	National Institute of Standards and Technology
NMEP	Net mean effective pressure
NMHC	Nonmethane hydrocarbon
NMOG	Nonmethane organic gas
NOx	Oxides of nitrogen, or nitrogen oxides
NTC	Negative temperature coefficient
OBD	Onboard diagnostics; *also* OBD II
OBDS	Onboard distillation system
OFP	Ozone-forming potential
ORVR	Onboard refueling vapor recovery
OSC	Oxygen storage material component; *also* Oxygen catalyst storage content; *also* Oxygen storage capacity
OVI	Open-valve injection
OXSW	Oxidation switch
PAS	Proportional ambient sampling
PCM	Programmable control module
PCV	Positive crankcase ventilation
PDA	Phase Doppler anemometry
p_F	Partial pressure (of fuel) (kPa)
PFI	Port fuel injection (or injected)
PGM	Platinum group metals

PIV	Particle image velocity (or velocimetry)
PLIEF	Planar laser-induced exciplex fluorescence
PLIF	Planar laser-induced fluorescence
POM	Acetal copolymer
POx	Partial oxidation reforming (or reactor)
ppm C_1	HC mole fraction in parts per million (ppm); based on HC as C_1H_x
Pt	Platinum
PWM	Pulse width modulation
PZEV	Partial zero emissions vehicle
QWS	Quick warm-up system
R&D	Research and development
RACV	Rotary air control valve
Reg-Neg	(Reformulated gasoline) Regulation–Negotiation Agreement
RFG	Reformulated gasoline
RMS	Root mean square
RMT	Remote mix tee
RON	Research Octane Number
ROV	Rollover valve
rpm	Revolutions per minute
RVP	Reid vapor pressure
SAIR	Secondary air system (monitor)
SAO	Smooth approach orifice
S/C	Steam-to-fuel ratio (in reformer)
SCV	Swirl control valve
SDIMEP	Standard deviation of indicated mean effective pressure
SHED	Sealed housing for evaporative determination
SI	Spark ignition (engine)
SMD	Sauter mean diameter
S_P	Mean piston speed (m/sec)
SULEV	Super ultra-low emissions vehicle; *also* SULEV II
SUV	Sport utility vehicle

T_{10}, T_{50}, T_{90}	Temperatures at 10, 50, and 90% distillation points, respectively
TAME	Tertiary amyl butyl ether
TBA	Tertiary-butanol; *also* t-butanol
TDC	Top dead center (of piston travel)
TE	Trapping efficiency (of hydrocarbon)
THC	Total hydrocarbon
TLEV	Transitional low emissions vehicle
TWC	Three-way catalyst
TWD	Total weighted demerits
T_{XX}	Termperature for XX% evaporation of the fuel
U'	Turbulence intensity (m/sec)
UEGO	Universal exhaust gas oxygen (sensor)
UFC	Underfloor catalyst
ULEV	Ultra-low emissions vehicle; *also* ULEV II
VOC	Volatile organic compound
VTEC	Variable valve timing and lift electronic control
VVA	Variable valve actuation
VVT	Variable valve timing/control monitor
VVT/L	Variable valve timing and lift (system)
WOT	Wide open throttle
WRAF	Wide-range air/fuel (sensor)
WRO_2	Wide-range oxygen (sensor)
WWMP	World Wide Mapping Point (Ford)
X_F	Fuel mole fraction
Y	Hydrogen-to-carbon ratio (of fuel)
ZEV	Zero emissions vehicle

δ	Boundary layer thickness (millimeters)
λ	Fuel/air equivalence ratio
τ	Characteristic diffusion time (milliseconds)
Φ	Fuel/air equivalence ratio

Index

Numbers followed by *f* or *t* denote figures or tables, respectively.

A/F. *See* Air/fuel
Acronyms list, 435–440
Air injection. *See* Secondary air injection
Air pump, 195, 245
Air-assist injector (AAI), 44–46, 47, 61, 68, 132
Air-forced injector (AFI), 44–46, 68, 132
Air/fuel (A/F) ratio, 52, 85, 175*f*
   calibration in PFI engine, 9–10
   controlling, 193–195
   during steady-state conditions, 100
   lean, to improve hydrocarbon oxidation, 159–160, 163
   mixture distribution diagnostics, 123*t*, 134, 135
   monitoring, 196–197
   onboard fuel reformers, 224, 225
   at PFI engine startup, 9
   and swirl, 58*f*
Air/fuel sweep conversions, 253*f*, 254*f*
Alcohol-fueled vehicles, 228–229
Alcohols, 208
Alkylperoxy radicals (RO_2), 105
Alliance of Automobile Manufacturers (AAM), 219
Alumina, 246, 249
Aromatics, 214*t*
ASTM distillation curve, 208, 210, 212, 232*f*
ATPZEV (Advanced Technology PZEV), Toyota, 431, 432*f*
Auto/Oil Air Quality Improvement Research Program (AQIRP), 218–219
Autothermal reforming, 225

Bag mini-diluter (BMD), 384–398
Bag sampling, 376–383
Benzene, 209*t*, 214*t*
BMW, 410–411

Calibration, catalyst light-off, 157–160
California, 208, 214, 241, 333, 336, 347, 354, 370, 371
California Air Resources Board (CARB), 333, 334, 356, 385
California Phase II reformulated gasoline (CP2-RFG), 12, 38, 40, 208, 209*t*, 214*t*, 217, 229, 336
Cam phasing, 61, 68
Camshaft position, 6
Canister (fuel system), 342–344
Canister bleed emissions, 344
Canister bypass system, 246
Carbon dioxide (CO_2), 105, 219, 380*f*, 381*f*
Carbon formation, 221–222
Carbon monoxide (CO), 105, 219, 348–349, 407
Catalyst, 221
   close-coupled (CC), 243, 246, 256, 257–259

443

Catalyst *(continued)*
    conversion efficiency, 1
    hydrocarbon trap, 269–281
        desorption rate, 273–274
        two-stage trap system, 279–281
        underfloor, 255
        zeolite, 270–272
    oxygen storage content, 352–354
    underbody, 259, 260f
    see also Three-way catalyst
Catalyst deactivation, 298–299
Catalyst light-off, 61, 228
    calibration
        air/fuel ratio, 159–160
        idle engine speed, 158
        spark retard, 159–160
Catalyst system
    PZEV standards, 255–264
    rapid warm-up designs, 242–246
    three-way catalyst, 241–242, 246–255
Catalyst system monitor, 351–354, 366t
Catalyst temperature, 5, 5f
    rapid warm-up designs, 242–246, 262f
Catalyst washcoat, 243, 246, 247, 249, 253, 255, 258, 259f
Catalytic converter, 173–174
    three-way system modeling, 283–317
Catalytic oxidation, 178–181
Central fuel injection (CFI) engine
    engine startup, 41–42
        comparison with PFI engine, 95, 96f, 97f
Ceramic substrate, 244
Ceria-zirconia, 246, 249
CFI. *See* Central fuel injection
Charge inhomogeneity, 43
Charge motion control valve (CMCV), 167, 168f

Charge motion control valves, 56–57
Chemical reactions, three-way catalytic converter system, 289–298, 300, 301
Chrysler, 167
Cinematography, for fuel spray analysis, 124
Clean Air Act Amendments (1990), 214
Close-coupled (CC) catalysts, 243, 246, 256, 257–259
Closed-valve injection (CVI), 14, 35–36, 53
    targeting, 52
    timing, 14, 45, 47–51
CMCV (charge motion control valve), 167, 168f
Coking, 222
Cold engine processes, diagnostics
    combustion, 142–145
    fuel delivery into cylinder, 130–134
    fuel spray characteristics, 121–127
    mixture distribution, 134–142
    wall wetting, 123t, 127–130
Cold start, 31, 83, 178f
    alcohol-fueled vehicles, 228–229
    cylinder pressure, 82–83
    PFI engine behavior, 80–86
    and POx fuel reformer, 226–227
    unaccounted fuel during, 84–86
    see also Engine startup
Cold-idle revolutions, 158
Cold-start emissions reduction control strategy monitor, 354–355
Cold-weather driveability, gasoline, 211–213
Combustion
    diagnostic techniques, 142–145
    port fuel injected engine, during engine startup, 7f, 20–22

Index

Combustion chamber
   crevices, and hydrocarbon emissions, 88–90
   liquid fuel impingement, 36–40
Comprehensive component monitoring (CCM), 354
Compression rings, 88, 89
Connectors, fuel system, 338
Conservation of mass equation, 306
Conservation of momentum equation, 306
Constant volume sampler (CVS), 369, 373f
   bag sampling, 376–383
   compared with bag mini-diluter, 385–387, 389f, 390, 394
   dilution air, 372–375
   dilution ratio optimization, 376, 377f, 378f
   exhaust dilution, 375–376
Coolant temperature
   after PFI engine shutdown, 5f
   during PFI engine startup, 5, 10, 11f, 17, 18f, 34f
   steady-state operation, 48, 49f, 91
Coordinating European Council (CEC), 215
Coordinating Research Council (CRC), 12, 212, 213
CP2-RFG (California Phase II reformulated gasoline), 38, 40, 336
Crank angle degrees (CAD)
   during PFI engine startup, 15–17, 82f, 104f
   IVO timing during PFI engine startup, 15–17
   warm engine, 102f
Crankcase, 39
Cranking speed, 55, 57, 81–82
CRC (Coordinating Research Council), 212, 213
CVI. *See* Closed-valve injection
Cylinder, 35, 82–83, 103
Cylinder blowby levels, 88–89, 91
Cylinder head gasket, 88

DaimlerChrysler, 61
Delphi, 63
Desorption rate (hydrocarbon trap), 273–274, 276–277
DI. *See* Driveability index
Diagnostic techniques, cold engine processes
   combustion, 142–145
   fuel delivery into cylinder, 130–134
   fuel spray characteristics, 121–127
   mixture distribution, 134–142
   wall wetting, 123t, 127–130
Diffusion coefficients, hydrocarbon fuels, 98
Diffusion time, hydrocarbon fuels, 98
Dilution air, 372–375
Dilution ratio optimization, 376, 377f, 378f
2,2 Dimethyl heptane, 209t
2,3 Dimethyl pentane, 209t
Diode laser, 136, 138f
Direct injection spark ignition (DISI) engine startup, 22–26
Distillation curve, gasoline, 208–211
Distillation temperatures
   T_{10}, 11–12, 210
   T_{50}, 12, 46, 56, 210, 211, 212, 214t
   T_{90}, 12, 46, 210, 214t
Driveability index (DI)
   and A/F ratio, 216
   calculations for, 12, 212–213
   and hydrocarbon emissions, 217, 218f
   importance to engine calibrator, 33
Droplet evaporation, 42–43
Droplet flow, 13–14

445

Droplet size, 12, 22, 55, 206
  and swirl, 57–58
  closed valve injection vs. open
    valve injection, 35–36, 50
  diagnostic techniques, 123t, 124–127, 132
  fuel injector types, 44–47, 62, 68
Dual cone fuel injector, 47
Dual-wall exhaust manifold, 244

E85 fuel, 231
EDI (evaporation driveability index), 213
Elastic scattering, 134
Electrically heated catalyst (EHC), 245, 255
Emissions research obstacles, 113
Engine acceleration, cold start, 81–82
Engine misfire monitor, 355–356
Engine process measurement
  combustion, 142–145
  fuel delivery into cylinder, 130–134
  fuel spray characteristics, 121–127
  mixture distribution, 134–142
  wall wetting, 123t, 127–130
Engine shutdown, port fuel injected engine
  four-cylinder engine at idle, 3f
  general engine behavior, 2–3
  impact on hydrocarbon emissions, 3f, 4
  role in emissions, 1–2
Engine speed, during a cold start, 81–83
Engine startup
  central fuel injection (CFI) engine, 41–42
  direct injection spark ignition engine, 22–26
  port fuel injected engine
    combustion, 20, 21f
  fuel mixture preparation, 10–20
  general behavior, 6–10
  impact on hydrocarbon emissions, 22
  initial conditions, 5–6
  role in emissions, 1–2
  see also Cold start
Equations
  tailpipe emissions dilution factor, 370–371, 375–376
  catalytic converter system modeling
    chemical reaction rates, 328–331
    conservation of mass, 306
    conservation of momentum, 306
    gas phase energy, 285
    gas phase species, 285
    oxygen storage capacity, 286
    surface energy, 286
    surface species, 286
ETBE (ethyl-tertiary butyl ether), 208
Ethanol (EtOH), 208, 228–229
Ethers, 208, 211
Ethyl-tertiary butyl ether (ETBE), 208
Evaporation driveability index (EDI), 213
Evaporative emissions, 369
  reducing, 336–345
  standards, 333–334
  testing, 334–336
  types, 334
Evaporative system monitor, 356–360, 366t
Exhaust dilution, 375–376
Exhaust emissions
  analyzer accuracy, 398–401
  bag mini-diluter, 384–398
  dilution factor equation, 370–371, 375–376
  tailpipe emissions limits, 371

Exhaust gas
    composition, 174, 175f
    mixing, 3
    temperature, with retarded ignition timing, 158f, 159, 160–162, 189–191
Exhaust gas ignition (EGI) system, 246
Exhaust gas recirculation (EGR) monitor, 360, 366t
Exhaust manifold, 173–174
    design, 112, 189, 190f
    dual-wall, 244
Exhaust valve closing (EVC), 15–16
Expansion ratio, 162

Fast-response flame ionization detector (FFID), 4, 48, 51, 102, 103, 104f, 123t
    combustion diagnostics, 143, 145
    fuel delivery diagnostics, 134–136
Federal Test Procedure (FTP), 11f, 159, 191, 215, 217, 218f, 348, 361, 365, 372
    cold start, 31, 79, 80–83, 85
    emissions, 256f, 262f, 264f
    onboard distillation system, 233, 334–335,
    stages, 67–68, 81
    three-way catalysts, 247, 248f, 252f, 254f, 255f
FFID. *See* Fast-response flame ionization detector
Flame ionization detector (FID), 123t, 127, 135–136, 335, 395, 396f
Flame kernel growth rate, 59
Flame propagation speed, 59–60, 99
Flame quench, 32, 35, 96–98
Flame wrinkling, 59–60
Flow velocities in-cylinder, 56–60
Ford, 411–413

Fraunhofer diffraction, 125
FTP. *See* Federal Test Procedure
Fuel, liquid, 32–43
Fuel absorption, into oil layers, 90–91
Fuel accounting, 84–86
Fuel cells, 221
Fuel distillation systems, 206, 229–233, 234
Fuel equivalence ratio, 16f, 18f, 21f
Fuel evaporation, 10, 11–16, 42–43
Fuel ignition, 2, 3f, 7f
Fuel injection
    amount during cold start and warm-up, 33
    at engine shutdown, 2–3
    determining fuel spray characteristics, 122–126
    droplet diameters, 12, 46, 47, 55
    during engine startup
        closed-valve injection, 13–14
        intake valve timing, 15–20
        minimum for stable combustion, 19f, 84f
        open-valve injection, 13–14
    at PFI engine startup, 6, 7f, 8, 12–14
    see also Closed-valve injection; Open-valve injection
Fuel injection targeting, 51–55, 66, 94
Fuel injection timing, 47–51, 92–93
Fuel injector, 206
Fuel injector heating, 55, 220
Fuel injectors
    air-assist injector, 44–46, 47, 61, 68, 132
    air-forced injector, 44–46, 68, 132
    dual cone injector, 47
    effect of timing on hydrocarbon emissions, 45
    effect of type on hydrocarbon emissions, 44–47
    multi-hole injector, 47, 50, 55, 68

Fuel injectors *(continued)*
  pencil stream injectors, 46
  pintle injector, 44–46, 55
  twelve-hole injector, 47
Fuel location, 94
Fuel mixture preparation, 10–20, 31–69, 124f
  defined, 32
  effects on hydrocarbon emissions, 32–43
  flow field effects, 14–20, 56–60
  "fuel compensation," 33–35
  and fuel injection, 12–14, 44–55
  fuel volatility, 11–12
  goals of, 32
  influenced by temperatures, 10–11
  strategies to improve, 60–64
Fuel noncombustion, causes
  fuel absorption, 86–87, 90–92
  fuel stored in crevices, 86–90
  partial burns, 98–100
  quenching, 96–98
  too-rich liquid fuel, 92–96
Fuel properties, 11–12
Fuel rail heating, 55
Fuel reformer systems, 206–207
  and alcohol-fueled vehicles, 228–229
  autothermal reforming, 225
  partial oxidation reforming (POx), 63, 223–224, 226–228, 229, 233
  steam reforming, 222–223
Fuel system, 336–342
Fuel system connectors, 338
Fuel system monitor, 360–362
Fuel system seals, 339
Fuel tank, 337, 338f, 339, 340–342
  plastic, 341
  steel, 340–341
Fuel transport, 122–126
Fuel vapor, 10, 33
  see also Evaporative emissions

Fuel vapor/air ratio, 33, 53–55
Fuel volatility, 11–12, 33, 36–37, 48–50, 50, 85, 216–218

Gas chromatography, 103
Gas phase energy equation, 285
Gas phase species equation, 285
Gasoline
  chemistry, 207–208
  cold-weather driveability, 211–213
  distillation curve, 208–211
  onboard fuel distillation, 229–231, 232
  reformulated gasoline (RFG), 214–215
  volatility, 208–211
  see also Fuel
Global consumption rate, 106–109

Heated exhaust gas oxygen (HEGO) sensor, 123t, 135, 414
Heavy aromatics, 40–41
Heavy ends, gasoline, 42, 43
HEGO (heated exhaust gas oxygen) sensor, 123t, 135, 414
Henry's law for a dilute system in equilibrium, 90
HEV (hybrid electric vehicle), 337
High octane fuel, 229–231
High-boiling-point tracer, 43
High-cell-density substrate, 243
High-driveability-index (high-DI) gasoline, 33, 35
Holography, for fuel spray analysis, 124, 125, 127
Honda, 61, 167, 413–416, 417f, 418f
Hybrid electric vehicle (HEV), 337
Hydrocarbon adsorber. *See* Hydrocarbon trap
Hydrocarbon consumption, 106

Index

Hydrocarbon conversion, 106
Hydrocarbon emission factors
    air/fuel ratio, 100, 163, 216
    ambient temperature, 215
    cold start and warm-up, 31, 35, 83, 64–69, 132
        need for over-fueling, 31, 32–33, 52–53, 83, 100, 205
    coolant temperature, 37
    driveability index (DI), 217, 218f
    exhaust manifold design, 189, 190f
    fuel, 35, 83, 140, 206, 215–219
    fuel absorption, 86–87, 90–92
    fuel injection targeting, 51–55, 132
    fuel injection timing, 37, 47–52
    fuel injector types, 44–47
    fuel stored in crevices, 86–90, 134, 143
    liquid fuel, 31–43, 92–95
    partial burns, 98–100
    piston ring-pack crevices, 89–90, 101
    quench layers, 96–98
    starting strategies, 40–41
    too-rich liquid fuel, 92–96
    vehicle design, 215
    wall wetting, 127
Hydrocarbon emissions, 407
    and closed-valve fuel injection, 14, 48
    at DISI engine startup, 22–25
    engine comparisons at startup
        CFI compared to PFI, 95, 96f, 97f
        DISI compared to MPI, 23–25
        DISI compared to PFI, 22, 23f
    and OBDS, 231–233
    and open-valve fuel injection, 13, 36
    and partial oxidation reformer (POx), 226–228
    at PFI engine cold start, 31, 79, 178f
    at PFI engine warm-up, 31
    and secondary air injection, 174, 175–176, 177, 178f, 181, 182f
    at shutdown, 3f, 4
    at startup, 7f, 8, 13, 14, 20, 22, 31
    and variable valve time (VVT) system, 14
Hydrocarbon oxidation
    chemical formula, at high temperatures, 105
    and cold-idle engine speed, 158
    consumption rates
        global consumption rate, 106–109
        post-flame, 109–112
    at low temperatures, 105
    and retarded ignition timing, 159–160
    and secondary air injection, 111–112
Hydrocarbon storage
    absorption into oil layers, 90–92
    in combustion chamber crevices, 88–90
Hydrocarbon transport
    cold engine, 103–104
    warm engine, 101–103
Hydrocarbon trap, 245, 269–281
    desorption rate, 273–274, 276–277
    two-stage trap system, 279–281
    underfloor, 255
    zeolite, 270–272
Hydrogen, 220, 225, 228
Hydrogen conversion, 187–188
Hydrogen peroxide (H_2O_2), 105
Hydroperoxyl radicals (HO_2), 105
Hydroxyl radicals (OH), 110

IACV (intake air control valve), 164–167
Idle engine speed, 158
Ignition timing, retarded. *See* Spark retard
Imaging, fuel droplets, 122–127
IMEP. *See* Indicated mean effective pressure
In-cylinder λ, during cold start, 53–55, 57, 67
In-cylinder equivalence ratio, φ, during cold start, 54–55, 57, 67
Indicated mean effective pressure (IMEP), 15, 20, 41, 60, 61, 99, 140
 and spark retard, 160–162, 167, 168f
Indolene, 39, 48–49, 51
Infrared imaging, 130, 136, 137f
Innova infrared detector, 335
Intake air control valve (IACV), 164–167
Intake manifold pressure, at engine shutdown, 3, 4, 8–9
Intake port wetting, 39–40, 94–95
Intake valve, 206
Intake valve closing (IVC), timing, 17
Intake valve opening (IVO), timing, 14–20
Intake valve temperature, during engine startup, 10–11, 34–35
In-use monitor performance ratio tracking, 348
Iso-butyl benzene, 209t
Iso-hexane, 209t
Iso-octane, 40, 48–50, 223, 224
 hydrocarbon oxidation, 107–108, 111
Iso-pentane, 48, 49f, 90, 91, 209t, 232f
Iso-propanol (IPA), 208
Iso-propyl benzene, 209t
IVC (intake valve closing), timing, 17

IVO (intake valve opening), timing, 14–20

Japan, 372, 399

Lanthanum, 250
Laser diagnostics, for fuel spray analysis, 123t, 124–127
Laser Doppler velocimetry (LDV), 123t, 126, 142
Laser extinction and absorption (LEA), 123t, 132
Laser-induced exciplex fluorescence (LIEF), 134, 140
Laser-induced fluorescence (LIF), 43, 46, 57, 110, 123t
 combustion diagnostics, 142–143
 fuel delivery diagnostics, 132, 134
 mixture distribution diagnostics, 138, 139–140
 wall wetting diagnostics, 128–130, 131f
LDV (laser Doppler velocimetry), 123t, 126, 142
LEA (laser extinction and absorption), 123t, 132
Leak detection, 358–360
Leidenfrost effect, 38
LEV (low emissions vehicle), 163, 164f, 219, 228, 349
LEV II standards, 333, 339, 348, 349, 352, 361, 365, 370
LIEF (laser-induced exciplex fluorescence), 134, 140
LIF. *See* Laser-induced fluorescence
Light alkanes, 40–41
Lincoln Navigator, 63–64, 65f
Liquefied petroleum gas (LPG), 36, 38–39
Low-boiling-point tracer, 43

Index

LPG (liquefied petroleum gas), 36, 38–39

MAF (mass airflow sensor), 196
Malfunction indicator light (MIL), 347, 349, 356
Manifold absolute pressure (MAP), 9, 54–55, 68, 81–83
Mass airflow sensor (MAF), 196
Mass flow controller, 378, 383, 397
Mathematical modeling, three-way catalytic converter system, 283–317
Mathematical nomenclature, 317–320
Methane, 97, 98, 110, 222
Methanol (MeOH), 11, 97, 208, 211, 228–229
Methylbenzene, 93, 209t
Methylcyclopentadienyl manganese tricarbonyl (MMT), 219
Methyl-tertiary butyl ether (MTBE), 208, 209t, 211, 215
Mid-boiling-point tracer, 43
Mie scattering, 123t, 134
Misfire, 40, 99
Mixture preparation. *See* Fuel mixture preparation
MMT (Methylcyclopentadienyl manganese tricarbonyl), 219
Modeling, three-way catalytic converter system
  catalyst behavior, 309–312, 313f, 314f, 315f, 316f
  chemical reactions, 289–298, 300, 301
  equations
    chemical reaction rates, 328–331
    conservation of mass, 306
    conservation of momentum, 306
    gas phase energy, 285
    gas phase species, 285
    oxygen storage capacity, 286
    surface energy, 286
    surface species, 286
  flow distribution, 304–309
  heat transfer, 302–304
  inlet flow distribution, 304–309
  mass transfer, 302–304
  multi-dimensional modeling, 286–289
  one-dimensional modeling, 284–286
  oxygen storage mechanism, 299–302
Molecular diffusion coefficients, hydrocarbon fuels, 98
MPI. *See* Multipoint injection
MTBE (methyl-tertiary butyl ether), 208, 209t, 211, 215
Multi-hole fuel injector, 47, 50, 55, 68
Multipoint injection (MPI) engine startup, compared to DISI engine, 23–25

National Institute of Standards and Technology (NIST), 399
Natural gas, 222
N-butane, 209t
N-decane, 209t
N-dodecane, 209t
Negative temperature coefficient (NTC), 105
New York Department of Environmental Conservation, 375
N-heptane, 108–109
Nissan, 62, 269–270, 416, 418f, 419–424
Nitrogen, 224

451

Nitrogen oxides (NOx), 175–176, 219, 291, 348–349, 407
NMHC (nonmethane hydrocarbon), 14, 62, 397f
NMOG (nonmethane organic gas) emissions, 40, 64, 233, 348–349
Nonmethane hydrocarbon (NMHC), 14, 62, 397f
Nonmethane organic gas (NMOG) emissions, 40, 64, 233, 348–349
N-pentane, 48, 232f
NTC (negative temperature coefficient), 105
N-undecane, 209t

OBD II (onboard diagnostics II) regulations, 196, 347
OBD II system, 196, 347, 348t
   catalyst system monitor, 351–354, 366t
   cold-start emissions reduction control strategy monitor, 354–355
   comprehensive component monitoring (CCM), 354
   diagnostic failure, 350–351
   engine misfire monitor, 355–356
   evaporative system monitor, 356–360, 366t
   exhaust gas recirculation monitor, 360, 366t
   fuel system monitor, 360–362
   in-use performance tracking, 365–366
   oxygen sensor monitor, 363–364, 366t
   secondary air system monitor, 364, 366t
   VVT control system monitor, 364–365, 366t
OBDS (onboard distillation system), 63–64, 65f, 68, 231–233

Octane, 207–208, 219
   high octane fuel, 229–231
Oil film sources, 35
Oil layer absorption, of hydrocarbons, 90–92
Oil temperature, 11f, 34f
Olefins, 105, 214t
Onboard diagnostics, 196–197
Onboard distillation system (OBDS), 63–64, 65f, 68, 231–233
Onboard fuel distillation, 229–233, 234
Onboard fuel reformers, 224, 225
Onboard refueling vapor recover (ORVR) system, 343, 344
Open-valve injection (OVI), 35–36, 53
   and hydrocarbon emissions, 13–14, 36, 45, 93
   targeting, 52
   timing, 47–51
ORVR (onboard refueling vapor recover) system, 343, 344
Over-fueling, 32–33, 52–53, 83, 100
OVI. *See* Open-valve injection
Oxidation. *See* Catalytic oxidation; Hydrocarbon oxidation; Thermal oxidation
Oxygen, 214t
Oxygen interference, 394–395
Oxygen sensor monitor, 363–364, 366t
Oxygen sensor signal, 352f
Oxygen storage capacity equation, 286
Oxygen storage content (catalyst), 352–354
Oxygenates, 207–208, 211, 213, 214, 215
Ozone-forming potential, 40

Index

Palladium, 241, 246–258
Partial oxidation reformer (POx), 63, 223–224, 226–228, 229, 233
Particle image velocimetry (PIV), 123t, 132, 141, 142f
PDA (phase Doppler anemometry), 123t, 125–126, 133
Pencil stream fuel injector, 46
PFI. *See* Port fuel injected
Phase Doppler anemometry (PDA), 123t, 125–126, 133
Pintle fuel injector, 44–46, 55
Piston
    fuel evaporation from, 14, 37–39, 94
    position at engine stop, 6, 57
    ring-pack crevices, 88, 89, 101
Piston strokes, 101–104
Piston wetting
    at engine startup, 14
    effect on hydrocarbon emissions, 37–39
    multihole injectors compared, 50
PIV (particle image velocimetry), 123t, 132, 141, 142f
Planar laser-induced fluorescence (PLIF)
    combustion diagnostics, 143, 144f
    fuel delivery diagnostics, 133, 134
    mixture distribution diagnostics, 140
Platinum, 241, 246, 247, 248f
Platinum group metals (PGM), 241, 243, 246–258
PLIF. *See* Planar laser-induced fluorescence
Port film, 46, 50
Port flow velocity, during engine startup, 14–15
Port fuel injected (PFI) engine
    behavior during a cold start, 80–86
    engine shutdown process, 2–4
    engine startup process, 5–22
Port wall film model, 33
Port wall temperature, during engine startup, 10, 11f, 34–35
POx (partial oxidation reformer), 63, 223–224, 226–228, 229, 233
Pressure, intake manifold, 3, 4, 8–9, 15
Prevaporized fueling, 41–42, 43, 46
Prevaporizer system, 220
Propane, 111
Pump module seal, 338
PZEV (partial zero emissions vehicle), 62, 68, 211, 242, 334, 339, 349
    BMW, 410–411
    catalyst system design, 255–264
    Ford, 411–413
    hydrocarbon trap, 269–281
    Nissan, 269–270, 416, 418f, 419–424
    system requirements, 407–410
    Toyota, 61–62, 63f, 424–428

Quench layers, 96–98, 113
Quenching distance, methane/air flames, 98

Raman scattering, 123t, 138–139
Rayleigh scattering, 123t, 138–139
Reformulated gasoline (RFG), 214–215
Reid vapor pressure (RVP), 11, 208–211, 214t
Remote mix tee (RMT), 378, 379f
Retarded ignition. *See* Spark retard
RFG (reformulated gasoline), 214–215
RFG Reg-Neg, 214–215
Rhodium, 241, 246, 247, 252–253, 259f

RVP (Reid vapor pressure), 11, 208–211, 214t

Sauter mean diameter (SMD), 12, 126
Sealed housing for evaporative determination (SHED), 335, 336f
Seals, fuel system, 339
Secondary air injection
    controlling quality, 193–195
    and engine fueling enrichment, 191.
    and hydrocarbon oxidation, 111–112
    monitoring, 196–197
    principles of, 174–181
    quantity, 191–193
    temperature and mixing, 181–188
    turbocharged engines, 198–199
    vee engines, 197–198
Secondary air injection system, 176–177, 199
Secondary air system monitor, 364, 366t
SHED (sealed housing for evaporative determination), 335, 336f
SMD (Sauter mean diameter), 12, 126
Sparger-type design, 112, 186, 187f
Spark ignition, dual, 167, 168f
Spark plug threads, 88
Spark retard, 40–41, 51, 56, 60, 63–64, 65f, 67, 68, 145
    to decrease cold-start emissions, 158f, 159–160
    effect on engine operation, 160–163
    enhancements, 163–167
    to raise exhaust gas temperatures, 56, 189–191
Speed flare, 7, 8–9 (speed surge in the startup process)
Steam reforming, 222–223

SULEV (super ultra-low emissions vehicle), 47, 54, 79, 85, 160, 163, 165f, 206, 217, 242, 370, 372
    Honda, 413–416, 417f, 418f
    Toyota, 428–432
SULEV II, 22, 349
Sulfur, 214t, 249–250, 251f
Surface energy equation, 286
Surface oxygen storage capacity equation, 286
Surface species equation, 286
Swirl, 57–58, 61, 62, 68
Swirl control valves, 163, 164f, 165f

T_{10}, 11–12, 210
T_{50}, 12, 46, 56, 210, 211, 212, 214t
T_{90}, 12, 46, 210, 214t
Tailpipe emissions dilution factor, 370–371, 375–376, 377f
Tailpipe emissions limits, 371
Tailpipe exhaust temperatures, 388f
TAME (tertiary amyl butyl ether TAME), 208
T-butanol (TBA), 208
Temperature
    ambient, definition of cold start, 31
    catalyst
        after PFI engine shutdown, 5f
        rapid warm-up designs, 242–246, 262f
        role in PFI engine startup process, 5
    coolant
        after PFI engine shutdown, 5f
        during PFI engine startup, 5, 10, 11f, 17, 18f, 34f
        steady-state operation, 48, 49f, 91
    exhaust gas
        during engine startup and warm-up, 34f

with retarded ignition timing, 158f, 159, 160–162, 189–191
fuel
    evaporation behavior, 11–12, 33
    intake valve, 10–11, 34–35
    oil, during PFI engine startup, 11f, 34f
    port wall, during engine startup, 10, 11f, 34–35
Tertiary amyl butyl ether (TAME), 208
Thermal oxidation, 178–181
    role of temperature and mixing, 181–191
Three-way catalyst, 241–242, 246, 269
    platinum group metal loads, 247–255, 256, 257–258, 259
    thermal durability, 251–252
Three-way catalytic converter, system modeling, 283–317
Toluene, 40, 209t
Top-land side clearances, 90
Total weighted demerits (TWD), 11
Toyota, 47, 63f, 228, 424–428, 428–432
Toyota hybrid ATPZEV system, 431, 432f
2,2,4 Trimethyl pentane, 209t
Tumble flow, 164, 165f
Turbocharged engines, secondary air injection, 198–199
Turbulence, in-cylinder, 32, 56, 57, 59, 83, 99, 101, 111
TWD (total weighted demerits), 11

U.S. Code of Federal Regulations (CFR), 335, 370, 372, 390, 400
U.S. Environmental Protection Agency (EPA), 217, 385
    evaporative emissions standards, 333, 334

RVP regulations, 209
UEGO (universal exhaust gas oxygen) sensor, 123t, 127–128, 134, 135, 196–197, 363–364, 414
ULEV (ultra-low emissions vehicle), 47, 64, 79, 160, 216, 217, 227, 243
ULEV II (ultra-low emissions vehicle II) standards, 333
Underbody catalyst, 259, 260f
Unical RF-A gasoline, 46
Universal exhaust gas oxygen (UEGO) sensor, 123t, 134, 135, 127–128, 196–197, 363–364, 414

Vapor dome, 336, 337
Vapor fueling, 43, 60
Vapor/air ratio, 33, 53–55, 67
Vaporized fuel delivery, 41–42, 43, 46
Variable valve time (VVT) system, 14–15, 56, 61–62, 63f, 68
Variable valve timing and lift electronic control (VTEC) system, 163, 413
Vee engines, 197–198
Volatile organic compound (VOC) emissions, 211
VTEC (variable valve timing and lift electronic control) system, 163, 413
VVT (variable valve time) system, 14–15, 56, 61–62, 63f, 68
VVT control system monitor, 364–365, 366t

Wall wetting
    diagnostic techniques, 123t, 127–130
    during engine warm-up, 94, 95
    during steady-state operation, 50
    effect on hydrocarbon emissions, 36–38, 51

Wall wetting *(continued)*
    at engine startup, 12, 14, 50, 66, 68
    and swirl, 57–58
Washcoat (catalyst), 243, 246, 247, 249, 253, 255, 258, 259*f*
Water condensation, 378*f*
Water electrolysis, 228
Wide-range oxygen (WRO$_2$) sensor. *See* Universal exhaust gas oxygen (UEGO) sensor

Xylene, 93, 209*t*

Zeolite, 245, 270–272
Zero Emissions Vehicle (ZEV), 334

About the Contributing Authors

Dr. Michael Akard is an Analytical Product Specialist at Horiba Instruments, Inc. in Ann Arbor, Michigan. He earned his B.S. in Chemistry from the University of New Mexico in 1990 and his Ph.D. in Analytical Chemistry from the University of Michigan, Ann Arbor, in 1994. Dr. Akard is the author of 11 articles in journals ranging from *Analytical Chemistry* to the *Journal of Chromatographic Science*. He has made 18 presentations at PITTCON, SAE, and Anachem conferences, and he holds a U.S. patent for a method and a product in high-speed gas chromatography.

Dr. Paul Andersen is Technical Director of Johnson Matthey's North American Technical Center. He earned his B.S. in Chemical Engineering from the University of Illinois—Urbana and his Ph.D. in Chemical Engineering from Northwestern University. Since joining Johnson Matthey in 1992, Dr. Andersen has worked on the development of various emissions control catalysts, including three-way catalysts, NOx adsorber catalysts, SCR catalysts, and oxidation catalysts.

Professor Choongsik Bae has worked in the Department of Mechanical Engineering, Korea Advanced Institute of Science and Technology (KAIST) since 1998. He received his B.S and M.S. degrees in Aerospace Engineering from Seoul National University in Korea. He received his Ph.D. in Mechanical Engineering from Imperial College, London, in 1993 and then worked as a Research Associate there. He later joined Chungnam National University and soon moved to KAIST. Professor Bae has been involved in various studies of experimental engine work and research on fuel spray, flow, and combustion in spark ignition and compression ignition engines. In 1997, he received the Arch T. Colwell Merit Award from SAE.

Dr. Todd Ballinger is a Staff Scientist in the Catalytic Systems Division of Johnson Matthey. He graduated from the University of Pittsburgh in 1993 with a Ph.D. in Physical Chemistry/Surface Science. After conducting post-doctoral research in catalysis and surface science at Texas A&M University and the U.S. Naval Research Laboratory, he joined the Catalytic Systems Division of Johnson Matthey in 1995. There, he has been involved in the development of advanced three-way catalysts and hydrocarbon traps/catalysts for automotive catalytic converters, as well as the development of advanced catalyst systems for achieving very low emissions from vehicles.

Mark Borland is a Senior Engineering Specialist in the Advanced Engine Systems Development Group of DaimlerChrysler Corporation. In his 13 years there, he has worked on the development of emissions systems and control algorithms for both gasoline- and diesel-powered vehicles. Most recently, Mr. Borland was part of the team responsible for the development of software and hardware concepts to meet tailpipe emissions on DaimlerChryler's first production PZEV vehicle. Mr. Borland earned his B.S. in Mechanical Engineering from the University of Wisconsin—Milwaukee and his M.S. in Applied Statistics from Oakland University in Rochester, Michigan.

Professor Wai K. Cheng is a Professor of Mechanical Engineering at Massachusetts Institute of Technology (MIT) and is the Associate Director of the Sloan Automotive Laboratory at MIT. His research interest lies in engine performance and emissions. Professor Cheng has made major contributions to the mixture preparation process in spark ignition engines, and he has authored more than 70 technical publications. He has received the Ralph R. Teetor Award in 1984, and the Oral Presentation Awards from SAE in 2002 and 2004. He is a Fellow of SAE.

Dr. James A. Eng is a Staff Research Engineer at General Motors Research and Development. He received his Ph.D. from Princeton University, working on hydrocarbon emissions and post-flame oxidation mechanisms from homogeneous charge spark ignition engines. During the past 10 years, Dr. Eng has worked in the areas of chemical kinetics, understanding the effects of fuels on engine performance and emissions, cold-start hydrocarbon emissions, and HCCI combustion.

Timothy Gernant is a Senior Product Development Engineer working on the calibration of onboard diagnostics at Ford Motor Company. He joined Ford in 1995 upon graduation from the University of Michigan, Ann Arbor, with a B.S. in Mechanical Engineering. Mr. Gernant continued his education at the University of Michigan, Ann Arbor, and received an M.S. in Automotive Engineering in 2000. He holds multiple patents relating to OBDII diagnostic systems.

About the Contributing Authors

Kathleen Grant has been an OBDII Calibration Engineer at Ford Motor Company for 10 years. She graduated from Wayne State University with a B.S. degree in Electrical Engineering and has worked with diagnostic systems and engine controls for more than 20 years. Ms. Grant has been awarded two U.S. patents relating to OBDII diagnostic systems. Previously, she was a project engineer at General Motors Corporation. Most recently, Ms. Grant has been calibrating OBD functionality on PZEV-emissions-level vehicles.

Christopher Hadre is the Core Design and Release Engineer for vapor canisters with the Evaporative Systems Group of DaimlerChrysler Corporation and has been a specialist in evaporative systems development for more than 5 years. He graduated with a B.A.Sc. in Mechanical Engineering from the University of Windsor in Canada in 1998.

Professor Matthew J. Hall is with the Department of Mechanical Engineering at the University of Texas—Austin, and he has more than 20 years of experience in automotive and vehicle research. He received his B.S. and M.S. in Mechanical Engineering from the University of Wisconsin—Madison and his Ph.D. from Princeton University. Professor Hall was a Post-Doctoral Fellow at the Combustion Research Facility of Sandia National Laboratories and at the University of California—Berkeley. His primary research interests center around combustion processes, with an emphasis on internal combustion engines. Professor Hall's focus is on experimental measurements studying engine performance, emissions, and flows, with a specialization in optical diagnostic techniques and sensors. He has received several awards from SAE, including the Ralph R. Teetor Award in 1993, the Arch T. Colwell Merit Award in 1985, the Horning Award in 1987, and the Myers Award in 1998.

Dr. David Lafyatis is a Technical Program Manager at Johnson Matthey. He earned his Ph.D. in Chemical Engineering from the University of Delaware and then completed a Post-Doctoral Fellowship at the Rijksuniversiteit Gent in Belgium, studying reaction kinetics in a transient reactor. For the past 9 years, Dr. Lafyatis has worked at Johnson Matthey in the area of exhaust aftertreatment technology.

Professor Ronald D. Matthews obtained his B.S. in Mechanical Engineering from the University of Texas, Austin, followed by three graduate degrees from the University of California—Berkeley, culminating in 1977 with a Ph.D. with a specialization in combustion. He joined the faculty of the Department of Mechanical Engineering at the University of Texas, Austin, in 1980, where he established its combustion and engine research program. Professor Matthews is Head of the General Motors Foundation Combustion Sciences and Automotive Research Laboratories on the University of Texas, Austin, campus. In 1980, he founded the SAE student chapter at the University of Texas, Austin, and has served as Faculty Advisor to the chapter since then. For more than 25 years, he has been involved in research in the areas of combustion, engines, emissions, and alternative fuels. Professor Matthews' research activities also include experimental work and numerical modeling of both fundamental combustion processes and combustion within engines. His present research focuses primarily on controlling hydrocarbon emissions from PFI spark ignition engines, the spark ignition process, engine friction, and alternative diesel fuels. Professor Matthews has received several SAE awards, including the Ralph R. Teetor Award in 1979, the Arch T. Colwell Merit Award in 1992, the Excellence in Engineering Education (Triple E) Award in 2002, the Phil Myers Award in 2002, and the Faculty Advisor Award in 1990, 1997, and 2002. In 1996 and again in 1998, the University of Texas, Austin, body of work on fractal engine modeling was nominated for the *ComputerWorld* Award and was selected for inclusion in the Smithsonian National Museum of American History, Permanent Research Collection on Information, Technology, and Society. Professor Matthews was elected a Fellow of SAE in 2002.

Kimiyoshi Nishizawa is a Senior Manager of the Powertrain Advanced Engineering Department No. 2 at Nissan Motor Co., Ltd. He graduated from Tokyo University in Japan and received his B.S. in Mechanical Engineering. His first job at Nissan was to design and develop components for gasoline engines. Later, he became an engine systems engineer, focusing on emissions reduction. Mr. Nishizawa's recent work has focused on developing the PZEV system for the Nissan Sentra CA for the U.S. market, which was launched in February 2000; a ULEV system for the Nissan Bluebird Sylphy for the Japanese market, which was launched in August 2000; and developing emissions systems to meet Japan ULEV standards and applying them to more than 80% of Nissan vehicles sold in Japan in March 2003.

Dr. Stephen Russ is a Technical Leader for Engine Combustion in the V-Engine Engineering Division of Ford Motor Company. He joined Ford after receiving his Ph.D. in Mechanical Engineering from the University of Minnesota, Minneapolis/Twin Cities, in 1993. Dr. Russ worked in the Ford Research Laboratory for six years, conducting research on engine combustion, emissions formation, and advanced diagnostics. For the past five years, he has led the development of several Ford V-engine programs. Dr. Russ has authored 20 SAE technical papers and has organized and chaired SAE technical sessions on engine combustion and emissions. He also has been awarded 11 U.S. patents for various engine technologies. Dr. Russ was selected to participate in the 1999–2000 SAE Industrial Lectureship Program and has given invited lectures at several universities.

Professor Tariq Shamim is an Associate Professor of Mechanical Engineering at the University of Michigan—Dearborn. He is a graduate of the University of Michigan—Ann Arbor, where he received both his Ph.D. in Mechanical Engineering and his M.S. in Aerospace Engineering. Professor Shamim received another M.S. in Mechanical Engineering from the University of Windsor in Canada and a B.S. in Mechanical Engineering from the NED University of Engineering and Technology in Karachi, Pakistan. His research and teaching interests are in the area of computational thermo-fluids, with major emphasis on combustion, emissions control, fuel cells, and thermal spray. Professor Shamim's research is supported by the National Science Foundation, U.S. Department of Energy, U.S. Department of Defense, and the automotive industry. He is actively involved with several professional organizations, including SAE, American Society of Mechanical Engineers (ASME), and the Combustion Institute. In 2004, he received the Ralph R. Teetor Award from SAE.

Jenny Spravsow has been employed by DaimlerChrysler since 1998 and currently holds the position of Product Development Engineer in the Advanced Evaporative Systems Group. She earned her B.S. in Mechanical Engineering from Lawrence Technological University in Southfield, Michigan, and her M.S. in Mechanical Engineering from Oakland University in Rochester, Michigan.

Glenn Zimlich is an OBD Calibration Technical Expert at Ford Motor Company. He joined Ford in 1990, after completing his Bachelor of Mechanical Engineering at the University of Detroit. He completed his Master of Mechanical Engineering from the University of Detroit in 1993. Mr. Zimlich has received multiple U.S. patents relating to onboard diagnostics.

About the Editor

Dr. Fuquan (Frank) Zhao currently is the Vice President of Product Engineering, and General Manager for the R&D Center at Brilliance Jinbei Automobile Corporation in Shenyang, China. In this position, Dr. Zhao is responsible for all activities related to vehicle product development.

Prior to this position, he was a Research Executive of Technical Affairs at Daimler-Chrysler Corporation. In this position, he was responsible for providing technical guidance and advice to all product team managers and engineers within DaimlerChrysler Corporation relating to engine development issues and advanced powertrain technologies. He represented the Chrysler Group in various consortium activities and served as a technical spokesman.

Dr. Zhao's other experience includes time as an Assistant Professor in Mechanical Engineering at Wayne State University, a Research Fellow at the Imperial College of Science, Technology, and Medicine in the United Kingdom, and a Postdoctoral Fellow at Wayne State University and the University of Hiroshima. Dr. Zhao received his B.S. in Mechanical Engineering from Jilin University of Technology (China) in 1985. He obtained his M.S. in Mechanical Engineering from the University of Hiroshima (Japan) in 1989 and his Ph.D. there in 1992.

Dr. Zhao is the principal author of more than 100 technical research papers on various subjects. He is the principal author of the book, *Automotive Gasoline Direct-Injection Engines*, published by SAE in 2002, and he is the leading editor of the book, *Homogeneous Charge Compression Ignition (HCCI) Engines*, published by SAE in 2003. Dr. Zhao also is the leading editor of the books

Direct Fuel Injection for Gasoline Engines and *Advanced Development in Ultra-Clean Gasoline-Powered Vehicles* published by SAE in 2000 and 2004, respectively. He has received several prestigious awards to recognize his technical achievements and his leadership in professional societies, including the 2002 SAE Forest R. McFarland Award. He served as the chair of the Combustion Committee for the SAE Fuels and Lubricants activity from 2002 to 2004. He is a Fellow of SAE. Currently, Dr. Zhao holds an honorary professorship from several leading Chinese Universities.